AGRICULTURE ISSUES AND POLICIES

LEGUMES

TYPES, NUTRITIONAL COMPOSITION AND HEALTH BENEFITS

Agriculture Issues and Policies

Additional books in this series can be found on Nova's website under the Series tab.

Additional e-books in this series can be found on Nova's website under the e-book tab.

Food Science and Technology

Additional books in this series can be found on Nova's website under the Series tab.

Additional e-books in this series can be found on Nova's website under the e-book tab.

AGRICULTURE ISSUES AND POLICIES

LEGUMES

TYPES, NUTRITIONAL COMPOSITION AND HEALTH BENEFITS

HIROTO SATOU
AND
REN NAKAMURA
EDITORS

nova publishers
New York

For permission to use material from this book please contact us:
Telephone 631-231-7269; Fax 631-231-8175
Web Site: http://www.novapublishers.com

Additional color graphics may be available in the e-book version of this book.

Library of Congress Cataloging-in-Publication Data

Legumes: types, nutritional composition and health benefits / editors: Hiroto Satou and Ren Nakamura.
p. cm.
Includes index.
ISBN: 978-1-62808-280-7 (hardcover)
1. Legumes. 2. Legumes--Composition. 3. Legumes--Health aspects. 4. Legumes as food. I. Satou, Hiroto. II. Nakamura, Ren.
SB177.L45L48 2013
633.3--dc23

2013021117

Published by Nova Science Publishers, Inc. † New York

CONTENTS

Preface		**vii**
Chapter 1	Legumes for Grazing and Health: The Case of *Bituminaria bituminosa* *María Pazos-Navarro, Mercedes Dabauza, Enrique Correal, David J Walker, José A del Río, Ana Ortuño, Pilar Méndez, Arnoldo Santos, Segundo Ríos, Vanessa Martínez-Frances and Daniel Real*	**1**
Chapter 2	Nutritional Characterization of Wild Legumes (*Lathyrus* and *Vicia* Genera) *Elena Pastor-Cavada, Julio E. Pastor, Rocío Juan and Javier Vioque*	**41**
Chapter 3	Legumes Leading the War against "Diabesity-the Obesity-Diabetes Epidemic": An Overview *Rocio Campos-Vega, B. Dave Oomah, Guadalupe Loarca-Piña and Haydé Azeneth Vergara-Castañeda*	**93**
Chapter 4	Soybean: Overview of the Nutritional Profile and the Implications for Nutrition and Health Effects *I. Mateos-Aparicio, A. Redondo-Cuenca and M. D. Tenorio-Sanz*	**125**
Chapter 5	Fermentation of Lesser Known Legumes *E. O. Farinde and S. B. Fasoyiro*	**159**
Chapter 6	Dark and Bright Facets of the Nutritional Value of Grass Pea (*Lathyrus sativus L.)* Seeds *Angela R. Piergiovanni, Venturino Bisignano, Giambattista Polignano and Lucia Lioi*	**179**
Chapter 7	Lentils (*Lens Culinaris* L.): Linking Whole Foods for Better Human Health *Dil Thavarajah, Pushparajah Thavarajah and Shiv Kumar*	**193**

Chapter 8	*Medicago truncatula*: A Model Organism to Study the Biology of Agriculturally Important Legume Crops *Mauricio Alberto Reynoso, Virginia Savy, Marcos Lancia, Karen Hobecker, Flavio Blanco and María Eugenia Zanetti*	**209**
Chapter 9	Antihypertensive Potential of Protein Hydrolysates from the Velvet Bean (*Mucuna pruriens*) *Maira Segura-Campos, Saulo Galicia-Martínez, Luis Chel-Guerrero and David Betancur-Ancona*	**225**
Chapter 10	Use of Herbicide (Roundup®): Potential Influence on Soil Metals Lability and on Radish Metal Uptake *Fabienne Baraud and Lydia Leleyter*	**241**
Chapter 11	Chemical and Structural Composition of Arabinogalactan Proteins (AGP) of the Mesquite (Prosopis spp) *Yolanda L. López-Franco, Faviola Bainori-Vasquez, Ana K. Sánchez-Castañeda, Karla Martínez-Robinson, Tomás Madera-Santana, Alma Rosa Toledo-Guillén and Jaime Lizardi-Mendoza*	**257**
Chapter 12	Common Bean as an Emerging Model Grain Legume *Virginia Dalla Via, Carolina Rípodas, Joaquín Clúa, Candela Narduzzi, Mauricio Reynoso, Mario Aguilar, María Eugenia Zanetti and Flavio Blanco*	**271**
Chapter 13	Health Benefits of Legumes *T. B. Ng and C. C. W. Ng*	**283**
Chapter 14	Current Update in Methodologies for Extraction and Analysis of Proteins and Isoflavones *Savithiry S. Natarajan and Devanand L. Luthria*	**295**
Index		**305**

PREFACE

In this book, the authors present topical research in the study of the types, nutritional composition and health benefits of legumes. Topics discussed include legumes (Bituminaria bituminosa) for grazing and health; nutritional characterization of wild legumes (lathyrus and vicia genera); legumes leading the war against "Diabesity-the obesity-diabetes epidemic"; soybeans nutritional profile and implications for nutrition and health effects; fermentation of lesser known legumes; dark and bright facets of nutritional value of grass pea (lathyrus sativus L.) seeds; lentils (Lens culinaris L.) and their link to better human health; Medicago truncatula as a model organism to study the biology of agriculturally important legume crops; antihypertensive potential of protein hydrolysates from Velvet bean (Mucuna pruriens); use of the herbicide Roundup on soil metals lability and on radish metal uptake; chemical and structural composition of arabinogalactan proteins (AGP) of mesquite (Prosopsis spp); common bean as an emerging model grain legume; the health benefits of legumes; and current update in methodologies for extraction and analysis of proteins and isoflavones.

Chapter 1 - Legumes have been used for centuries for different purposes as human food (grain legumes such as lentils, chickpeas, etc.), animal feed (pasture and forage legumes) and as sources of secondary metabolites with health, medical or nutraceutical benefits (pharmaceuticals, cosmetics and chemicals). Due to this relevance, some of these species are the subject of breeding programmes for selecting hybrids with desirable traits. To aid in these programmes, Plant Biotechnology offers multiple methods for conservation and commercial propagation of valuable plants for different objectives, such as the pharmaceutical industry or pasture production.

Bituminaria bituminosa (L.) C.H. Stirton (syn. *Psoralea bituminosa* L., *Fabaceae, Psoraleeae*; Stirton 1981) is a perennial legume widely distributed in the Mediterranean Basin and Macaronesia. The plants are grazed by small herbivores, but are mainly used for hay to feed milking goats, mostly in the Canary Islands. Moreover, *B. bituminosa* plants are a source of secondary metabolites of pharmaceutical and medicinal interest, such as furanocoumarins (FCs, psoralen and angelicin) and pterocarpans (bitucarpin A and B). So, their possible physiological functions and pharmacological properties are being analysed.

Our research group began to search for alternative perennial pasture legumes for low-rainfall Mediterranean environments at the end of the 1980's. *B. bituminosa* showed desirable traits (tolerance to cold and/or drought), that draw it as a good option. So, the first step was to select new germplasm with different traits to increase genetic variability for starting a breeding programme, with the most promising varieties, accessions and natural hybrids to deliver cultivars in the minimum time. To achieve this objective, conventional crossing between cultivars and wild accessions, and techniques for interspecific crossing breeding were carried out, in order to obtain new hybrid lines with a combination of desirable traits. Then, for a commercial approach, studies based on seed production of *B. bituminosa* selected cultivars have to be carried out and also research on the alternative uses of this species, based on its content of bioactive components with pharmaceutical potential.

As a result of this breeding programme, the authors' research group has selected several hybrid and non-hybrid lines showing high levels of expression of, or a combination of desirable traits such as high forage quality, tolerance to biotic or abiotic stresses, or high contents of FCs. Therefore, the development of biotechnological tools such as micropropagation or *in vitro* plant regeneration protocols allowed the possibility of selecting somaclonal variants, obtaining pollen-derived haploid plants for optimising plant breeding programmes, and regenerating genetically transformed plants to control the biosynthesis of FCs. Moreover, these techniques offer attractive approaches for the large-scale propagation and conservation of germplasm resources of these plants.

In the present chapter, the authors describe this species, show its potential uses and the strategies that may be followed to carry out a traditional breeding programme in *B. bituminosa* and the tools developed to assist in it.

Chapter 2 - In several regions of the world, legumes play an important role in human and animal nutrition as they are a cheap source of protein and other nutrients. In recent years have been shown that the consumption of proteins, certain lipids, polyphenols, or bioactive peptides contained in legumes have a beneficial effect on health.

In addition, since the "green revolution," a large amount of the world's phytodiversity has been lost because local crops have been substituted by other genetically uniform crops, which have higher yields. However, the diversification of cultivars is necessary to adapt cultivars to soil and water conditions, or to limit phytosanitary risks. Hence, knowledge of local plants could provide useful data for the possible development of future crops.

Vicia and *Lathyrus* are major genera of the tribe *Fabeae* and have wild species in Southern Spain. A total of 46 different wild species have been studied. To the authors' knowledge, this is the first time that a deep nutritional characterization of wild populations of *Lathyrus* and *Vicia* species has been made.

In this work the fatty acid composition of the seed oil, the polyphenol concentration and the antioxidant activity of the polyphenols extracted from the seeds, the amino acid composition and nutritional quality of seed proteins have been studied in 15 Lathyrus species and in 28 *Vicia* species (31 taxa).

The main seed oil fatty acids ranged from $C_{14:0}$ to $C_{20:0}$. All studied species showed higher amounts of total unsaturated fatty acids than saturated ones. Among unsaturated fatty acids, the most abundant were linoleic, oleic and linolenic acids. Among studied species, the $\omega6/\omega3$ ratio was variable, ranging from 1.7 in *V. articulata* to 17.3 in *V. faba*, there being nine species in which the $\omega6/\omega3$ ratio was below 5.

Polyphenolic contents ranged from 1.9 mg/g meal in *V. narbonensis* to 29.2 mg/g meal in *L. sphaericus*. The highest specific polyphenols antioxidant activity (referred to as catechin equivalents in the extracts) was observed in the extracts from *Lathyrus annuus*, *Vicia parviflora* and *Vicia tenuifolia*. On the contrary, the highest total antioxidant activity (referred as to extract volume) was observed in *L. aphaca* y *V. sativa*. Results show that studied

Lathyrus and *Vicia* species are rich in phenolic compounds with higher antioxidant activity than phenolics of widely consumed legumes such as soy, chickpea or lupin.

Protein contents in studied species ranged from 17.7% in *Lathyrus sativus* to 32% in *V. pyrenaica*. Among essential amino acids the most abundant were Leu, Lys, Phe, Thr and Val. Also, all species contained Lys above FAO recommendations. On the contrary, all species were limiting in Trp ranging from 0.5% to 1.0% content and were also limiting in sulphur containing amino acids, Met and Cys. In vitro protein digestibility (IVPD) in studied legumes ranged from 76.0% in *L. pratensis* to 86.5% in *L. annuus*. Nutritional parameters, such as protein efficiency ratio, corresponded to high quality proteins. Protein digestibility corrected amino acid score was similar to the observed in other commercial legumes such as lentil or peanut.

In conclusion, results confirm the interest of studying wild populations of *Lathyrus* and *Vicia* species as source of seeds with good nutritional characteristics and contribute to revalorize the cultivation of these legumes.

Chapter 3 - The rates of obesity and diabetes are worrying. The worldwide prevalence of type 2 diabetes mellitus (T2DM) is rising rapidly primarily due to the surge in obesity, combined with ageing populations and a trend towards urbanization. Overweight and obesity are key features of diabetes and prevention of excessive weight gain is a fiscally prudent global health priority. The new term, 'diabesity' has been coined to describe people burdened with obesity and type 2 diabetes. Individuals with diabesity are at increased risk of multiple comorbidities (including chronic and degenerative diseases) and represent a huge economic burden on health services around the world. Some adjuvants can assist in weight loss and reduction in blood glucose. In this regard, there is strong evidence that a diet high in whole grains, like legumes, is associated with lower body mass index, waist circumference and reduced risk of overweight, obesity and various associated chronic and degenerative diseases. Therefore, diets high in whole grains and legumes can help reduce weight gain and elicit weight loss with energy controlled diets high in cereals and legumes. Legume consumption has been highlighted as the perfect way to improve and/or promote glycemic control, food intake and body weight in obese, glucose-intolerant and diabetic subjects. The mechanism of action that could explain those effects is linked to the concomitant presence of bioactive compounds like glycoproteins and fiber, among others. This chapter explores the role of legumes, particularly their bioactive compounds on the war against obesity and diabetes.

Chapter 4 - Soybean represents an excellent source of high quality protein, it has a low content in saturated fat, contains a great amount of dietary fiber and the isoflavone content makes soybean singular among other legumes. Most of the studies have been focused on the positive effect of soybean protein over hyperlipidemia and the prevention of cardiovascular disease. Besides, isoflavones focused many attention due to they are associated with the prevention and the treatment of different chronic diseases. Some studies have shown also the role of the soy dietary fiber and the galacto-oligosaccharides in different diseases such as hypercholesterolemia, diabetes and obesity.

Soybean is a source of functional ingredients interesting from the nutritional point of view and because their potential health effects. Indeed, soybeans and soy foods are widely promoted and eaten based on assumed relationships between its consumption and beneficial health effects. This chapter is focused on the latest results of research on health properties and uses of soybean as well as an overview of the nutritional profile of this legume.

Chapter 5 - Legumes belong to the family *Leguminosae*. Legumes are major sources of plant protein and oil. Some of the major and well known legumes are cowpea, peas, clover, lentils, lupins, mesquite, carob, peanut, locust bean, groundnut, wisteria, and kenturkey coffee tree. Some minor and lesser known legumes are pigeon pea, jack bean, sword bean, gram pea, moth bean, mungo bean, winged bean, velvet bean, kidney bean, lima bean, African yam bean, black gram and bambara groundnut. The lesser legumes have been reported to be less accepted either due to their long cooking time or their anti-nutrient content. This had been improved through the use of fermentation technologies, hence the theme fermentation of lesser legumes. This chapter focused on the importance of legumes, some examples of lesser legumes, nutrient composition and anti nutrient content of some of the lesser legumes , methods of fermentation of some lesser legumes with emphasis on lima bean, pigeon pea and African yam bean. Other areas discussed are the effects of fermentation on the anti nutrient content of these lesser legumes and the micro organisms involved in their fermentation process. Cooking is a unique and compulsory precursor of fermentation to processing these lesser legumes. Fermentation improved the nutrient composition and reduced the anti-nutrient contents of the lesser legumes reviewed. Protein, fat, and ash content of lima bean were increased from 16.81%, 1.30% and 2.41% in the raw lima bean to 17.52%, 2.17% and 2.68% respectively at 72 h of fermentation. Minerals (Mg, P and K) also increased with fermentation time in lima bean. Mg increased from 8,100 ppm in the raw lima bean to 8200 ppm at 72 h of fermentation, P and K also increased from 8,400 ppm, 2,100 ppm in the raw lima bean to 12,800 ppm, 6,900 ppm respectively at 72 h of fermentation. . Tannin, phytate, cyanide and trypsin inhibitors were reduced from 11.0 mg/kg, 17.0 mg/kg, 1.87 mg/kg and 6.32 mg/kg in the raw lima bean to 6.01 mg/kg, 5.52 mg/kg, 0.80 mg/kg and 1.0 mg/kg respectively at 72 h of fermentation. In fermented pigeon pea seeds, phosphorus content was increased to 0.51 mg/100g in comparison to 0.44 mg/100g in the raw seeds. There was no significant difference in the sodium, potassium, calcium and magnesium contents of the raw pigeon pea seeds and the fermented seeds. Tannin and phytate in raw seeds were significantly reduced from 4.23 mg/100g and 2.20 mg/100g to 1.19 mg/100g and 1.65 mg/kg by fermentation. Fermentation reduced the flatulence potential of African yam bean, α-galactoside was reduced to the range of 22 – 39 %. The microorganisms involved in the fermentation process were found to be mainly *Bacillus species* and were common to almost all the lesser legumes reviewed. They include *Bacillus subtilis, B. plumilus, B. licheniformis,* and *B. megatarium. Bacillus species* are able to break down protein, carbohydrate and fat, which are the major constituents of these legumes. Fermented lesser legumes have their nutritional quality and acceptability improved over the starting material.

Chapter 6 - Grass pea (*Lathyrus sativus* L.) is an annual food, feed and fodder crop cultivated in temperate regions of Asia, Europe, America and East Africa. This pulse has several advantageous traits such as: high yielding, nitrogen fixation, extensive tolerance of drought and water-logging, adaptation to a wide range of soil types, resistance to insects and pests. Grass pea is a popular source of vegetable proteins in subsistence farming systems, mainly in some developing countries. However, an extensive use for food and manufacturing feedstock is limited by the presence in seeds of appreciable quantities of β-N-oxalyl-L-α,β-diaminoproprionic acid (β-ODAP), in addition to other antinutritional factors typical of pulses (i.e.: digestive enzyme inhibitors, lectins, phytic acid, etc.). ODAP is a non protein amino acid associated with neurolathyrism, a neurological disorder that manifests in humans and animals when grass pea seeds became a staple food for long periods. As a consequence of the reputed grain toxicity, a significant drop of grass pea cultivation took place in the second half of 20th century at worldwide level. Presently, it is considered a model for sustainable agriculture, so efforts are underway for the reintroduction in North and Latin America, Australia and Southern Europe. The aim of these initiatives is to counter the effect of climate change on agricultural productivity and to provide an alternative crop in the soils overexploited by continuous cereal cultivation. The purpose of this review is to give an overview of the most recent literature dealing with nutritional and nutraceutical quality of grass pea seeds.

Chapter 7 - The failure to link current agricultural production with human nutritional needs has led to the development of unhealthy food systems, which cause both malnutrition and chronic diseases. Malnutrition is responsible for the deaths of 30% of children under 5 years of age and accounts for 11% of the global disease burden. Poor diets and nutrition resulting from unhealthy food systems also create huge economic burdens for healthcare systems with negative consequences on child development and long-term direct impacts on sustainable development. Therefore, sustainable agricultural systems are essential for providing food energy and sufficient daily nutrients for humans, and in so doing support health and general well-being. Cool season food legumes, especially lentils (*Lens culinaris* L.), have the potential to provide adequate amounts of protein, prebiotic carbohydrates, and essential micronutrients; a 50 g serving can provide 3.7-4.5 mg iron (Fe), 2.2-2.7 mg zinc (Zn), 22-34 μg of selenium (Se), 50-250 μg of beta-carotene, and 50-300 μg of folates. Unlike other grains, lentils are very low in phytic acid (2.5-4.4 mg g^{-1}), which binds Fe and Zn and thus renders these nutrients poorly bioavailable. The present chapter provides an overview of current global lentil production, global malnutrition issues, and the promise of pulse crops, particularly lentil, as a whole food solution to combat global malnutrition issues. In addition, the superior mineral bioavailability profiles and genetic potential of lentil for further enrichment of bioavailable mineral micronutrients will be briefly discussed with respect to developing sustainable food systems.

Chapter 8 - *Medicago truncatula* (barrel medic) is a species native to the Mediterranean that has been adopted as a model organism for temperate legumes due to its relatively small size, short generation time, small diploid genome and the capacity to reproduce by self- and out-crossing. It is a close relative of alfalfa (*M. sativa*), which is used as forage, but it is also a model organism for other legume crops used for human consumption such as pea (*Pisum sativum*), lentil (*Lens culinaris*) and faba bean (*Vicia faba*). As many other legumes, *M. truncatula* establishes a nitrogen fixing symbiosis with soil bacteria called *Rhizobia*, which accounts for a significant proportion of the nitrogen incorporated into the ecosystems. In addition, it has the capacity to interact with fungi to form mycorrhiza, an ancient root endosymbiosis important for plant mineral nutrition. In the past decade, a significant amount of genetic and molecular tools has been developed to study the biology of this model species: the sequencing of nearly its complete genome, the generation of genome-wide expression datasets, the development of techniques for stable and hairy root transformation, and the establishment of mutant collections for genetic studies. This compendium of tools allowed the dissection of the genetic networks that govern symbiotic association with bacteria and fungi. The knowledge generated in this model legume is crucial to improve crop productivity in agriculture.

Chapter 9 - Lifestyle-related diseases are becoming increasingly more prevalent worldwide, and arterial hypertension is among the most common. High blood pressure is a risk factor for cardiovascular disease, but can be controlled in many ways from inhibiting enzymes that promote it to mitigating oxidation processes. Angiotensin I-converting enzyme (ACE) controls blood pressure by converting angiotensin I into angiotensin II, a pressor hormone that causes peripheral vasoconstriction, sodium reabsorption in the kidneys and increased body fluid retention. Produced by decomposition of angiotensin II, angiotensin III controls aldosterone production and raises blood pressure. Inhibiting ACE is one of the most effective methods for suppressing increased blood pressure. Recent studies on the biological activities and potential health benefits of peptides from food proteins have shown that these peptides may provide specific health benefits, such as lowering blood pressure. Alkaline extraction is a method for protein extraction from legumes. The present chapter describes the inhibitory activity of ACE in enzymatic hydrolysates from *Mucuna pruriens* protein concentrates. The concentrates were prepared by wet fractionation, and enzymatically hydrolyzed using different enzymes (Alcalase®, Flavourzyme®, pepsin, pancreatin and the sequential system Alcalase®-Flavourzyme®) for different lengths of time (5–120 min). Their biological potential was evaluated based on the IC_{50} value (i.e. the hydrolysate concentration required to produce 50% ACE inhibition). Hydrolysis of *M. pruriens* protein concentrates with commercial and gastric enzymes generates extensive hydrolysates. These can be used to produce low molecular weight peptides, which may possess hypotensive and antihypertensive properties. Protein hydrolysates with these properties have commercial potential as 'health-enhancing ingredients' in the manufacture of functional foods aimed at preventing and/or treating cardiovascular diseases or oxidative stress, while also providing added value to this legume.

Chapter 10 - Important quantities of metals can be released from soils, accumulate in plants and then enter the food chain. Thus, the consumption of vegetables grown on contaminated soils has become one of the major routes of human exposure to potentially toxic metals. It is then crucial to better understand the various processes controlling metal mobilization / immobilization, uptake, and sequestration by plants, for better risk assessment. The aims of this work were to determine the potential influence of a commercial herbicide (Roundup®) on metals lability (mobility) and phytoavailability. Metals lability was estimated by single chemical extraction (HCl or EDTA) and phytoavailability by determination of the metals quantities actually taken by plants grown the on studied soils. Radish, an important, common vegetable in direct contact with soil, known as a useful indicator of metal phytoavailability, as a moderate accumulator (Davies and Hougton, 1984; Baraud and Leleyter 2012) was selected as a test plant for this purpose.

Chapter 11 - Arabinogalactan proteins (AGPs) are commonly found in higher plants. The AGPs are usually located on cell membranes and gum exudates. The AGPs characterized so far have a protein content of less than 10% with more than 90% carbohydrate content. As a consequence of their protein component, these macromolecules can be classified in two groups according to their amino acids composition: classical AGPs and non-classical AGPs. Furthermore, structures similar to glycosylphosphatidylinositol (GPI) lipid anchor have been reported in the classical AGP type molecules.

Mesquite gum (MG) is known to be comprised of macromolecular complexes of type II AG and proteoglycans (arabinogalactan-proteins, AGPs) containing of 3-7% of protein.

The existence of lipid components and hydroxyproline and proline amino acids representative of AGPs, have been not documented for MG. This report is focused on the chemical and structural characterization of AGPs of mesquite (Prosopis spp). The obtained data corresponds to a physicochemical composition similar to that reported previously for MG. Lipid composition determination demonstrated the presence of C16:0, C18:0 and C22:0 fatty acids. With the EDX analysis, it was possible to detect traces of phosphorus that could be related to the GPI anchor. Additionally, alanine, serine, glycine, threonine and hydroxyproline were detected in MG; these amino acids are part of classical AGPs. The results suggest that the AGP of Sonorean mesquite can be considered classical AGPs.

Chapter 12 - Common bean (*Phaseolus vulgaris*) is a legume plant original from America. It was domesticated independently in the Mesoamerican and Andean geographical centers. From there, it was disseminated to the rest of the world, becoming the most important grain legume in terms of human alimentation. Beans from this plant are the most important source of proteins in the diet of undeveloped and developing countries, representing more than 50% of the grains consumed around the world. Common bean, as well as many other legumes, is able to establish a symbiotic association with soil bacteria, allowing atmospheric nitrogen to be incorporated into ecosystems and soils, having a crucial impact in agriculture. In the last years, many genetic tools have been developed for this species, including sequencing of the genome, developing of transformation and reverse genetic techniques, germoplasm resources and databases of expression libraries. The existence of two genetic pools allowed establishing an experimental model to study genetic affinity between legumes and rhizobia, contributing to understand the genetic bases of nitrogen fixation efficiency.

Chapter 13 - Legumes exhibit a variety of health benefits. Fenugreek seeds lower blood cholesterol and glucose levels. They are rich in fiber and saponins, and thus suppress accumulation of fat and upregulate the low density lipoprotein receptor. They also manifest anticancer and antidiabetic activities.Peanuts have beneficial effects on cholesterol level, hypertension, cancer, inflammation, oxidative stress, and vascular reactivity. Faba beans protect against DNA damage induced by hydrogen peroxide, inhibit the actions of pancreatic lipase and angiotensin converting enzyme, and suppress the proliferation of a variety of cancer cells besides exerting alpha-glucosidase activity. Adzuki bean demonstrate hydrogen peroxide, 2,2'-diphenyl-1-picrylhydrazyl, and superoxide anion scavenging activities. Kidney beans exhibit antimutagenic activity. Garden pea manifests angiotensin converting enzyme inhibitory and antioxidant activities. Mung beans have antidiabetic activity. Chickpea has an abundance of the essential fatty acids linoleic and oleic acids, vitamins and plant sterols. Bowman inhibitors possess anticarcinogenic properties.Soybean protein exerts a cardioprotective action and soybean phospholipids demonstrate hypolipidemic activity. Soybean is characterized by an abundance of phytoestrogens which alleviate postmenopausal symptoms and prevent osteoporosis and cardiovascular and malignant diseases. Although controversies regarding some of the aforementioned health benefits exist such as the anticancer action of isoflavones,in general consumption of legumes will promote health.

Chapter 14 - Legumes such as soybeans are staple diet for a significant proportion of the Asian population. In the past several decades, soybeans and its derived food products have become popular globally because of its protein, oil, and other bioactive phytochemicals content. This short communication summarizes recent advances in analytical methodologies used for the extraction and analysis of proteins and isoflavones from soybean and soy products.

In: Legumes
Editors: Hiroto Satou and Ren Nakamura
ISBN: 978-1-62808-280-7

Chapter 1

LEGUMES FOR GRAZING AND HEALTH: THE CASE OF *BITUMINARIA BITUMINOSA*

María Pazos-Navarro[1], Mercedes Dabauza[1], Enrique Correal[1], David J Walker[1], José A del Río[2], Ana Ortuño[2], Pilar Méndez[3], Arnoldo Santos[4], Segundo Ríos[5], Vanessa Martínez-Frances[5] and Daniel Real[6,7,8,9]

[1]Department of Natural Resources, IMIDA, Murcia, Spain
[2]Department of Plant Biology, Faculty of Biology, University of Murcia, Espinardo, Murcia, Spain
[3]Unidad de Producción Animal, Pastos y Forrajes Instituto Canario de Investigaciones Agrarias (ICIA), Tenerife, Islas Canarias, Spain
[4]Unidad de Botánica, ICIA, Tenerife, Islas Canarias, Spain
[5]Centro Iberoamericano de la Biodiversidad (CIBIO), Universidad de Alicante Carretera San Vicente del Raspeig, Alicante, Spain
[6]Department of Agriculture and Food Western Australia, WA, Australia
[7]Future Farm Industries Cooperative Research Centre, The University of Western Australia, Crawley, WA, Australia
[8]School of Plant Biology, The University of Western Australia, Crawley, WA, Australia
[9]Centre for Legumes in Mediterranean Agriculture, The University of Western Australia, Crawley, WA, Australia

ABSTRACT

Legumes have been used for centuries for different purposes as human food (grain legumes such as lentils, chickpeas, etc.), animal feed (pasture and forage legumes) and as sources of secondary metabolites with health, medical or nutraceutical benefits (pharmaceuticals, cosmetics and chemicals). Due to this relevance, some of these species are the subject of breeding programmes for selecting hybrids with desirable traits. To aid in these programmes, Plant Biotechnology offers multiple methods for conservation and commercial propagation of valuable plants for different objectives, such as the pharmaceutical industry or pasture production.

Bituminaria bituminosa (L.) C.H. Stirton (syn. *Psoralea bituminosa* L., *Fabaceae, Psoraleeae*; Stirton 1981) is a perennial legume widely distributed in the Mediterranean Basin and Macaronesia. The plants are grazed by small herbivores, but are mainly used for hay to feed milking goats, mostly in the Canary Islands. Moreover, *B. bituminosa* plants are a source of secondary metabolites of pharmaceutical and medicinal interest, such as furanocoumarins (FCs, psoralen and angelicin) and pterocarpans (bitucarpin A and B). So, their possible physiological functions and pharmacological properties are being analysed.

Our research group began to search for alternative perennial pasture legumes for low-rainfall Mediterranean environments at the end of the 1980's. *B. bituminosa* showed desirable traits (tolerance to cold and/or drought), that draw it as a good option. So, the first step was to select new germplasm with different traits to increase genetic variability for starting a breeding programme, with the most promising varieties, accessions and natural hybrids to deliver cultivars in the minimum time. To achieve this objective, conventional crossing between cultivars and wild accessions, and techniques for interspecific crossing breeding were carried out, in order to obtain new hybrid lines with a combination of desirable traits. Then, for a commercial approach, studies based on seed production of *B. bituminosa* selected cultivars have to be carried out and also research on the alternative uses of this species, based on its content of bioactive components with pharmaceutical potential.

As a result of this breeding programme, our research group has selected several hybrid and non-hybrid lines showing high levels of expression of, or a combination of desirable traits such as high forage quality, tolerance to biotic or abiotic stresses, or high contents of FCs. Therefore, the development of biotechnological tools such as micropropagation or *in vitro* plant regeneration protocols allowed the possibility of selecting somaclonal variants, obtaining pollen-derived haploid plants for optimising plant breeding programmes, and regenerating genetically transformed plants to control the biosynthesis of FCs. Moreover, these techniques offer attractive approaches for the large-scale propagation and conservation of germplasm resources of these plants.

In the present chapter, we describe this species, show its potential uses and the strategies that may be followed to carry out a traditional breeding programme in *B. bituminosa* and the tools developed to assist in it.

1. Introduction

The legume family (*Fabaceae*) grows in a wide range of agro-ecological conditions, and the plant morphologies vary from herbs to giant forest trees; including major grain legumes, oilseed, forage, ornamental and medicinal crops, and agro-forestry species (Lewis et al. 2005). This family has been divided into 3 major subfamilies, with unequal distribution: *Papilionoideae, Caesalpinioideae* and *Mimosoideae* (Singh et al. 2007).

Legumes are economically important in world trade because of their uses as human food and animal feed and in other commercial applications (pharmaceutical products, soap, paints, resins, coatings, cosmetics and chemicals among others) (Singh et al. 2007). The *Fabaceae* are second only to the *Gramineae* in agricultural importance. While cereals are an excellent source of energy in the form of carbohydrates, legumes are the major source of dietary protein for a large proportion of the human population. Legumes also provide essential amino acids and minerals required by humans (Grusak 2002). They are also important forages in temperate (e.g. alfalfa, clover) and tropical (*Stylosanthes* spp., *Desmodium* spp.) regions.

Legume biomass is also used as fodder, as cover crops and as green manure. A wide range of annual and perennial forage legumes are used in Mediterranean areas because of the different environmental conditions and farming systems. Annual (*Trifolium subterraneum* L.and *Medicago* spp.) or biennial (*Hedysarum coronarium* L.) legumes (Lombardi et al. 2001; Roggero et al. 2001) are established as pure or mixed swards to increase fodder production or the nutritive value of pastures. Perennial legumes provide numerous agro-environmental benefits due to continuous soil cover. In Mediterranean areas, summer drought favours plants with adaptations to protect meristematic tissues and for dehydration delay, dehydration tolerance and summer dormancy. Drought problems for legumes are likely to worsen with the projected rapid expansion of water-stressed areas of the world (Graham and Vance 2003).

There is an absence of drought-tolerant herbaceous perennial forage legume and herb options other than lucerne (*Medicago sativa* L.) for environments with Mediterranean-like climates, common in extensive areas of southern Australia, the Mediterranean basin and Chile (Dear et al. 2001; Li et al. 2008). Clearly, there is great deal of material available from dry areas in the Mediterranean basin and elsewhere. For example, there is a great diversity of perennial legumes in the southernmost parts of Italy; hence, in the future other semi-arid parts of the Mediterranean basin should be explored (Coks 2003). Alternative perennial legumes have the potential to complement and expand the feed base for grazing livestock and provide biologically-fixed N and an opportunity to control weeds in cropping rotations (Real et al. 2011). These novel species would assist in combating soil degradation attributable to deep drainage, by using more out-of-season rainfall compared with predominantly-winter-growing annual pastures and cropping systems, and be sufficiently persistent to incrementally dry the soil profile over successive years (Dear and Ewing 2008; Sandral et al. 2006). Although there is a great number of legume species available from dry areas in the Mediterranean basin (Cocks 2003), frequently, those which are adapted (Mediterranean flora) have not been selected and the selected commercial varieties are not adapted (Lorenzetti and Falcinelli 1987).

Among their multiple characteristics, legumes also synthesise secondary compounds that protect the plant against the attack of pathogens and pests (Dixon et al. 2002; Ndakidemi and Dakora 2003). These compounds also have health-promoting activities relevant to humans. So, the moderate consumption of legumes may help to prevent strokes, cardiovascular, Parkinson's, Alzheimer's and Huntington's diseases, liver ailments and even some forms of cancer (Singh 2005, 2007; Maurich et al. 2004, 2006). They have a blood cholesterol-reducing effect (Andersen et al. 1984) and are also included in the diet of insulin-dependent diabetics (Jenkins et al. 2003). Certain legumes, however, produce anti-nutritional factors, such as trypsin inhibitors and phyto-hemaglutinins (Gupta, 1987) and allergens, the latter being a severe problem in peanut (Spergel and Fiedler 2001). Genomics approaches, including metabolomics and proteomics, are essential to understanding the metabolic pathways that produce these health-promoting or anti-nutritional secondary compounds and for enhancing or eliminating, respectively, their production in the plant.

Our botany and agronomy research group (ICIA-Tenerife and IMIDA-Murcia) has worked on the screening of perennial legumes well adapted to dry Mediterranean conditions prevalent in the Canary Island and South-east mainland Spain agricultural areas. After screening for drought tolerance between native shrubby and herbaceous perennial legumes (species from genera *Chamaecytisus*, *Teline*, *Medicago*, *Colutea*, etc), we finally ended up selecting *Bituminaria bituminosa* (L.) C.H. Stirton, a multipurpose perennial legume widely distributed in the Mediterranean basin and Macaronesia (Correal et al. 1994, 2006; Méndez 1993; Méndez and Fernández 1992; Méndez et al. 2000; Pascual et al. 1991; Ríos et al. 1993, 1992; Sánchez et al. 1996). Thus, *B. bituminosa* is well adapted to high temperature and low rainfall and has a high retention of leaves when moisture stressed, therefore providing valuable feed over summer (Real et al. 2011; Real and Verbyla 2010). Other authors reported that *B. bituminosa* was different from other species evaluated; it showed a very high vigour both in spring and in summer. It has been shown to be a robust species that can survive in dry areas (Pagnotta et al. 2003). The *B. bituminosa* plants are grazed by cattle in the range of northern Israel (Sternberg et al. 2006) and used as hay to feed milking goats in the Canary Islands. In addition, it could be used for the phyto-stabilisation of former mining areas contaminated by heavy metals (Walker et al. 2007). Moreover, this species also synthesises bioactive compounds which have current and potential uses for modern medicine, such us pterocarpans and furanocoumarins (FCs, psoralen and angelicin) (Martínez et al. 2010).

Due to the large morphological and genetic diversity found in *B. bituminosa* and its potential relevance for Mediterranean pasture-crop systems and the pharmaceutical industry, our research group in Spain, in collaboration with Australian researchers, is carrying out a breeding programme and we have been able to select lines with high forage quality, tolerance to biotic or abiotic stresses or high contents of FCs. These plants are valuable for the pharmaceutical or livestock industries, which are interested in developing new or cheaper drugs or new drought-tolerant perennial legume cultivars for Mediterranean-like climates. To aid in this programme and to evaluate different agronomic traits (stress tolerance or FC content), hydroponic culture, controlled-freezing chamber and field studies have been carried out. We have evaluated the physiological response of *B. bituminosa* to heavy metals (Martínez-Fernández et al. 2011, 2012) and have related FC accumulation to indicators of stress (water relations and proline) (Walker et al. 2012). We have also developed *in vitro* culture systems (micropropagation and plant regeneration by organogenesis or somatic embryogenesis) and genomic tools for (i) mass propagation of selected genotypes (germplasm conservation, animal feeding purposes or large-scale production of FCs), (ii) production of synthetic seeds by encapsulation of breeding line somatic embryos, (iii) genetic engineering to study the FC biosynthesis pathway and (iv) genetic characterisation of the diversity within *B. bituminosa* breeding programmes (Pazos-Navarro et al. 2011, 2012, 2013)

The potential role of *B. bituminosa* in extensive grazing systems in Mediterranean climatic zones and as a source of pharmaceutical products will be discussed in this chapter.

2. Geographic Distribution and Taxonomy

2.1. Botanical History

B. bituminosa has been known from ancient times due to its bituminous odour and trifoliate leaves. It was mentioned for the first time in Herodotus's work, according to Ríos (present chapter). Later, it was described by Dioscórides (1st century AD) in his famous book

Materia Medica, according to the first copy known, the Codex Vindobonensis (Bizancio, 512 AD) – where it is mentioned as *Trifolium asphaltites.* In the same way, it was reported and described by several botanists from the XVI and XVII centuries, such as P.A. Mattioli (1544), A. Laguna (1555) and J. Gerard (1597, 1633), sometimes together with a drawing with flowers (Gerard) or without flowers (Mattioli, Laguna).

Likewise, it was described in several works of pre-Linnean botanists (XVII century and the first half of the XVIII century) such as Bauhin, Morison or Lobelius, most of them indicated by Linnaeus (Hortus Cliffortianus, 1753). When Linnaeus published his famous work *Species Plantarum* (1753), he described for the first time the genus *Psoralea* with 16 species, but these are currently, distributed in several genera: *Bituminaria, Cullen, Cyamopsis, Dalea, Indigofera, Otholobium, Pediomelum, Psoralea* and *Rhynchosia* (Jarvis 2007). Among the described species was *Psoralea bituminosa* currently included in the *Bituminaria* genus. This species had been previously mentioned, as *Trifolium*, under polynomic description, in the *Hortus Cliffortianus*. This was the first scientific description of *Psoralea bituminosa* according to the new nomenclatural system proposed by Linnaeus. In the same work (1753), the distribution of this species was mentioned as "Habitat in Siciliae, Italiae, Narbonae collibus maritimus", all belonging to the Mediterranean region.

The genus *Bituminaria* Heist, ex Fabricius (Enum., ed. 2: 165, 1759) is based on *Psoralea bituminosa* L., Sp. Pl. 763 (1753), lectotipified by Jafri in Jafri & El-Gadi, Fl. Libya 86: 39 preserved in the Linnaean herbarium (LINN) with the number N. 928.19, according to Jarvis (2007: 776). Medik (1787) established the genus *Aspalthium,* which is considered by Stirton as a synonym of *Bituminaria.* Several authors, such as Stirton or Lewis et al. (2005), enclose this accepted genus in the Family *Fabaceae*, subfamily *Papilionoideae* and Tribe *Psoraleeae*.

Stirton (Bothalia, 317-318, 1981) accepted the genus *Bituminaria* Heist. ex Fabricius (*Psoralea* pro parte), which is considered to be related to *Pediomelum* Rydb. from North America. He recognized two subgenera (*Bituminaria* and *Christevenia*), each of them with only one species, respectively: *B. bituminosa* (L.) C.H. Stirton based on *Psoralea bituminosa* L. and *Bituminaria acaulis* (Stev.) C.H. Sirton based on *Psoralea acaulis* Stev. apud Hoffm, in Comm. with type from "iberia occidentalis" [sic], West Georgia (Turkey). These two species are also the only ones recognized in the Flora Iberica by S. Talavera (Vol. VII(I), 1999), who mentions the paternity of the latter as *B. acaulis* Hoffm., a species present in central-western Asia, reaching Anatolia.

Stirton (Bothalia, 317-318, 1981) reported the extreme variability of *B. bituminosa*, suggesting that it should be investigated. He proposed the possible differentiation of *Psoralea morisiana* Pignatti & Metlesics, from Cerdeña (Italy), later accepted and combined as *Bituminaria* by Greuter (1986), as *B. morisiana* (Pignatti and Metlesics) Greuter.

S. Talavera (Vol. VI I(I), 1999), did not mention *B. morisiana*, although he reported the existence of populations in the Iberian Peninsula, with some differences that could be related to *Psoralea plumosa* described by Rch. in the Fl. Germ. Excursion 869 (1832), which must be a synonym of *Psoralea palaestina* Gouan, Ill. Observ. Bot.: 51 (1773) and, in any case, included in *B. bituminosa* variability.

Another species recently combined by Greuter (Willldenowia, 16 (1): 108, 1986) was *Psoralea flaccida* Nábelek, as *Bituminaria flaccida* (Nábelek) Greuter, described for Tunisia and also mentioned by Boulos (1999) for Egypt, but not mentioned in the Flora Iberica.

Rencently a new species (*B. basaltica*) has been described by Minissale et al. (2013) for the islet of Filicudi (Sicily, Italia).

2.2. Geographic Distribution

After Linnaeus, several works reported more information about the distribution of *B. bituminosa*, not only in European Mediterranean countries but also those of Mediterranean Africa (Flora Libyca, Flora of Morocco, Flora of Egypt,…) and the near East. The present known distribution for *B. bituminosa* includes not only Mediterranean areas, but also southern areas of temperate Europe and the northern border of Saharo-Arabian areas. It is not present in the northwest of the Iberian Peninsula but does grow in the Macaronesian islands of Madeira (Portugal) and the Canaries (Spain).

B. bituminosa, which is widely distributed, is a polymorphic taxon well adapted to different habitats, regarding both altitude and climatic conditions; thus, *B. bituminosa* is present between latitudes 44° and 29ºN and longitudes 15ºW and 40ºE; its presence in high-elevation areas may extend up to 1300 m in the Mediterranean basin, and in the Canary Islands and Morocco up to 1800-2000 m. For this reason several species, subspecies, varieties and forma have been described, most of them not accepted currently.

R. Maire in 1927 proposed the new variety *atropurpurea* (as *Psoralea bituminosa*), characterised by its dark-violet flower and located in the Moroccan High Atlas; although this species has not been recognized in flora catalogues of Morocco, such as Fenanne et al. (1999), Fennane and Tatou (2005) or Valdés et al. (2002), it may be because the intense colour of its petals is not exclusive and can be observed also in other localities along its distribution. Nevertheless, according to our own observations, it is possible that in other parts of Morocco, such as the Anti Atlas Mountains, well-differentiated cryptic taxa could exist, but have not yet been described, or were observed by Jahandiez and Maire (1934, 1941) but not recognized by the previously-mentioned authors.

Jahandiez and Maire (1934) not only accepted the var. *atropurpurea* (*Psoralea bituminosa*) but also the var. *latifolia* Moris reported at Chaouen (Rif) by Emberger and Maire. Later, these authors (1941) reported other varieties and forma in the Catalogue of the Moroccan Flora:

- *Psoralea bituminosa* var. *typica* Fiori f. *decumbens* S. et Ma., nom. nudum,
- *P. bituminosa* var. *angustifolia* (Guss.) Strobl. (*P. b.* f. *linearifolia* S. et Ma., nom nudum),
- *P. bituminosa* var. *rotundata* Maire and,
- *P. bituminosa* var. *laxa* Maire et Weiller.

It is not known if these taxa have been recognized and combined into *B. bituminosa*. R. Negré (1961), in his work about the little flora of arid regions of western Morocco, did not distinguish varieties within *P. bituminosa*. Jafri (1980), in his Flora of Libya, did not mention varieties for *P. bituminosa,* although he pointed out the variability of the studied material, indicating the high diversity not only for leaves but also for stipules, calyx, fruit beak, etc.; however, Ozenda (1991), did not report the presence of *P. bituminosa* in his “Flora and vegetation of Sahara" but only *Cullen plicata* as *Psoralea plicata*; however, none of the current flora accept the mentioned "infra-taxa".

For Egypt, L. Boulos (1991) listed two species: *B. bituminosa* and *B. flaccida*. He considered that the latter could be only an extreme variation of the former, having the terminal lobe of lowers leaves from 0.5 to 1.5 cm and floriferous heads with 3-8 flowers (10-25 for the typical plants).

2.3. Taxonomy

We are listing some of the taxa described as *Bituminaria* or *Psoralea* (belonging to *Bituminaria*) not only at the specific level but also the infraspecific one. Those in bold type have been accepted by different authors.

2.3.1. Mediterranean-South temperate Europe-East Asia

- ***Bituminaria acaulis*** (Steven) C.H. Stirton.
- *B. bituminosa* ssp. *pontica* (A. P. Khokhr.) Zernov (Northen Caucasus and Crimea). This subspecies was published as *Psoralea pontica* and combined later to the subspecific rank. If were to be recognized as a taxon, the rest of the infrataxa should be included in *B. bituminosa* ssp. *bituminosa*.
- ***B. bituminosa*** (L.) C.H. Stirton var. ***bituminosa.***
- *B. bituminosa* var. *atropurpurea* [as *Psoralea*] Maire (Morocco, High Atlas).
- *B. bituminosa* var. *brachycarpa* (Felmann) A. Danin, as *Psoralea*, presents small flowers and fruits. It is located in Israel and probably in Palestine.
- *B. bituminosa* var. *communis* (Webb) [*B.b.* var. *bituminosa*].
- *B. bituminosa* var. ***hulensis*** Felmann. It is located in wet areas of Hula Valley (Israel) and probably in Palestine.
- *B. bituminosa* var. *latifolia* Moris (1837-1859) [reported by Muñoz and Correal 1998].
- *B. bituminosa* var. *palestina* (Gouan) Ralf L. Jahn.; plants very hairy, located in Palestine.
- *B. bituminosa* var. *prostrata* Zoh.; prostrate plants located in Palestine.
- *B. bituminosa* f. *angustifolia* Grenier and Godron (1847-1850) (Muñoz and Correal 1998).
- Pau (1916), according to Coca (2003), divided the species (*P. bituminosa*) into two infraspecific taxa: f. *latebracteata* and f. *parvibracteata.* Fiori (1923-1929) proposed the var. *typica* from Italy. More recently, Arnot (1985) reported a partial revision without recognizing infraspecific taxa, probably due to the limited amount and diversity of the studied material.
- ***Bituminaria flaccida*** (Nábelek) Greuter, Willdenowia 16 (1) 108, 1986 [described from Tunisia according to the Index Kewensis as *Psoralea*] (in Muñoz and Correal 1998)
- ***Bituminaria morisiana*** (Pignatii & Metlesics) Greuter, syn.: *P. morisiana* Pignatii and Metlesics.
- *Psoralea taurica* Ledeb ex Nym. [*Bituminaria bituminosa* according to Stirton]

2.3.2. Macaronesian area

Madeira archipelago (Portugal)

- *B. bituminosa* (L.) C.H. Stirton var. *bituminosa*. No infraspecific taxa have been described from these islands.

Canary Islands (Spain)

- *B. bituminosa* (L.) C.H. Stirton var. *bituminosa*
- *B. bituminosa* var. ***crassiuscula*** (Méndez et al. (1991)
- *B. bituminosa* var. ***albomarginata*** (Méndez et al. (1991) [Cited by Webb & Berhtel., 1842, as var. *palaestina*]

Some investigations are currently in progress to differentiate between Canarian populations (var. *bituminosa)* and the Mediterranean ones. So, new names, not only for the common *B. bituminosa* in the Canary Islands but also for some populations with very narrow leaflets of the dry southern areas (especially on Tenerife Island), should be proposed. These considerations are supported by molecular studies such as those of Coca et al. (2003), Muñoz et al. (2000) and Pazos-Navarro et al. (2011). These studies point out genetic differences between populations of *B. bituminosa* from the western Mediterranean basin and those of Macaronesia – particularly those of the Canary Islands.

In general, the different subspecies, varieties and forma described for the *B. bituminosa* complex may belong to variations related to the typical variety (*bituminosa*). It is possible that some taxa, previously described, might have enough characteristics to be accepted as true taxa; however, further investigations on the species distribution area, particularly in the northwest African region, are necessary.

3. Genetic Variability Studies of *Bituminaria bituminosa*

Until recently, *B. bituminosa* was considered a 'genomic orphan' species (Varshney et al. 2009), with almost no genomic resources available. The few genetic studies developed for this species have shown genetic and morphological differences between *B. bituminosa* accessions from Mediterranean basin and Macaronesian locations.

Firstly, large morphological differences found between *B. bituminosa* accessions at different locations in the Canary Islands gave origin to the description of two new varieties, *albomarginata*, from very dry habitats, and *crassiuscula,* from 2000 m altitude (Méndez et al. 1990-91), beside the common variety *bituminosa*, which is present throughout the Canary Islands, and several ecotypes related to particular geographic locations of the Canary Islands.

Later, Muñoz et al. (2000), using PCR and morphological analysis, separated accessions of the Canary Islands from those of mainland Spain. Molecular analysis also identified differences within Canarian and peninsular accessions, the Canarian accessions being more diverse than those from mainland Spain. These initial results reflected the large intraspecific diversity of the species.

A second study by Juan et al. (2005), based on AFLP analysis, confirmed previous results published by Muñoz et al. (2000); thus, mainland Spain and Mediterranean populations of var. *bituminosa* were separated in one cluster, and populations of the Canary Islands in a second cluster containing the varieties *albomarginata*, *crassiuscula* and *bituminosa*.

Walker et al. (2006) determined the 2C nuclear DNA content of *B. bituminosa* for populations from the Canary Islands, mainland Spain and Sardinia, being in the range of 1.00 – 1.09 pg (2n = 2x = 20). Although no significant differences were found, there was true intraspecific variation in DNA content, related to the geographic origin of the population, being lower for the Canarian populations.

In a recent study using DNA sequencing technologies, Pazos-Navarro et al. (2011) established an efficient method to develop molecular markers (SSR) for *B. bituminosa*. The results showed clear differences between the three varieties (*albomarginata*, *bituminosa* and *crassiuscula*). Moreover, differences between Canarian and Mediterranean *bituminosa* accessions were observed and a new botanical division within this variety could be evaluated. Other authors had already pointed out these differences, based on genetic, morphological and physiological analyses (Muñoz et al. 2000; Coca et al. 2004). Additionally, results obtained by Pazos-Navarro et al. (2011) showed evidence of cross-pollination between the three varieties in the IMIDA-Murcia *B. bituminosa* breeding programmes.

After these genetic studies, *B. bituminosa* can no longer be considered a genomic orphan species, having now a large (albeit incomplete) repertoire of expressed gene sequences (432,306 mRNA molecules) that can serve as a resource for future genetic studies. Moreover, the expressed gene repertoire discovered in this work could be useful for follow-up experiments targeting biochemical pathways and/or important agronomic traits.

4. USES OF *BITUMINARIA BITUMINOSA*

4.1. Traditional Livestock Use in the Canary Islands

The presence of livestock in the Canary Islands has been known since pre-Hispanic times (Álvarez et al. 1977). Aboriginal people arrived with goats, sheep, pigs, dogs and cats, plus some cereal and legume species (Morales et al. 2009). Cattle were introduced after the Spanish conquered the islands, partly from Africa, but mostly from the Iberian Peninsula (Capote et al. 2004).

All the aboriginal island communities carried out intense livestock activity from the middle of the 1st millennium BD until the annexation by the Castilian Spanish (XV century), the herds of goats, sheep and pigs being one of the main contributors to the economy (Pérez 1963). The grazing was conducted throughout the whole of the Canary Islands, moving livestock from coastal areas to the top of the high-altitude areas according to the season of the year, so that an optimal utilisation of the grazing resources could be obtained along a vertical transhumance.

Tedera, the local name for *B. bituminosa*, has traditionally been used as a forage plant in the Canary Islands, where it grows spontaneously from sea level to over 2000 m, both in natural and altered environments, preferably in disturbed soils of waysides and roads but also in abandoned farmlands with slightly-compacted soils of poor quality. Tedera is usually not cultivated as a forage crop but collected from natural populations and offered to livestock as hay. The first known reference in the islands to the word “tedera” is from a seventeenth century document (Corrales and Corbella 2001). Subsequently, other authors cite this species as forage of interest in the islands (Pérez and Sagot 1867; Bolinaga 1915). Mr. Bolinaga, the Higher Gardener of La Orotava Botanic Garden (Tenerife Island) wrote a press note about the interest in cultivating tedera on the best soils, like alfalfa, but as a rainfed crop. He also indicated that many seeds were collected annually for this purpose, not specifying by who or where. There are also references to tedera seed collected from La Palma, ordered by the Agriculture and Industrial Plants Office of the United States Department of Agriculture (1928) and carried out by David Fairchild (1930) (Francisco Ortega et al. 2012). In his article, Fairchild refers to the importance of goats in the Canarian agriculture, tedera together with other endemic shrub legumes like tagasaste (*Chamaecytisus proliferus* ssp. *proliferus* var. *palmensis*) and gacia (*Teline stenopetala* ssp. *stenopetala*) being the main feed resource for small ruminants on La Palma Island. Currently, there are no data about the presence of tedera in archaeological excavations carried out in the Islands, as is the case for the ruderal flora, involuntarily introduced by the pre-Hispanic population, such as *Malva parviflora*, *Amaranthus* sp, *Solanum nigrum* and *Chenopodium murale*, which established as weeds on cropland. Though this does not mean that tedera was not present in the Islands before human settlement because archaeological research has focused mainly on human feeding (Morales et al. 2009). However, because grazing was one of the most-important activities of the aboriginal population and tedera is one of the few species remaining green during long periods of the summer and autumn, it is probable that it was used in pastures.

There are no data available on how tedera arrived to the Canary Islands, but its colonisation of each island could have occurred at different times; thus, considering that some of the oldest rocks in the islands are present in Fuerteventura –more than 20 million years-some populations could be very old (pers.com. A. Santos).

Several studies of the aboriginal language have proposed that the word "tedera" is of indigenous origin, deriving from the Berber language (Wölfel 1965). It is also interesting to note that the term Tarada (which sounds like Tedera) is the name given by Tuaregs to *Cullen plicatum* (syn. *Psoralea plicata*) (Benchelah et al. 2000), a species with some morphological similarities to *B. bituminosa*, and which is appreciated by camels, gazelles and mouflons. Guinea (1947) also refers to *Cullen plicatum* as excellent fodder, besides being used as a substitute for tea in central Sahara and Algeria. All these facts suggest that knowledge of this plant as a fodder species could have been brought along by the first inhabitants of the islands. Other opinions point out the Hispanic origin of the term (Alvar 1959; Llorente 1987); thus, considering that it is a bitumen plant and therefore could burn easily, we cannot discard the hypothesis that the plant name was a combination of two Spanish words, "tea" (resin-impregnated wood used to illuminate or to start a fire) and "asphalt" (bitumen).

More recent technical reports on the livestock development possibilities on the Islands (E.D.E.S. 1969, Ramos-Rodríguez and de la Puente-Jiménez 1983) considered tedera as an excellent fodder for ruminants, usually in the form of hay, but also grazed directly or mixed with cereal straw or other herbs. These reports also mention that on La Palma Island, tedera is sown in early autumn, with the arrival of the first rains, but with no tillage or control of seeding rates. In the 1990s, tedera was also used by the Nature Conservation Institute of Tenerife to control soil erosion and, although the project was abandoned, even today its effectiveness for controlling soil losses can be seen in some areas.

4.2. Current Research into the Use of *B. bituminosa* as Livestock Feed in the Canary Islands

In 1980s, the pasture and forage research team of the ICIA (Canarian Institute for Agricultural Research) started prospecting the variability of *B. bituminosa* and other forage legumes of interest present in the archipelago, and, at the same time, collecting germplasm and plant material of native pasture legumes (Guma et al. 2010). As a result of this, two endemic varieties of *B. bituminosa* were described: var. *albomarginata* and var. *crassiuscula*, which together with var. *bituminosa* are the three infraspecific taxa accepted for the moment (Méndez et al. 1990-91).

Populations of var. *albomarginata* thrive in environments having low rainfall (< 300 mm) but high relative humidity, because it is usually present in areas not far from the sea (300-600 m above sea level, on the potential areas of the thermophiles forest). The var. *bituminosa* is broadly distributed across the Canary Islands (350-800 mm), with significant differences between biotypes. The var. *crassiuscula* is adapted to annual precipitation of around 500 mm, part of it as snow. Its natural populations are located to 1800-2200 m above sea level on the potential areas of the pine forest. In the Canary Islands, tedera populations are adapted to a broad range of soils, from deep sandy to stony shallow soils; thus, it can be found on soils with pH from 4.7 to 8.5, textures from sandy to clay loam (according to USDA nomenclature -USDA, 1978-) and electrical conductivities of 0.5-4.6 dS m^{-1}. Mycorrhizal studies with the three varieties showed that var. *albomarginata* was the most dependent on the mycorrhizal symbiosis, but without significant differences from the other two varieties (Jaizme et al. 2001).

Agronomic trials to evaluate the forage production, nutritional value and plant-livestock interface of Canarian tederas focused mainly on populations of var. *bituminosa* (common-tedera) and var. *albomarginata* (albo-tedera). These were done using milking goats, initially measuring the quantity and quality of hay produced by common-tedera and albo-tedera in comparison with lucerne hay (Méndez et al. 2000); later, the seasonal grazing value and grazing recovery of the three Canarian tedera varieties were evaluated (Ventura et al. 2009; Méndez et al. 2011). Variety *crassiuscula* is considered an important genetic resource, but not a good forage species since it is less productive, less persistent and does not grow well in the warm climatic conditions prevailing in the majority of the Canary Island livestock farming areas. Work on sheep has been done by Western Australia (WA) colleagues and work with beef cattle by Israeli colleagues.

Because of its adaptation to dry climatic conditions, the Canarian tedera is a herbaceous perennial tolerant to drought and has a relatively-high net photosynthetic rate and, hence, good potential as a forage legume (Pn:14-18 μmol m^{-1} s^{-1}): var *albomarginata* has the highest water-use efficiency (WUE) of the three varieties, with a value close to 5 mmol CO_2 mol^{-1} H_2O (Méndez et al. 2000).

Dry matter yields of common-tedera cultivated during 4 years under irrigation in experimental plots at the ICIA (Tenerife) were respectively 7.5, 13.5, 7.7 and 5.5 Tm ha^{-1} of hay, with a mean water consumption of 1,600 m^3 ha^{-1} in a warm area receiving 350 mm of annual rainfall; the seeding rate was around 35 kg ha^{-1} (Méndez 2009; Méndez et al. 2008). The *albomarginata* pure variety took longer to yield enough biomass for a first cutting and so it could not be exploited until the second year, and it was cut once per year; hence, forage yields of *albomarginata* were about 50% lower than those of common-tedera; however, its edible biomass yield (leaves and stems with a diameter less than 5 mm) was 10 – 20% higher than common-tedera, and it needs less water to produce similar amounts of forage. Additionally, it was observed that natural hybrids with one *albomarginata* parent were more productive and persistent than common-tedera (unpublished data, P. Méndez).

When the forage yields (green matter, GM) of individual plants were compared, the highest values were for mainland Spain populations (2058 kg GM plant^{-1}) and for some Canarian tedera of var. *bituminosa* (1700-1900 kg GM plant^{-1}) (±30-35 % Dry Matter, DM), followed by *crassiuscula* accessions and other ecotypes of var. *bituminosa* (1100-1400 kg GM plant^{-1}), var. *albomarginata* and common-tedera (1000-1090 kg GM plant^{-1}) (±25% DM). The lower yields corresponded to ecotypes related to var. *bituminosa* and to some ecotypes of var. *albomarginata* (350-900 kg GM plant^{-1}).

The nutritive value of Canarian common-tedera, evaluated by Álvarez et al. (2004, 2007) and Ventura et al. (2004), gave the following results: Crude Protein content between 15-20%, *in vivo* Dry Matter Digestibility (DMD), Organic Matter Digestibility (OMD) and Crude Protein Digestibility (CPD) of 60.0, 61.6 and 51.4%, respectively, an energy content of 0.74 MFU (milk fodder units), high contents of Ca, P and K and a daily intake of 0.54 kg DM animal^{-1}, which corresponds to an intake of 31 DMD kg^{-1} W^{0,75}; the *in vitro* methods for Protein Ruminal Degradability give higher values (72.2%). A diet based on common-tedera hay had a positive influence on milk and cheese production of Canarian goat breeds (Álvarez et al. 2007; 2008).

During palatability trials with Canarian milking goats, the animals preferred common-tedera hay over alfalfa hay, and the amounts consumed of tedera were significantly higher than for alfalfa; furthermore, alfalfa yields under irrigation were 2.5-times higher than tedera yields, but alfalfa consumed 4-times more water than tedera (Méndez 2000). Moreover, the three Canary tedera varieties (*bituminosa*, *albomarginata* and *crassiuscula*) had crude protein and digestible energy contents slightly higher than a medium-quality alfalfa; however, when tedera was cut in summer and offered fresh to goats, it was rejected, but not if offered as hay (Ventura et al. 2000; 2009).

To study the effect of goat grazing, in a grazing trial (Méndez et al. 2011a) common-tedera had a good recovery when goats consumed less than 90 % of its edible biomass. The results suggest a use for tedera as a perennial legume for grazing that could maintain and even extend the grazing period in areas of semi-arid Mediterranean climate, it being possible to get milk yield and quality values similar to those obtained with alfalfa (Méndez et al. 2011b).

The furanocoumarins psoralen and angelicin are present in the leaves, stems and fruits of this species with the total concentration varying from 100 to 8000 mg kg^{-1} (Innocenti et al. 1997; Martínez et al. 2010; Méndez et al. 2001; Tava et al. 2007; Zobel et al. 1991); however, angelicin is not detected in the pure Canarian *albomarginata* variety from Lanzarote Island, or it is only present in traces in some individuals (less than 8 mg kg^{-1}) (Méndez and Fernández 1990, Méndez et al. 2001; 2006). The predominance of angular FCs (angelicin) could be related to phylogenetically-new taxa and to more-advanced biosynthetic pathways (Berembaum and Feeny 1981); this, if relevant, could indicate that var. *albomarginata* is the most-ancestral group.

4.3. Traditional Medicine

In the oldest medicinal plant treaties is described, quite accurately and in detail, a trifoliate and smelly legume called, by Herodoto of Halicarnaso (5th century BC), *Triphyllon* (Forster 1942; Rivera et al. 2012) and later, by Dioscórides (1st century AD), *Trifolium asphaltites*. These classical authors emphasized two unequivocal features of the plant: (1) the presence of trifoliate leaves and (2) the strong smell of asphalt or bitumen when it is collected.

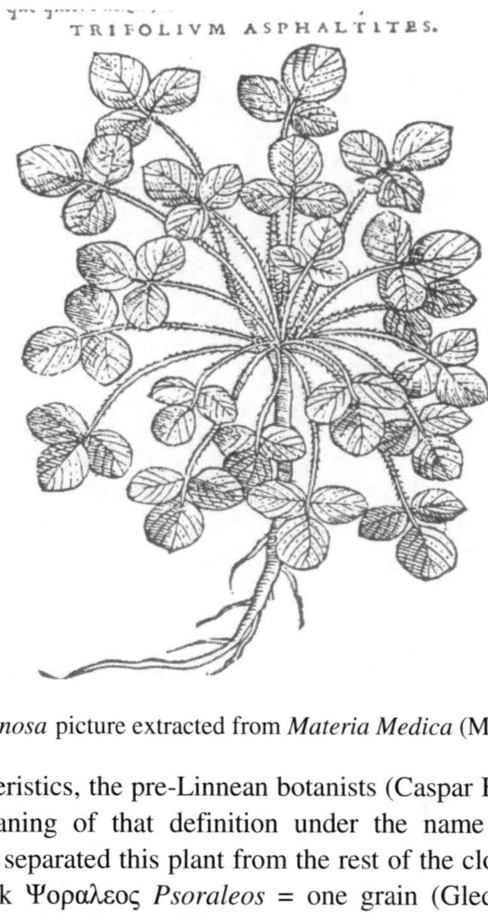

Figure 1. *Bituminaria bituminosa* picture extracted from *Materia Medica* (Mattioli 1544).

Just for these characteristics, the pre-Linnean botanists (Caspar Bauhin, Tournefort, Ray, etc.) maintained the meaning of that definition under the name *Trifolium bituminosum redolens*. Later, Linnaeus separated this plant from the rest of the clovers, creating the genus *Psoralea* (from the Greek Ψοραλεος *Psoraleos* = one grain (Gledhill 2008), keeping the reference to the asphalt smell in the specific epithet *bituminosa*. Finally, in relatively-recent times, Stirton (1981), because it has an indehiscent legume, segregated it into a new genus, *Bituminaria* (*bitumen, bituminis* = tarry, clammy, adhesive, smelling of tar (Gledhill 2008), thus highlighting doubly the smelly characteristic present in the Mediterranean strains of this species.

Knowledge of this medicinal plant has been accurate since the times of Dioscórides, which is not always common. There are many references confusing it with other medicinal species. The Italian translation of *Materia Medica* (Mattioli 1544) presents a picture in which appears, perfectly-drawn, a non-flowering rosette of *B. bituminosa* (Figure 1). In this translation and in later editions, such as the Castilian edition of Andres Laguna (1566), it appears next to other clovers under the name *Trifolium asphaltites*, always differentiated by its smell and medicinal virtues from other trifoliate "simples" (*Trifolium sensu stricto*).

From an ethnobotanical point of view, it is considered that the greater the number and types of popular names that a plant collects, the more traditional knowledge exists or existed about it (Rivera and Obon 1991).

In Spain, the tradition of medicinal plants is directly influenced by the great medical and pharmaceutical development achieved during the Muslim times in Al-Andalus (centuries VIII-XV). In the case that concerns us, this is shown by the many names under which this "simple" was compiled in the Spanish Moorish books preserved (Asín Palacios 1943; Bustamante et al. 2007). The relationship between Arabic, Mozarabic or Romance names and those registered in Castilian and Catalan ethnobotanical works (Table 1) is very evident, many of them being mere translations of the previous oldest. Among the names recorded, 60% of them describe one or two of the defining characteristics in *Bituminaria*: trifoliate leaves and bituminous odour (Table 1). Some names allure to the smell to goats or to its similarity to another plant which also has an intense odour, the rue (*Ruta graveolens*). There is also a coincidence in the medicinal uses between rue and *B. bituminosa*, especially those related to female physiology.

However, two of the most-common names in Spain, "angelote" (angel-like) and "cecinegra" (black-browed), have a different meaning, probably corresponding to a close observation of its corolla, which presents a black spot in the keel. This detailed popular knowledge highlights the importance of this medicinal plant. Similarly, another name, "contraruda" (Green et al. 2008), although not kept frequently, reveals the great value that our traditional medicine placed on *B. bituminosa* as an antidote.

Despite being a plant present in medicine since ancient times, its absence in modern Pharmacopoeias and Phytotherapy treatises is striking (Vanaclocha and Cañigueral 2003; Cañigueral pers.com.). This may be due to the phototoxicity caused by the furanocoumarins (FCs) present in the plant and, probably, the absence of pharmacological studies and clinical trials (that could support the many uses it had in the past), which exclude this plant from our modern medicines. But its enormous potential as a source of active compounds to cure different diseases is supported throughout history by its presence from Dioscórides to Hispano-Arab physicians (Bustamante et al. 2007) and in Arabic and Latin texts, which were propagated across Europe through translations into Italian (Mattioli 1544), Castilian (Laguna 1566) and other European languages. These medicinal uses remain in the Spanish tradition, as evidenced by over 30 medicinal uses reported in recent ethnobotanical works (Table 2).

Since ancient times, its more-referenced therapeutic uses are as vulnerary applications and for different processes related to women's health (menstruation, abortion and post-partum phase), although the diversity of therapeutic indications is very large (Table 2). Some less-frequent therapeutic indications cited since Dioscórides are for fluid retention or as treatment of certain nervous diseases (epilepsy, hysteria). There exists also a curious coincidence between the doctrine of Dioscórides and Ayurvedic medicine; thus, in Europe, the administration of seeds (beans) - crushed or swallowed - to combat fevers is advised, something that is currently done in India with seeds of *Psoralea corylifolia*. Recent studies have shown a significant antimicrobial activity of *B. bituminosa* flower extract against *Mycobacterium phlei* and a moderate activity against *Staphylococcus aureus* and *Candida albicans* (Miro, 1981). From ancient references, veterinary uses (mostly abortive) and some magical uses are reported also (Forster 1942; Rivera et al. 2012).

Table 1. Popular names of *Bituminaria bituminosa* acquired in different languages

Popular names of *Bituminaria bituminosa*		Emphasizes names							References
	English meaning	stinky smell	good smell	leaves morph.	medicinal uses	heliotropism	other analogies	unknown	
Names of the classics greco-roman tradition									
trifolium asphaltites	bituminous clover	x		x					Laguna (1566)
asphaltio	like asphalt	x							Laguna (1566)
triphyllon	three leaves			x					Rivera et al. (2012)
Names of Al-Andalous tradition (arabic and romances)									
buṭdah, budah, yarbah buṭdah	stinky, foul smell	x							Bustamante et al. (2007; 2011a; 2011b)
aṭrifulūn, ṭarīfulun	three leaves			x					Bustamante et al. (2007; 2011a; 2011b)
ğasdat harrān	small fig						x		Bustamante et al. (2007; 2011a; 2011b)
puṭda, puda, yerba puṭda	stinky, foul smell	x							Asín Palacios (1943)
ṭūrnah šūl	sunflower-like					x			Bustamante et al. (2007; 2011a; 2011b)
yarbah rubdah	stinky, rue-like smell	x							Bustamante et al. (2007; 2011a; 2011b)
šgîriya	unknown							x	Bellakhdar (1997)
Botanical prelinaean names									
trifolium bitumen redolens	bituminous clover	x		x					Tournefort (1700)
trifolium bituminosum	bituminous clover	x		x					James (1747)
Names in Spanish languages									
Castilian names									
angelote, angelota	angel-like						x		Font i Quer (1985); Reyes Prosper (1915); Lázaro Ibiza (1906); Obón and Rivera (1991); Rivera et al. (1994; 2008); Bellakhdar (1997); Cano and Ríos (1996)
cecinegra, cejinegra	black-browed						x		Font i Quer (1985); Reyes Prosper (1915); Obón and Rivera (1991);

Table 1. (Continued)

Popular names of *Bituminaria bituminosa*		Emphasizes names							References
	English meaning	stinky smell	good smell	leaves morph.	medicinal uses	heliotropism	other analogies	unknown	
									Rivera et al. (1994; 2008); Cano and Ríos (1996)
cecinegra, cejinegra	black-browed						x		Font i Quer (1985); Reyes Prosper (1915); Obón and Rivera (1991); Rivera et al. (1994; 2008); Cano and Ríos (1996)
contrarruda	antidote				x				Verde et al. (2008)
hierba cabruna, yerba cabruna, hierba cabrera	goat-scented herb	x							Font i Quer (1985); Reyes Prosper (1915); Lázaro Ibiza (1906); Bellakhdar (1997); Pascual (1978)
hierba de las quemaduras, hierba de los quemaos	heal burns herb				x				Fajardo et al. (2007); Verde et al. (2008)
hierba de los granos	pimple's herb				x				Molero et al. (1992)
hierba teresa	teresa herb							x	Verde et al. (2008)
higueruela	small fig						x		Font i Quer (1985); Rivera et al. (2008); Bellakhdar (1997)
pestosa, mata pestosa, hierba pestosa, apestosa	stinky, foul smell	x							Pers. Com. S.Ríos; Obón and Rivera (1991); Rivera et al. (1994; 2008); Cano and Ríos (1996)
ruda, rudón, ruda basta, ruda de las cabras	rue-like herb	x							Obón and Rivera (1991); Rivera et al. (1994; 2008); Verde et al. (2008)

Popular names of *Bituminaria bituminosa*		Emphasizes names							References
	English meaning	stinky smell	good smell	leaves morph.	medicinal uses	heliotropism	other analogies	unknown	
trébol bituminoso	bituminous clover	x		x					Font i Quer (1985); Reyes Prosper (1915)
trébol hediondo, trébol heliondo, helionda	bituminous clover	x		x					Font i Quer (1985); Reyes Prosper (1915); Lázaro Ibiza (1906); Bellakhdar (1997); Pascual (1978)
Catalan names									
cabruna, herba cabruna	goat-scented herb	x							Font i Quer (1985); Moll (2003)
camabruna, calambruna, calabrulla, cabrulles, cabruja	corruption derived from cabruna?						x		Font i Quer (1985); Parada et al. (2002)
girasol, girasol de marge, girasol bord	sunflower-like					x			Font i Quer (1985); Bellakhdar (1997); Pellicer (2000)
herba bruna	brown herb						x		Font i Quer (1985)
pedrenca	rocky soil ecology?						x		Font i Quer (1985); Bellakhdar (1997)
trèvol pudent, herba pudenta	stinky, foul smell	x							Font i Quer (1985); Bellakhdar (1997); Pellicer (2000)
Others European names									
psoralée bitumineuse (fr.*)	bituminous clover	x							Bonnier and Layens (1964)
trevullu malu, trivozu malu, truvullu malu (sard.*)	bad clover	x							Congia (1998)
trifoglio bituminoso (it.*)	bituminous clover	x		x					Congia (1998)
stinking trefoil (eng.*)	bituminous clover	x		x					James (1747)

*Note: (fr. French, sard. Sardinian Languages, it. Italian, eng. English).

Table 2. Medicinal uses of *Bituminaria bituminosa* reported by numerous authors

Medicinal uses	Plant parts used	Form of administration	References
Immune system, allergies			
anti-inflammatory	entire plant	infusion	Mulet (1991)
blood cleansing	entire plant	infusion	González-Tejero (1989)
Respiratory apparatus			
pleurisy	seeds/fruits (legumes)	decoction	James (1747)
Infectious and contagious diseases			
antiseptic	root	decoction	Mulet (1991); Villar et al. (1987)
antimicrobial	flowers	experimental data	Miró (1981)
Digestive system			
stimulant	fresh leaves	infusion	Bellakhdar (1997); Villar et al. (1987)
phlegmatic tumors and others	fresh leaves	cataplasm	Bustamante et al. (2004)
Kidney and urinary disorders			
urine retention	fresh leaves, seeds/fruits (legumes)	infusion, decoction	James (1747); Gómez de Ortega (1784); Laguna (1566)
dropsy	fresh leaves, seeds/fruits (legumes)	infusion, decoction	James (1747); Gómez de Ortega (1784); Laguna (1566)
Proctology			
antihemorroidal	fresh leaves, root	cataplasm	Mulet (1991)
Cardio-vascular system			
hemostasis	leaves and twigs	decoction	Mulet (1991)
aphrodisiac	seeds/fruits (legumes)	decoction	Alliota et al. (2003); Rivera et al. (2012)
Nervous system/fevers			
coral gout or epilepsy	seeds/fruits (legumes)	infusion	Fournier (1948); James (1747); Laguna (1566)
hysteria	seeds/fruits (legumes)	decoction	Gómez de Ortega (1784)
antispasmodic	seeds/fruits (legumes)	decoction	Fournier (1948);
palsy or paralysis	seeds/fruits (legumes)	pressed oil	Gómez de Ortega (1784)
chills	seeds/fruits (legumes)	milled and ingested	Dioscórides
antipyretic	seeds/fruits (legumes)	decoction	Alliota et al. (2003); Fournier (1948); Rivera et al. (2012)
Women´s diseases			
promote postpartum	seeds/fruits (legumes)	infusion	Alliota et al. (2003); James (1747); Rivera et al. (2012)
abortion	seeds/fruits (legumes)	infusion/ decoction	Alliota et al. (2003); Bellakhdar (1997); Rivera et al. (2012); (Villar et al. (1987); Verde et al. (2008)
provoke menstruation	seeds/fruits (legumes)	infusion/ decoction	Alliota et al. (2003); Fournier (1948); James (1747); Gómez de Ortega (1784); Laguna (1566); Rivera et al. (2012)

Medicinal uses	Plant parts used	Form of administration	References
Muscular system			
rib pain	leaves and fruits (legumes)	infusion	Gómez de Ortega (1784); Laguna (1566)
rheumatism	entire plant	decoction	Alliota et al. (2003); Rivera et al. (2012)
Skin problems			
treat wounds	flowers, leaves and twigs	decoction, unguent	Bellakhdar (1997); Cano and Ríos (1996); Font i Quer (1985); Mulet (1991); Obón and Rivera (1991); S. Ríos (Pers. Com.); Rivera et al. (1994; 2008)
burns	leaves fried in olive oil	unguent	Fajardo et al. (2007); Verde et al. (2008)
removal of verrucas	entire plant	cataplasm, powder	Bustamante et al. (2004)
treat sores	fresh leaves with salt	cataplasm	Bustamante et al. (2004)
Antidotes			
against snake and other venoms	entire plant	decoction	Dioscórides; James (1747); Gómez de Ortega (1784); Laguna (1566)
Other uses			
Veterinary			
sheep diseases	leaves and twigs	decoction	Verde et al. (2008)
Pastures and fodder crop			
forage	entire plant	fresh and dried (Heno)	Pellicer (2000); S. Ríos (Pers. Com.); Rivera et al. (2008)
Magical			
was used by the Magians	ash of entire plant	strewing over sacrificial victims	Forster (1942); Rivera et al. (2012)

The fruits or the whole plant are often used for medicinal purposes, being used less frequently the flowers or roots. For internal administration, the infusions or decoctions are the most-classical form of administration. Externally, poultices and baths with fresh plant material or powders have also been used, but ointments are less frequent (Table 2).

In conclusion, *B. bituminosa* presents a rich medicinal and pharmaceutical history, from the beginning of Western medicine until relatively-recent times, which has remained alive in the Mediterranean folk medicine and particularly in the Spanish tradition. Its absence from modern pharmacopoeias may be due to the phototoxicity of its FCs, or simply the lack of clinical trials that confirm its uses; however, its high content of FCs (psoralen and angelicin mainly) could ensure the return of this plant to the drug industry in the near future.

4.4. Potential Use for Modern Medicine

A study of the composition of *B. bituminosa* (Tava et al. 2007) revealed the presence of different terpenic compounds, such as isopentenyl alcohol, furanmethanol tetradecanol, hexadecanol and caryophyllene, phenolic compounds (vanillic acid ethyl ester, plicatin B, etc.) and furanocoumarins such as angelicin and psoralen. Other studies (Maurich et al. 2006;

Pistelli et al. 2003) have shown the presence of the pterocarpans, erybraedin C and bitucarpin A.

4.4.1. Furanocoumarins

Furanocoumarins are a sub-group of phenolic compounds included in the coumarins group. Coumarins are lactones with the basic structure of 1,2-benzopyrone and an oxygenation at the C-7 position. The FCs can be grouped into the linear type, where the furan ring is attached at carbons 6 and 7, and angular type, where this ring is attached at carbons 7 and 8 of the coumarin structure, and these are sub-divided into linear FCs (psoralens), such as psoralen (1), and angular ones (angelicins) like angelicin (2) (Figure 2). Linear FCs have been identified in a great variety of plant families, and the highest concentrations have been found in the *Umbelliferae* (*Apiaceae*), *Rutaceae*, *Leguminosae* (*Fabaceae*) and *Moraceae*. The angular FCs are distributed less widely and confined principally to the *Apiaceae* and *Leguminosae* (Bourgaud et al. 1989). In the case of *B. bituminosa,* different authors have described the presence of FCs, among them psoralen and angelicin (Innocenti et al. 1991, 1997; Martínez-Fernández et al. 2011; Méndez et al. 2001; Tava et al. 2007; Walker et al. 2012).

4.4.2. Therapeutic Activity of furanocoumarins

The FCs have been used in folk medicine for a long time. The Indian sacred book "Athara Veda" describes the treatment of leukoderma (vitiligo) using a poultice from a plant now known as *Psoralea corylifolia,* and the ancient Egyptians used *Ammi majus* for the same disorder. The first FC, 5-methoxypsoralen (5-MOP), was isolated in 1838 by Kalbrunner from bergamot oil.

These compounds are a therapeutically-important sub-type and have various clinical applications. The most-outstanding property of FCs is their great ability to sensitise cells to visible light, sunlight and, especially, near-ultraviolet (UV) light. This results in strong toxicity, mutagenicity and possibly carcinogenicity. The mechanism of action is well known. After intercalation into the double helix of the DNA and molecular complexing, the light-activated FCs react with the pyrimidine bases, especially thymine. The FCs are typical phototoxic compounds that produce photodermatitis in combination with UV light exposure, as well as cytotoxic and mutagenic disorders (Gupta and Anderson 1987; Stern 2007). At the molecular level, FCs bind to cellular constituents such as proteins and lipids, damaging lysosomes, leading to the formation of reactive oxygen species and contributing to the formation of novel antigens by covalent modification of proteins. The FCs are well known for interfering with drug metabolism, in particular with cytochromes P450 (CYP). Since the discovery of FCs and other structurally-related biomolecules, extensive research on different aspects of therapeutic interest has been carried out due to the singular behaviour of these compounds when irradiated with UV light (Curini et al. 2006). These properties have made possible the use of these molecules in PUVA therapy (psoralen plus UV-A radiation) or extracorporeal photopheresis for the treatment of many skin diseases such as vitiligo or psoriasis, and also for several autoimmune diseases (systemic lupus erythematosus, Crohn's disease, type 1 diabetes mellitus or multiple sclerosis) as well as for the treatment of cutaneous T-cell lymphoma, solid organ transplant rejection and graft versus host disease. Importantly, these compounds - alone or in combination with other drugs - represent promising candidates to develop new therapies or improve the existing ones (Hönigsmann et al. 2008; Scarisbrick et al. 2008).

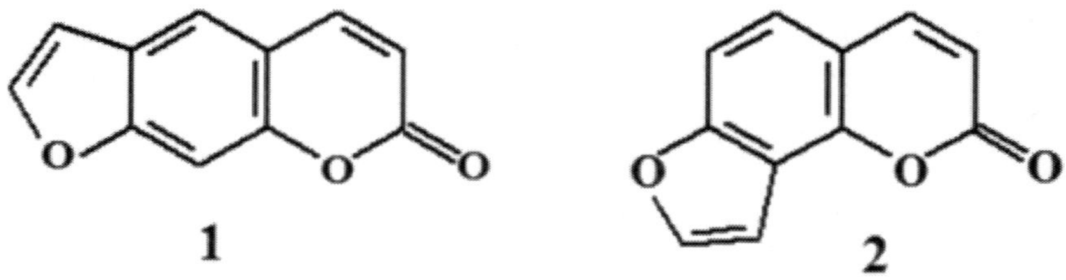

Figure 2. Structures of furanocoumarins present in *B. bituminosa*: psoralen (1) and angelicin (2).

4.4.3. Parts Used for Medicinal Purposes

Although FCs are found in all plant parts, the maximum concentration is found in the fruits (legumes containing one seed), buds and leaves, especially in the younger leaves, since FC synthesis is thought to occur here and in the outer skin of the plant. Factors that can bring about an increase in the concentration of FCs in various plant structures include seasonal changes, with FC levels increasing in the summer; disease-causing pathogens; mechanical damage to plants, including air pollutants and other environmental conditions; and plant size (Hagemeier et al. 1999; Hamerski and Matern 1998; Vasconsuelo and Boland 2007).

4.4.4. Furanocoumarins as phytoalexins and allelochemicals

The FCs are recognised as potent phytoalexins. Most plants accumulating these compounds possess a highly-inducible biosynthetic pathway that can be triggered by various biotic and abiotic stresses. Since FCs are strongly phototoxic, their presence in plants has been demonstrated to be a protective mechanism against phytopathogenic microorganisms and herbivores. So, when roots of *Glehnia littoralis* (*Apiaceae*) were inoculated with *Pseudomonas cichorii* the production of four linear FCs (psoralen, xanthotoxin, bergapten and demethylsuberosin) was induced as phytoalexins. Cells of *Petroselinum crispum* (*Apiaceae*) treated with elicitors derived from different species of the genus *Phytophthora* expressed defence-associated genes and the production of FC phytoalexins. Similar results were observed in *A. majus* cells treated with fungus elicitors (Heathpagliuso et al. 1992; Masuda et al. 1998).

Many phototoxic plants grow preferentially in sunlit habitats and hence maximise the exposure of herbivores to photoactivating wavelengths. Plant chemistry can have deleterious effects on insect parasitoids, including reduction in body size, increased development time and increased mortality. The linear FCs psoralen, 5-MOP (bergapten) and 8-methoxypsoralen (8-MOP, xanthotoxin) have considerable biological activity against herbivores, including humans. Increased dietary concentrations of each FC significantly decreased insect larval weight, extended generation time and induced higher mortality. Xanthotoxin was the most toxic, followed by psoralen and bergapten. At low concentrations they have proven deadly to many insect species, thus providing significant protection against insect herbivores. Even arthropods that have developed some tolerance to these compounds can show reduced performance upon ingestion, including delayed development that may be associated with antifeeding responses and reduced survival (Li et al. 2007; Stevenson et al. 2003).

5. BREEDING

The breeding technique for *B. bituminosa* was established in 2007 by Real et al. which - combined with the molecular tools developed by Pazos-Navarro et al. (2011) - permitted the identification of hybrids, either natural or the result of hand-crosses between elite material.

The genetic variability already observed in *B. bituminosa* can be summarised in some of the following desirable plant traits: a) a broad soil and climatic adaptation and hence, a large potential seed market for commercial varieties; b) a wide seed yield range to select lines with high seed production, retention and germination (Correal et al. 2008); differences found were mostly of genetic origin, but biotic (pollinators, ants, birds) and environmental factors (rain and temperatures) should be considered; c) a large variability in its FC content (psoralen and angelicin), with higher levels in var. *bituminosa* than in var. *albomarginata,* makes it an ideal source for the extraction of these compounds for pharmaceutical purposes (Del Río et al. 2010), and also for screening plant material with lower FC contents for animal production.

Due to the potential advantages of *B. bituminosa* for Mediterranean-like areas of southern Australia, an agreement led by Dr. D. Real was established on July 2008 for the research, breeding and commercialisation of *B. bituminosa* between three Spanish institutions (ICIA-Tenerife, IMIDA-Murcia, CIBIO-Alicante) and the national research consortium Future Farm Industries Co-operative Research Centre (FFI CRC) of Australia. This agreement benefited from the previous breeding experience of the IMIDA-Murcia and ICIA-Tenerife with *B. bituminosa*, providing native accessions and natural hybrids for the start of an Australian breeding programme on tedera, to cross the best parent plants in order to combine elite material into one or several cultivars, selecting in the breeding process homozygous plants with the combination of desirable characters.

Since the agreement was signed, the Spanish team has done work on the following topics: germplasm characterisation; drought, cold and heavy metals tolerance; *in vitro* propagation; molecular markers, root nodule bacteria and breeding for high secondary compounds content for pharmaceutical purposes.

To simplify the description of the different *B. bituminosa* varieties, we shall use the following common names:

a. *B. bituminosa* var. *albomarginata* (Canary Islands): albo tedera
b. *B. bituminosa* var. *crassiuscula* (Canary Islands): Teide tedera
c. *B. bituminosa* var. *bituminosa* (Canary Islands): tedera
d. *B. bituminosa* var. *bituminosa* (Mediterranean basin): Mediterranean tedera.

Two Spanish research teams - ICIA-Tenerife and IMIDA-Murcia - started in the early 1980s the work on characterisation, screening and breeding of tedera; in the 2000s, a third team -CIBIO-Alicante- joined also, and from this work important results were obtained:

- Accessions of tedera from the Mediterranean basin are generally of low forage value and persist mostly for two years under the conditions of the southeast of mainland Spain.

- Accessions of tedera from the Canary Islands are generally more productive, of better forage quality and perennial, but they do not tolerate the low winter temperatures prevalent at high altitude in Mediterranean environments.
- A future research objective should be to produce hybrids of higher yield, quality, persistence and tolerance to cold winters.

In order to differentiate between Canarian and Mediterranean accessions, a large number of morphological and behavioural characteristics were used, such as habitat, growth habits, persistence and stress tolerance. Some of them are summarised as follows:

5.1. Mediterranean Tedera Populations (var. *bituminosa*)

- They grow in autumn-winter with a rosette of leaves; some accessions tolerate cold winters (with absolute minimum temperatures as low as -10ºC).
- They have long stems and narrow leaves during flowering at the end of spring and they shed leaves in summer.
- They are present mostly in alkaline soils but also in acid soils.
- They are present in environments with mean rainfall ranging from 300-1000mm.
- They are biennial (short lived), but exceptionally, 1-5% of perennial plants were found in certain populations like Calnegre and Llano Beal, both accessions from Murcia.

5.2. Canarian Tedera Populations.

5.2.1. Tedera or common tedera (var. bituminosa)

- It is traditionally harvested from native stands and consumed as hay by milking goats.
- It is present in alkaline and acid soils and in areas with 300-1000mm of mean annual rainfall.

5.2.2. Teide tedera (var. crassiuscula)

- It grows at high altitudes (3ºC minimum in coldest months of growth) in sub-humid habitats (500-700 mm rainfall), but becomes dormant in winter, escaping cold temperatures.
- It does not survive at 1000 m in NW Murcia (1-3ºC minimum in coldest month) (Muñoz and Correal 1999).
- It presents low perennity when cultivated in Tenerife and Murcia.

5.2.3. Albo tedera (var. albomarginata)

A drought-tolerant plant; “green” in summer; native to coastal, semi-arid habitats having 150-300 mm annual rainfall

- It is a dwarf shrub presenting a cushion growth habit, a high leaf/shoot ratio, woody basal stems and high perennity.
- It is only native to Lanzarote but related populations are present also in Fuerteventura, Gran Canaria, Tenerife and La Gomera in the geologically oldest areas.
- It does not tolerate frost.

On Tenerife Island, where one ecotype close to albo-tedera is present in dry coastal areas, common tedera in intermediate-altitude zones and Teide-tedera in high-altitude areas, P. Méndez compared the agronomic performance of populations of the three varieties at the ICIA field station and observed vigorous tedera plants growing between plant rows in the second year of cultivation. Additionally, the first tedera cultivar from Tenerife, introduced into Murcia in the early 1980s, performed as a vigorous hybrid when compared to lower-yielding Mediterranean populations of peninsular Spain.

Something similar to what happened in Tenerife was observed at the IMIDA-Murcia after cultivating populations of the three tedera varieties during several years (in the 1990s and 2000s); however, in the case of Murcia, Mediterranean populations were present and new, natural hybrids between Canarian and Mediterranean tedera were identified. Natural hybrids were recognised by the presence of a recombination of morphological and physiological traits (vigour, perennity and tolerance to abiotic stresses) from the different varieties co-existing in the same field station of La Alberca.

From breeding work carried out during the 2000s (Juan et al. 2003; 2004; Real et al. 2008), more results were obtained and can be summarised as follows:

- *B. bituminosa* is an inbreeding species that also outcrosses.
- The three tedera varieties produce self-pollinated seed, but seed production is increased when pollinators visit the flowers.
- Environmental conditions and pollinators have an effect on legume production.

In addition, the Spanish teams worked during the 2000s on screening and selection for specific plant traits such as (1) tolerance to cold winters and dry summers, (2) growth adaptation to mild winters and long, dry summers and (3) high pharmaceutical compounds production. In this sense, several interesting results were obtained. The different botanical varieties have differing frost tolerance. The least tolerant is albo-tedera (var. *albomarginata*) that is burnt by mild frost of short duration; however, the plant is often able to recover. The Teide-tedera (var. *crassiuscula*) is present at the top of the Cañadas area, growing at altitudes of 2000 m above sea level close to Teide Mountain in Tenerife, and some populations are under snow for several weeks each year. The Mediterranean tedera (var. *bituminosa*) from high-altitude sites is the most frost-tolerant, being able to survive through severe winters, as was the case of populations La Perdiz (hybrid P6-E7P-F23A12) and Llano del Beal (hybrid P3-E7P-F28A15) from Murcia and Mikropolis from Greece, during a field trial carried out in El Chaparral-Cehegín (Murcia), at 800 m altitude, during the years 2009-2011, with minimum temperatures between -4 and -7°C. Additionally, natural hybrids between Mediterranean tedera and Canarian tederas (albo-tedera, Teide-tedera and common tedera) were also tested and some were frost-tolerant; however, controls of Canarian pure albo-tedera and common tedera were frost-sensitive, as was seen during their first introduction into peninsular Spain in the 1980s and 1990s. Some morphological characters were observed in the cold-tolerant accessions evaluated at El Chaparral-Cehegín, such as leaves with intense green colour and sunken leaf nerves. Currently, two crossing progenies of potential cold-tolerant breeding lines are being evaluated: one between Spanish Mediterranean tedera accessions La Perdiz x Llano del Beal, and another between Greek accession Mikropolis x La Palma Island common tedera accession.

Tedera varieties differ also in their drought tolerance and associated mechanisms, albo-tedera being the most drought-tolerant and able to remain green during the summer with minimal leaf drop. The other tedera varieties are also drought-tolerant but they tend to drop more leaves, in order to reduce evapo-transpiration, under severe drought conditions (Correal et al. 2003). In a field trial at Purias-Lorca (Murcia), the tolerance to long, dry summers of 20 tedera breeding lines was tested. Accessions retaining a good proportion of leaves during the summer were selected; two were natural hybrids of Mediterranean tedera populations -Llano del Beal and Calnegre- from Murcia with albo-tedera (P1-E7P-F28A3 and P7-E33OMV-F19A2), and a third one was a hybrid between albo-tedera and Mediterranean tedera (P5-E7P-F45A3). In all cases, summer drought tolerance was related with an albo-tedera parent (Famara-Lanzarote) and close ecotypes (Vinamar-Fuerteventura, Teno-Tenerife), which could be detected in the hybrid by the presence of morphological traits like woody stems, prostrate growth and perennial habit.

Advanced plant material with high psoralen and angelicin contents (from progenies F5-F6) has also been obtained, because of the pharmaceutical potential of these bioactive components. Analytical work was carried out to check the contents of FCs in accessions and elite material, the aim being to develop new cultivars as pasture for grazing animals with low content of these metabolites due to their possible anti-nutritive effect.

On the other hand, we have also evaluated other traits such as seed production, retention and germination rate. From these studies, several results were obtained:

- Canarian tedera populations like Famara (albo-tedera) and Boca de Tauce (Teide-tedera) retain legumes for longer than mainland-Spanish populations (Mediterranean tedera), which shed legumes when the stems are still green.
- Ants harvest large amounts of pods during late spring and early summer.
- Mean pod weight for 10 populations was 2.6 g/100 pods, and the mean weight of one pod was 26 mg.
- Weight of 100 legumes ranged from 1 to 6 g, the highest being for accession Galilea from Israel, followed by Canarian varieties (2.5-3.0 g).
- Seed production of high-FC clones at IMIDA-Murcia La Alberca field station was:
    - Clone P6A6 (psoralen): in 2011 (1st year) 90 g plant^{-1}; in 2012 (2nd year) 198 g plant^{-1}; seed production ranged from 137- 270 g plant^{-1}.
    - Clone B4P11R6 (angelicin): in 2011 (1st year) 40 g plant^{-1}; in 2012 (2nd year) 109 g plant^{-1}; seed production ranged from 52 – 255 g plant^{-1}.
- Seeds germinate better after scarification - 87% for scarified vs. 21% for non-scarified - and at constant temperatures.

Until 2007, when Real et al. developed a hand-crossing technique, we had to rely on natural hybrids (F2), which was a rather-inefficient and slow process because we had to follow a pedigree selection process in which the best-performing individuals of the 25-30 plants of an F3 progeny were selected by particular screening criteria, and planted again in families of plant rows in an F4 and so on. However, since 2008, we have used hand crossing to recombine plant traits of some of the best breeding material, such as the already-mentioned cold-tolerant breeding lines.

Thus, our breeding programme has consisted mostly of following natural crossings with hybrid vigour, progeny selection and recurrent selection of successive plant progenies, and ultimately, in hand-crossings between prominent tedera varieties and accessions, plus progeny selection and breeding according to particular plant traits. Beside the already-mentioned desirable tedera plant traits like persistence and high forage yield under environmental stresses such as drought and cold, we did some screening for high legume-seed retention, high seed production and high seed germination.

No toxicity has been reported for *B. bituminosa*. There are only reports that some accessions from the Mediterranean basin are unpalatable due to their smell. Also, goats reject the plants when the FC content is higher than 0.5-1.0 %; this, together with the rapid excretion, prevents toxicity problems (Ramos et al. 1998). After the ingestion by goats of 10 mg kg^{-1} xanthotoxine (8-methoxypsoralen), rapid metabolism and excretion were observed. So, human toxicity resulting from the ingestion of the milk or its derivates is improbable (Ivie 1987; Pangilinan et al. 1992). Condensed tannins expressed as catechin equivalents ranged from 0.1% for var. *albomarginata* to 0.5% for var. *bituminosa*, and total phenols expressed as tannic acid equivalents ranged from 0.9% for var. *albomarginata* to 1.4% for var. *bituminosa* (Ventura et al. 2012).

As a result of this breeding programme, our research group has selected several breeding lines showing high forage yield, tolerance to abiotic stresses or high FCs content. Some of the results obtained in this breeding programme may be compared with those obtained by ICIA-Tenerife or WA: for example, tedera (1) is able to grow all-year-round if not limited due to extreme drought during summer or very-low temperatures during winter, (2) forage yields (Méndez et al. 2006) ranged from 400-2000 g GM $plant^{-1}$ (20-25% Dry Matter content) in experiments carried out at ICIA-Tenerife with the three tedera varieties, the more-productive accessions being Mediterranean tedera from mainland Spain (e.g. Llano del Beal) and tederas from the Canary Islands (e.g. Tenteniguada), (3) forage yields ranged from 200-450 g DM $plant^{-1}$ in experiments carried out at IMIDA-Murcia (Correal et al. 2003), Mediterranean accessions also being more productive. Moreover, (4) in both locations (Tenerife and Murcia) natural hybrids between accessions of Mediterranean tedera and Canarian tederas gave forage yields of 1-4 kg DM $plant^{-1}$, that is, 5-8 times higher than those obtained from native accessions. Finally, (5) similar results (3 Mg DM ha^{-1}) were obtained with hybrids in WA, where large-scale tedera trials are being carried out to evaluate its productivity under sheep grazing.

6. Tools to Assist the *Bituminaria bituminosa* Breeding Programme

Once breeding lines have been selected, it is necessary to develop tools in order to assist in germplasm conservation, evaluation of tolerance to stresses and FC content evaluation.

Plant biotechnology, in its broadest sense, offers multiple methods, including plant tissue culture, hydroponic culture, cell biology and molecular biology, for the evaluation, germplasm conservation and commercial propagation of selected breeding lines valuable for the pharmaceutical industry (large-scale production of interesting secondary metabolites), or for pasture production (Wink 1999).

6.1. Hydroponic Culture

The cultivation of plants in hydroponic culture (i.e., in aerated nutrient solution) is a useful tool for the study of the effect of external factors, individually or in combination, on all aspects of plant growth: net biomass production, physiology, biochemistry and anatomy (Figure 3). It also gives rise to clean roots, which can be analysed easily. We have employed it to study the response of *B. bituminosa* to external stresses: salinity, waterlogging (hypoxia), water deprivation, heavy metals and elevated temperature.

Teakle and Real (2010) grew *B. bituminosa* accessions in saline (200 mM NaCl) and/or stagnant nutrient solution, to simulate the conditions in waterlogged, saline soils of pastures in southern Australia. After 27 days of exposure to these conditions, root and shoot growth declined (relative to control conditions of non-saline, aerated solution) in the order: stagnant, non-saline solution > aerated, saline solution > stagnant, saline solution. There was variation among the accessions; for example, the accession most sensitive to "waterlogging" was the only one which did not develop root aerenchyma. The results indicate that it will be possible to select elite plants with elevated tolerance of transient waterlogging in saline soils. Unpublished, preliminary work showed that exposure of *B. bituminosa* to moderate salinity (90 mM NaCl) produced increases of up to 30% in the tissue levels of the FCs, angelicin and psoralen.

Hydroponic assays in which polyethylene glycol (PEG) was included in the nutrient solution in order to lower its water potential were performed in order to examine the physiological response of *B. bituminosa* to water deprivation. They showed that, as expected, the root hydraulic conductivity of water decreased in the presence of PEG while the root:shoot biomass ratio increased; this is a common phenomenon which minimises the water loss/uptake ratio. The two accessions analysed did, however, differ in their accumulation of proline (an indicator of water deprivation) and the results - when combined with those of pot and field assays - gave valuable insights into the drought tolerance of this species (Martínez-Fernández et al. 2012).

The effects on plants of heavy metals present in nutrient solution, singly or in combination, can be studied in detail, since commercially-available computer programs permit the chemical speciation of the metals to be determined. Thus, Martínez-Fernández et al. (2011) studied the effects of high nutrient solution zinc (Zn) concentrations on the growth and physiology of two contrasting populations of *B. bituminosa* (one, "C2", from a Zn-contaminated site and another from a "clean" site). Although the growth of the plants of population C2 was more Zn-resistant, the two populations had similar tissue levels of Zn and organic acids (which may detoxify intracellular Zn) as well as similar water relations. These authors also found that exposure to elevated cadmium (Cd), copper (Cu) or Zn disrupted plant water relations and increased the tissue accumulation of angelicin and psoralen, which protect stressed plants against infection and feeding.

Hydroponic culture of plants does have certain drawbacks. For example, the size and age of plants that can be grown are usually limited. Walker et al. (2012) found that plants of *B. bituminosa* grown hydroponically in the vegetative stage increased their leaf FC accumulation when the day/night temperature regime was changed from 22/17 ºC to 33/22 ºC. However, in the field, mature plants from the same accessions had lower leaf FC levels in early summer (mean maximum temperature of 30 ºC) than in the autumn (20 ºC), attributable to greater FC accumulation in seeds and roots in early summer, where their protective effects help to ensure the survival of the population or individual plant.

6.2. *In vitro* Plant Tissue Culture

The development of biotechnological tools by *in vitro* tissue culture systems, initiated from cells, tissues and organs of *B. bituminosa*, offers an attractive approach for industrial purposes: large-scale propagation, conservation of germplasm, the possibility of selecting somaclonal variants, obtaining anther or microspore-derived haploid plants and regenerating genetically-transformed plants to control the biosynthesis of FCs, and other bioactive compounds in a short span of time, for industrial uses.

Since no previous work on *in vitro* culture had been carried out with this species, we firstly established micropropagation protocols for several clones of tedera and inter-varietal hybrids resulting from crosses among var. *bituminosa* (either Mediterranean or Canarian) and the other two varieties (*albomarginata* and *crassiuscula*) selected in the breeding programmes for their potential as forage or as sources of FCs. By using apical and nodal explants cultured on a modified Murashige and Skoog (MS) medium (1962) with organic additives and plant growth regulators (PGRs), differences among genotypes were observed not only in the frequency of sprouted shoots but also in the rooting frequency of shoots. The micropropagated plants were established *ex vitro* in the field and the FC levels were equal to or even higher than those in the non-micropropagated mother plants. Moreover, they maintained the original characteristics of the mother plants, showing that the micropropagation protocols established are a good method for the multiplication of selected plants (Pazos-Navarro et al. 2012).

Plant regeneration from explants by *de novo* shoot or embryo formation via organogenesis (Pazos-Navarro et al. 2013) or somatic embryogenesis (unpublished data) has been evaluated. A high frequency of plant regeneration via organogenesis using different explants of tedera has been achieved (Pazos-Navarro et al. 2013). Three explant types from *in vitro* plants were used: leaflets, petiole-leaf-attachment (PLA) and petioles (Figure 3). For callus induction we use a modified MS medium supplemented with organic additives and different concentrations and combinations of plant growth regulators (PGRs): Naphthalene-acetic acid (NAA) or Indole-acetic acid (IAA; 0.5, 5 μM) combined with Benzylaminopurine (BA; 5, 10, 20 μM). For shoot development, 0.5 μM NAA or IAA was combined with 5 μM BA. After 4 weeks of culture on callus induction medium, 100% organogenic calli was obtained independent of the explant type and culture medium assayed; the aspect of the calli was compact and green (Figure 3). However, on the shoot development medium, the average number of developed shoots per callus depended on the explant type and the previous callus induction medium. For leaflet and PLA explants, the previous culture on 5 μM IAA + 10 μM

BA induction medium was better, whereas for petioles 0.5 μM NAA + 10 μM BA was better. Independent of the explant type, the developed shoots rooted on 10 μM NAA + 1 μM GA_3. The *ex vitro* acclimatisation was successful, achieving 96.4 % survival in the greenhouse (Figure 3) and 100 % in the field (Figure 3).

The FC content (i.e., psoralen, angelicin and total content) was examined along the plant regeneration process, from organogenic calli, *in vitro* regenerated shoots and *ex vitro* plants (grown in the greenhouse and grown in the field for 1 or 4 months), and compared with cultivated plants in the experimental field. The accumulation of FCs depended on the level of cellular organisation (the lowest FC content was in calli), the developmental stage of the plants and their exposure to environmental factors, being higher in *ex vitro* plants in the field, as Diwan and Malpathak (2010) observed in *in vitro cultures* of *R. graveolens* (another FC producer species).

The plant regeneration via embryogenesis has also been achieved from different explants of tedera and albo-tedera (unpublished data). The results show that explant type is a key factor in the frequency of embryogenesis and in the number of embryos per callus.

Figure 3. a. Micropropagation of *Bituminaria bituminosa* plants from nodal segments, b. organogenic callus with multiple shoots, c. regenerated plants form organogenic calli, d. acclimatisation of plants, e. plants in hydroponic culture, and f. micropropagated plants growing in the experimental field. Bar: 20 mm (a, c); 5 mm (b); 50 mm (d); 100 mm (e).

Genetic characterisation of *B. bituminosa*, elicitation experiments involving abiotic and biotic factors and cryopreservation of meristems are currently in progress, in collaboration with several research groups from Western Australia or Spain.

6.3. Screening of Plant Cold Tolerance Using a Controlled-freezing Chamber

For young plants grown in pots, the use of a controlled-freezing chamber is a simple way to screen their cold (freezing) tolerance and hence select lines appropriate for cultivation in frost-prone areas. The plants are placed in the chamber (set at ambient temperature) and then the temperature is reduced slowly (1-4ºC $hour^{-1}$) to the desired temperature (for *B. bituminosa*, between -5 and -10ºC); the plants are then removed and replaced in their original place of cultivation. They are monitored for up to one month, while visual symptoms of freezing damage (leaf and shoot meristem necrosis) appear. The populations or breeding lines showing least damage can be selected for planting at the chosen field site. The "selection temperature" is chosen according to the harshness of the winter at this site.

For plants already growing in the field, measurement of the freezing-induced electrolyte leakage from detached leaves, using a conductivity meter, is a simple method of assessing their cold tolerance before the onset of winter. This leakage arises due to damage caused to the cell plasma membrane by freezing, particularly due to loss of intracellular water to ice on the leaf surface, down a water potential gradient (Xin and Browse 2000). The leaves are misted with distilled water (to allow the formation of ice crystals on the surfaces) before being placed in the controlled-freezing chamber. The temperature is then reduced slowly, with a period of one hour at -1ºC to allow ice formation, to the desired temperatures (for *B. bituminosa*, between -2 and -12ºC); the temperature is maintained at each of these temperatures for 30 minutes and then a sample of the leaves is removed and stored at 4ºC. After thawing-out and development of membrane damage (typically 24 hours), the leaves are incubated in water and the electrical conductivity (EC) of the "extract" measured; the leaves are then heated to 90 ºC before the EC (assumed to represent the maximum value, after destruction of the membranes) is measured again. Expression of the post-thawing EC as a percentage of the maximum permits calculation of the temperature at which 50% membrane damage occurred; this LT_{50} value can be used to compare plants, populations and acclimation. The performance of these assays at different points of the plant growth cycle (for example, in early winter, when the plants are acclimated to cold, and in spring, following de-acclimation) allows correlation of the *in vivo* cold tolerance of the leaves with physiological parameters measured at the same time. In the case of *B. bituminosa*, Walker et al. (2010) demonstrated a relationship between cold tolerance and osmotic adjustment in the leaves, related to the accumulation of soluble sugars and amino acids.

Discrepancies may arise between the apparent cold tolerance determined with plants grown in pots or detached leaves and the behaviour of plants in the field during winter. This is because it is impossible to recreate natural conditions (which represent a combination of different, simultaneous stresses) for pot-grown plants (which may also be subjected to freezing tests after little or no acclimation). Also, for "quick" (24 hours) assays with detached leaves, membrane injury is assessed but damage which takes longer to develop (such as oxidative stress and necrosis) is not. However, freezing-chamber assays accelerate the screening process.

CONCLUSION

Thanks to this multidisciplinary research group (IMIDA-ICIA-CIBIO-MU-WA) and other international groups (Italy, Israel, Greece and Turkey), *B. bituminosa* has been shown to be an important legume with two potential uses: (1) forage and (2) pharmaceutical. In this sense, breeding lines have been selected showing desirable traits such as high forage yield, tolerance to abiotic stresses or high FC content. However, efforts are still needed to collect germplasm from cold Mediterranean environments, to identify the core collection in each Mediterranean region for the important breeder's plant traits and to develop molecular markers for key plant traits. So, as future priorities, we need to (1) establish a Mediterranean breeding programme to develop high-yielding and quality tedera cultivars adapted to cold environments, (2) foster international cooperation for the efficient management of genetic resources and human knowledge, (3) develop new genetic tools in order to select new individuals for pharmaceutical purposes and (4) search for new germplasm from different origins, to incorporate into plant breeding programmes so as to enrich the lines already obtained.

REFERENCES

Alliota, G; Piomelli, D; Pollio, A, Touwaide, A. *Le Piante Medicinalle del Corpus Hippocraticum*. Geurini e Associati. Napoli, 2003.

Alvar, M. *El español hablado en Tenerife*. CSIC. Madrid, Spain, 1959.

Álvarez, J. Leyenda erudita sobre la población de Canarias con africanos de lenguas cortadas. *Anuario de Estudios Atlánticos*, 1977, 23, 51-81.

Álvarez, S; Fresno, M; Darmanin, N; Briggs, HR; Méndez, P; Castro, N. Native forages Assessment for the improvement of milk production in goats. *Journal of Animal Veterinary Advances,* 2007, 6 (1), 94-98.

Álvarez, S; Fresno, M; Méndez, P. In: Hopkins A, Gustaffson T, Bertilsson J, Dalin G, Nilsdotter-Linde N, Spórndly E. *Biodiversity and Animal Feed. Future Challenges for Grassland Production. 22nd General Meeting European Grassland Federation*. Uppsala-Sweden; 2008, 72.

Asín Palacios, M. *Glosario de voces romances. Registradas por el botánico anónimo hispano-musulmán (Siglos XI-XII).* CSIC/Escuela de Estudios Árabes. Madrid-Granada; 1943.

Bellakhdar, J. *Mèdecine arabe ancienne et savoirs populaires. La pharmacopée marocaíne traditionelle.* Ed.Dumas. Paris, 1997.

Bolinaga, J. *Pastos para secano*. La Región; 1915, 6, 7-9.

Bonnier, G; Layens, G. *Flore complète portative de la France, de la Suisse et de la Belgie.* Ed. Librairie Générale. Paris; 1964.

Bourgaud, F; Allard N; Guckert A; Forlot P. Natural sources of furocoumarins (psoralens). In: *Psoralens, Past, Present and Future of Photochemoprotection and other biological activities*. John Libbey Eurotext, Montrouge; 1989, 219-229.

Bustamante, J; Corriente, F; Tilmatine, M. Abulhayr Al 'Išbīlī (s.V/XI) "*Kitā*bu 'Umdati *Ţţabīb Fī Ma 'Rifati Nnabāt Likulli Labīb*" (*Doctor Basical Book for Botany Knowledge for all expert*). Vol. II. CSIC Madrid, 2007.

Bustamante, J; Corriente, F; Tilmatine, M. Abulhayr Al 'Išbīlī (s.V/XI) "*Kitā*bu 'Umdati *Ţţabīb Fī Ma 'Rifati Nnabāt Likulli Labīb*" (*Doctor Basical Book for Botany Knowledge for all expert*). Vol. III. T.1. CSIC Madrid, 2010.

Bustamante, J; Corriente, F; Tilmatine, M. Abulhayr Al 'Išbīlī (s.V/XI) "*Kitā*bu 'Umdati *Ţţabīb Fī Ma 'Rifati Nnabāt Likulli Labīb*" (*Doctor Basical Book for Botany Knowledge for all expert*). Vol. III. T.2. CSIC Madrid, 2010.

Cano, F; Ríos, S. 8. Etnobotánica. In: Ríos, S. & Alcaraz, F. Editors. *Flora de las riberas y zonas húmedas de la cuenca del río Segura.* Ed. Serv. Publ. Universidad de Murcia, 1996.

Capote, J; Tejera, A; Amills, M; Argüello, A; Fresnol, M; López, JL. Influencia histórica y actual de los genotipos canarios en la población caprina americana (FAO) *Animal Genetic Resources Information*, 2004, 35, 49-60.

Cianfrogna, JA; Zangerl, AR; Berenbaum, MR. Effects of furanocoumarins on feeding behavior of parsnip webworms *Depressaria pastinacella. Journal of Chemical Ecology*, 2002, 28, 1365-1375.

Coca, B; Ríos, S; Juan, A; Méndez, P; Santos, A; Correal, E. Germoplasma de tedera (*Bituminaria bituminosa*): caracterización y descriptores. In: García B editor. *Pastos y Ganadería Extensiva.* Salamanca (Spain): Sociedad Española para el Estudio de los Pastos. CSIC, 2004, 123-127.

Cocks, PS. The adaptation of perennial legumes to Mediterranean conditions. In: Bennett SJ editor. *New perennial legumes for sustainable agriculture*. University of Western Australia Press, 2003, 224.

Congia, P. *Dizionario Botanico Sardo.* Ed. Zonza. Cagliari, 1998.

Corrales, C; Corbella, D. *Diccionario histórico del español en Canarias*. ICE. Colección Documentos de la Gomera, Spain; 2001.

Correal, E; Hoyos, A; Ríos, S; Méndez, P; Real, D; Snowball, R; Costa, J. Seed production of *Bituminaria bituminosa*: size, production, retention and germination capacity of the legumes. In: *Proceedings 12th Meeting FAO-CIHEAM Inter-regional Cooperation Research and Development Network on Pastures and Fodder crops*. Elvas (Portugal): CIHEAM; Options Méditerranéennes, Series A, 2008, 79, 379-383.

Correal, E; Moñino, I; Méndez, P; Ríos, S. Valoración agronómica de siete poblaciones de *Bituminaria bituminosa*: efecto de la sequía estival y siega sobre su persistencia y producción de hoja. In: Robles AB, editor. *Pastos, Desarrollo y Conservación (SEEP).* Junta de Andalucía, Consejería de Agricultura y Pesca. 2003, 35-40.

Correal, E; Robledo, A; Ríos, S; Rivera, D. Mediterranean dryland sheep-cereal systems. In: Lloveras J, González-Rodríguez A, Vázquez-Yañez O, Piñeiro J, Santamaría O, Olea L, Poblaciones MJ, editors. *Grassland Science in Europe*. EGF. Badajoz; 2006, vol 11, 14-26.

Curini, M; Carvotto, G; Epifano, F; Giannone, G. Chemistry and biological activity of natural and synthetic prenyloxycoumarins. *Current Medicinal Chemistry*, 2006, 13, 199-222.

Dear, BS; Ewing, MA. The search for new pasture plants to achieve more sustainable production systems in southern Australia. *Australian Journal of Experimental Agriculture*, 2008, 48, 387-396.

Dear, BS; Moore, GA; Hughes, SJ. Adaptation and potential contribution of temperate perennial legumes to the southern Australian wheat belt: A Review. *Australian Journal of Experimental Agriculture*, 2003, 43, 1-18.

Del Río, JA; Ortuño, A; Pérez, L; Bennet, RG; Real, D; Correal, E. Furanocoumarin content in *Bituminaria bituminosa* varieties and *Cullen* species. In: Porqueddu C, Ríos S, editors. *13th Meeting of the FAO-CIHEAM Sub-Network on Mediterranean Pastures and Fodder Crops. The Contributions of Grasslands to the Conservation of Mediterranean Biodiversity*. Alicante (Spain). Zaragoza: CIHEAM/CIBIO/FAO/SEEP; Options Méditerranéennes, Series A: Mediterranean Seminars; 2010, 92, 67-70.

Dioscórides, P. In: Laguna, A. *Pedacio Dioscórides Anazarbeo. Acerca de la Materia Médica Medicinal y de los venenos mortíferos.* (ed. facsimil). Biblioteca de Clásicos de la Medicina y de la Farmacia Española. Madrid; 1566.

E.D.E.S. *Estudio de las posibilidades de desarrollo agrícola en la isla de El Hierro*. Informe (inédito). 1969.

Fajardo, J; Verde, A; Rivera, D; Obon, C. *Etnobotánica en la Serranía de Cuenca. Las plantas y el hombre.* Diput. de Cuenca; 2007.

Font i Quer, P. *Plantas medicinales. El Dioscórides renovado*. Ed. Labor. Barcelona; 1985.

Forster, E. Trees and plants in Herodotus. *The Classical Review,* 1942, 56 (2), 57-63.

Fournier, P. *Le livre des plantes médicinales et vénéneuses de France*. Ed. Paul Lechevalier. Paris, 1948.

Gómez de Ortega, C. *Continuación de la Flora Española o Historia de las plantas de España, que escribía Joseph Quer.* T.VI. Madrid; 1784.

González-Tejero, Mª. *Investigaciones etnobotánicas en la provincia de Granada.* Tesis doctoral (inéd.) Facultad de Farmacia. Universidad de Granada; 1989.

Graham, PH; Vance, CP. Legumes: importance and constraints to greater use. *Plant Physiology*, 2003, 131, 872-877.

Guinea, E. 1947. Catálogo razonado de las plantas del Sahara español. *Anales del Jardín Botánico de Madrid*, 8(1), 357-442.

Guma, IR; Santos, A; Reyes, A; Padrón, MA; Méndez, P; González, R. Perennial forage legumes endemic to the Canary Islands: Collection and *ex situ* conservation. *Genetic Resources and Crop Evolution*, 2010, 58, 181-187.

Gupta, AK; Anderson, TF. Psoralen photochemotherapy. *Journal of American Academy of Dermatology*, 1987, 17, 703-734.

Hagemeier, J; Batz, O; Schmidt, J; Wray, V; Hahlbrock, K; Strack, D. Accumulation of phthalides in elicitor-treated cell suspension cultures of *Petroselinum crispum. Phytochemistry*, 1999, 51, 629–635.

Hamerski, D; Matern, U. Elicitor-induced biosynthesis of psoralens in *Ammi majus* L. suspension cultures. *European Journal of Biochemistry*, 1998, 171, 369–375.

Heathpagliuso, S; Matlin, SA; Fang, NB; Thompson, RH; Rappaport, L. Stimulation of furanocoumarin accumulation in celery and celeriac tissues by *Fusarium oxysporum* f sp apii, *Phytochemistry*, 1992, 31, 2683-2688.

Hönigsmann, H; Szeimies, RR; Knobler, R. Photochemotherapy and Photodynamic Therapy. In: *Fitzpatrick Dermatology in General Medicine*. McGraw-Hill Professional, New York, NY, USA; 2008; 2249-62.

Innocenti, G; Cappelletti, EM; Caporale, G. Furocoumarin contents in the vegetative organs of cultivated *Psoralea* species. *International Journal of Pharmacognosy*, 1991, 29 (4), 311-316.

Innocenti, G; Piovan, A; Filippini, R; Caniato, R; Cappelletti, EM. Quantitative recovery of furanocoumarins from *Psoralea bituminosa*. *Phytochemical Analysis*, 1997, 8, 84-86.

Jaizme, MC; Méndez, P; Flores, H. Efecto de la micorrización sobre la tedera (*Bituminaria bituminosa*) en las primeras fases de cultivo. In: *Biodiversidad en Pastos*, 2001, 181-187.

James, R. *Pharmacopeia Universalis: or a new universal English dispensatory*. London; 1747.

Juan, A; Coca, B; Crespo, MB; Ríos, S; Correal, E. Datos preliminares sobre variabilidad molecular poblacional en el agregado de *Bituminaria bituminosa* (*Leguminosae*). In: de la Roza B, editor. *Producciones agroganaderas: gestión eficiente y conservación del medio natural.* Sociedad Española para el Estudio de los Pastos. XLV Reunión Científica. Gijón, Asturias (Spain), 2005, Vol II, 971-977.

Juan, A; Moñino, I; Correal, E; Crespo, MB; Coca, B. 2003. Producción de frutos en poblaciones canarias e ibéricas de *Bituminaria bituminosa*. In: Robles AB, editors. *Pastos, Desarrollo y Conservación (SEEP).* Junta de Andalucía, Consejería de Agricultura y Pesca. 2003, 41-46.

Juan, A; Moñino, I; Correal, E; Méndez, P; Crespo, MB. 2004. Comparación de las tasas de fructificación de *Bituminaria bituminosa* (*Leguminosae*) bajo condiciones de cultivo en Canarias y la Península Ibérica. In: García B, editor. *Pastos y Ganadería Extensiva.* Salamanca (Spain): Sociedad Española para el Estudio de los Pastos. CSIC, 2004, 111-115.

Laguna, A. *Pedacio Dioscórides Anazarbeo. Acerca de la Materia Médica Medicinal y de los venenos mortíferos*. (ed. facsimil). Biblioteca de Clásicos de la Medicina y de la Farmacia Española. Madrid; 1566.

Lázaro Ibiza, B. *Botánica descriptiva. Compendio de la Flora Española*. T.I. Madrid, 1906.

Lewis, G; Schrire, B; Mackinder, B; Lock, M. *Legumes of the world.* Royal Botanical Gardens, Kew, Richmond, Surrey, UK (Ed); 2005.

Li, GD; Lodge, GM; Moore, GA; Craig, AD; Dear, BS; Boschma, P. Evaluation of perennial pasture legumes and herbs to identify species with high herbage production and persistence in mixed farming zones in southern Australia. *Australian Journal of Experimental Agriculture*, 2008, 48, 449-466.

Li, X; Schuler, M; Berenbaum, M. Molecular mechanisms of metabolic resistance to synthetic and natural xenobiotics. *Annual Review of Entomology*, 2007, 52, 231-253.

Llorente Maldonado de Guevara, A. *El léxico del tomo I del "Atlas lingüístico y etnográfico de las islas Canarias"*. Universidad de Extremadura; 1987.

Machado, A. La biodiversidad de las Islas Canarias. Capítulo 7. In: Pineda FD, de Miguel JM, Casado MA, Montalvo J, editors. *La diversidad biológica de España*. Madrid (Spain): Pearson Educación, S.A., 2002; 89-99.

Macías Hernández, A. El sector pecuario en la economía canaria, 1400-1970. Algunas notas para su estudio. *Reunion SEEP, Tenerife;* 1995.

Macías Hernández, A. La economía de los primeros isleños. In: Bethencourt Massieu, A. editor, *Historia de Canarias,* Las Palmas de Gran Canaria, 1995.

Martínez, S; Correal, E; Real, D; Ortuño, A; Del Río, JA. *Bituminaria bituminosa*: A source of furanocoumarins of pharmaceutical interest. In: Awaad AS, Govil JN, SinghVK editors. *Encyclopedia of Medicinal Plants. Drug Plants I.* Houston, TX, USA.: Stadium Press LLC, 2010, 307–322.

Martínez-Fernández, D; Walker, DJ; Romero, P; Correal, E. The physiology of drought tolerance in Tedera (*Bituminaria bituminosa*). In: Porqueddu C, Ríos S, editors. *13th Meeting of the FAO-CIHEAM Sub-Network on Mediterranean Pastures and Fodder Crops. The Contributions of Grasslands to the Conservation of Mediterranean Biodiversity.* Alicante (Spain). Zaragoza: CIHEAM/CIBIO/FAO/SEEP; Options Méditerranéennes. Series A: Mediterranean Seminars; 2010, 92, 155-159.

Martínez-Fernández, D; Walker, DJ; Romero-Espinar, P; Flores, P; Del Río, JA. Physiological responses of *Bituminaria bituminosa* to heavy metals. *Journal of Plant Physiology*, 2011. 168, 2206-2211.

Martínez-Fernández, D; Walker, DJ; Romero, P; Martínez-Ballesta, MC; Correal, E. The response of the leguminous fodder plant *Bituminaria bituminosa* to water stress. *Journal of Agronomy and Crop Science*, 2012, 198 (6), 442-451.

Masuda, T; Takasugi, M; Anetai, M. Psoralen and other linear furanocoumarins as phytoalexins in *Glehnia littoralis*. *Phytochemistry*, 1998, 47, 13-16.

Maurich, T; Iorio, M; Chimenti, D; Turchi, G. Erybraedin C and bitucarpin A, two structurally related pterocarpans purified from *Bituminaria bituminosa*, induced apoptosis in human colon adenocarcinoma cell lines MMR- and p53-proficient and –deficient in a dose-, time-, and structure-dependent fashion. *Chemico-Biological Interactions*, 2006, 159, 104-116.

Méndez, P. Forage potential of Canary Island Legumes. In: FAO REUR Technical Series 28. *Management of Mediterranean Shurblands and Related Forage Resources.*Chania-Greece. 1993; 141-144.

Méndez, P. El heno de tedera (*Bituminaria bituminosa*): un forraje apetecible para el caprino. In: *III Reunión Ibérica de Pastos y Forrajes*. Bragança-A Coruña-Lugo; 2000, 411-414.

Méndez, P. La tedera. Un forraje conocido pero poco valorado. *Agropalca,* 2009, 4, 19.

Méndez, P; Álvarez, S; Fresno, M. Nutritive value and yield under irrigation of tedera hay (*Bituminaria bituminosa*) an alternative forage for warm semi-arid to sub-humid climate. In: Hopkins A, Gustaffson T, Bertilsson J, Dalin G, Nilsdotter-Linde N, Spórndly E. *Biodiversity and Animal Feed. Future Challenges for Grassland Production. 22nd General Meeting European Grassland Federation*. Uppsala-Sweden; 2008, 52.

Méndez, P; Álvarez, S; Morales, E; Bermejo, L. Quantity and quality of goat milk from a herd grazing on a cultivated pasture with tedera (*Bituminaria bituminosa*) and barley as main constituent. *Advances in Animal Biosciences*, 2011b, 2 (2), 495.

Méndez, P; Bermejo, L; Morales, E; Álvarez, S. Effect of goat grazing on tedera pasture (*Bituminaria bituminosa*). *Advances in Animal Biosciences*, 2011a, 2 (2), 375.

Méndez, P; Díaz, E; Rivero, R. Contenido en cumarinas del forraje verde de tedera (*Bituminaria bituminosa*). In: Crespo MB, Juan A, Ríos S, editors. *Biodiversidad en pastos. Actas de la XLI Reunión Científica de la SEEP* – Foro Iberoamericano de Pastos, CIBIO, Alicante, España; 2001, 335-339.

Méndez, P; Fernández, M. Interés forrajero de las variedades de *Bituminaria bituminosa* (L.) Stirton (tedera) de Canarias. *Pastos,* 1990, nº extraordinario, 264-271.

Méndez, P; Fernández, M. La producción forrajera de las islas Canarias. *El Campo. Boletín de Información Agraria*; 1992, 124: 66-70.

Méndez, P; Fernández, M; Santos, A. Variedades de *Bituminaria bituminosa* (L.) Stirton (*Leguminosae*) en el archipiélago canario. *Pastos*, 1990-91, vol XX-XXI, 157-166.

Méndez, P; Peters, J; Morales, D; Jiménez, MS. Características del intercambio gaseoso de la tedera (*Bituminaria bituminosa*). In: Consellería de Agricultura, Ganadería e Política Agroalimentaria. *III Reunión Ibérica de Pastos y Forrajes.* Bragança-A Coruña-Lugo; 2000, 91-96.

Méndez, P; Santos, A; Correal, E; Ríos, S. Agronomic traits as forage crop of nineteen populations of *Bituminaria bituminosa.* In: Lloveras J, González-Rodríguez A, Vázquez-Yañez O, Piñeiro J, Santamaría O, Olea L, Poblaciones MJ, editors. *Grassland science in Europe,* EGF, Badajoz; 2006, vol.11, 300-302.

Minissale, P; Brullo, C; Brullo, S; Del Galdo, GG; Sciandrello, S. *Bituminaria basaltica* (Fabaceae), a new species from Italy. *Phytotaxa,* 2013, 98 (1), 1-15.

Miró, M. *Screening fitoquímico en plantas superiores. Detección de actividad antiinflamatoria y antibacteriana.* Memoria doctoral (ined.). Facultad de Farmacia. Universidad de Granada; 1981.

Molero, J; Pérez Raya, F; Valle, F. (Eds.). *Parque Natural de Sierra Nevada. Paisaje, Fauna, Flora, Itinerarios.* Ed. Rueda. Madrid; 1992.

Moll, M. *Medicina popular Menorquina. Segles XVI-XXI.* Ed. Documenta Balear. Mallorca; 2003.

Morales, J; Rodríguez, A; Alberto, V; Machado, C; Criado, C; Rando, JC. The impact of human activities on the natural environment of the Canary Islands (Spain) during the pre-Hispanic stage (3rd–2ndCentury BC to 15th Century AD): an overview. *Environmental Archaeology,* 2009, 14(1), 27-36.

Mulet, P. *Estudio etnobotánico de la provincia de Castellón.* Excma. Diput. de Castelló; 1991.

Muñoz, A; Correal, E. *Bituminaria bituminosa* (L.) Stirton, leguminosa de interés forrajero en la Cuenca Mediterránea: I. Situación taxonómica, distribución y autoecología. In: *Actas XXXVIII Reunión Científica de la SEEP* Soria (Spain); 1998, 87-91.

Muñoz, A; Correal, E. *Bituminaria bituminosa* (L) Stirton, leguminosa de interés forrajero en la Cuenca Mediterránea: II. Comportamiento frente a frío y selección de material tolerante. In: *Actas XXXIX Reunión Científica de la SEEP* Almería (Spain); 1999, 257-262.

Muñoz, A; Ortiz-Dorda, J; Correal, E. Morphological and molecular characterization of *Bituminaria bituminosa* accessions from south-east Spain and the Canary Islands. In: CIHEAM, *Cahiers Options Méditerranéenne,* 2000, 45, 103-107.

Obón, C. & Rivera, D. *Las plantas medicinales de nuestra Región.* Ed. Regional de Murcia; 1991.

Pagnotta, MA; Bennet, PS; Cocks, PS; Ewing, M; Howieson, J; Moore, G; Norman, H; Nutt, B; Porqueddu, C; Russi, L; Snowball, R. Evaluation of perennial legumes collected in Italy: First year data. In: Bennett SJ editor. *New perennial legumes for sustainable agriculture.* University of Western Australia Press, 2003; 224.

Parada, M; Selga, A; Bonet, Mª; Valles, J. *Etnobotànica de les Terres Gironines.* Dip. de Girona, 2002.

Pascual, H. *Leguminosas de la Península Ibérica y Baleares.* INIA/Ministerio de Agricultura. Madrid; 1978.

Pascual, MJ; Correal, E. *Especies leguminosas de interés económico.* Pisa Ediciones, Valencia. 1991, 94 pp.

Pazos-Navarro, M; Dabauza, M; Correal, E; Hanson, K; Teakle, N; Real, D; Nelson, M. Next generation DNA sequencing technology delivers valuable genetic markers for the genomic orphan legume species, *Bituminaria bituminosa. BMC Genetics,* 2011, 12, 104.

Pazos-Navarro, M; Del Río, JA; Ortuño, A; Correal, E; Dabauza, M. Micropropagation from apical and nodal segments of *Bituminaria bituminosa* and the furanocoumarin content of propagated plants. *Journal of Horticultural Science & Biotechnology,* 2012, 87(1), 29-35.

Pazos-Navarro, M; Del Río, JA; Ortuño, A; Romero-Espinar, P; Correal, E; Dabauza, M. Plant regeneration from different explant types of *Bituminaria bituminosa* and furanocoumarin content along plant regeneration stages. *Plant Growth Regulation,* 2013 (in press) DOI 10.1007/s10725-013-9784-9.

Pellicer, J. *Costumari Botànic. (2).* Ed. del Bullent. Valencia, 2000.

Pérez, J. La ganadería canaria. Notas Histórico- Etnográficas. *Anuario de Estudios Atlánticos,* 1963, 9, 237.

Pérez, V; Sagot, P. *De la végétation aux îles Canaries. Des plantes des pays tempérés et des plantes des régions intertropicales et Physionomie générale de leur agricultura.* París, 1867.

Pistelli, L; Noccioli, C; Appendino, G; Bianchi, F; Sterner, O; Ballero, M. Pterocarpans from *Bituminaria bituminosa* and *Bituminaria morisiana. Phytochemistry,* 2003, 64, 595-598.

Ramos, G; Frutos, P; Giráldez, FC; Mantecón, AR. Los compuestos secundarios de las plantas en la nutrición de los herbívoros. *Archivos de Zootecnia,* 1998, 47, 597-620.

Ramos-Rodríguez, E; de la Puente-Jiménez, JL. *Posibilidades de desarrollo ganadero con base territorial en las islas Canarias.* Informe de la Agencia de Desarrollo Ganadero de la Consejería de Agricultura, Ganadería y Pesca de la Junta de Canarias. (inédito). 1983.

Real, D; Albersten, T; Snowball, R; Howieson, J; Revell, C; Ewing, M; Correal, E; Méndez, P; Ríos, S. *Bituminaria bituminosa* var. *albomarginata* (Lancelot trefoil), a novel perennial forage legume for low-rainfall Mediterranean environments in Western Australia. In: *XXI International Grassland Congress and VIII International Rangeland Congress*, Hohhot, China, 2008, Vol II, 452.

Real, D; Correal, E; Méndez, P; Santos, A.; Ríos, S; Sternberg, M; Dini-Papanastasi, O; Pecetti, L; Tava, A. *Bituminaria bituminosa* C.H. Stirton. Grassland Species Profiles, CIAT/FAO. www.fao.org/ag/AGP/AGPC/doc/GBASE/new_species/tedera/bitbit. htm. 2009.

Real, D; Foster, K; Correal, E; Kidd, D; Méndez, P. Techniques for artificial hybridization in the self-pollinated forage legume *Bituminaria bituminosa* and the cross-pollinated forage legume *Hedysarum coronarium.* In: XXI *International Grassland Congress and VIII International Rangeland Congress*, Hohhot, China, 2008, Vol II, 347.

Real, D; Li, GD; Clark, S; Allbersten, TO; Hayes, RC; Denton, MD; D'antuono, MF; Dear, BS. Evaluation of perennial forage legumes and herbs in six Mediterranean environments. *Chilean Journal of Agricultural Research*, 2011, 71(3), 357-369.

Reyes Prosper, E. *Las estepas de España y su vegetación.* Madrid, 1915.

Ríos, S; Correal, E; Robledo, A. First screening of the more interesting pasture legumes present in matorral areas of south-east Spain. In: *Proceedings IVth. Int. Rangeland Congress*. Montpellier, Francia Edit. AFP, 1993. Vol. 1, 417-421.

Ríos, S; Robledo, A; Correal, E. Perennial pasture legumes of south-east Spain. FAO Bull. *Herba*, 1992, 5, 68-76.

Rivera, D; Alcaraz, F; Verde, A; Fajardo, J; Obón, C. *Las plantas en la cultura popular. Vol. 9*. Ed. Soc. Medit. de Historia Natural/CAM. Murcia, 2008.

Rivera, D; Matilla, G; Obón, C; Alcaraz, F. *Plants and humans in the Near East and Caucasus. Vol. 2. The Plants: Angiosperms*. Plants and humans/Edit.um. Murcia, 2012.

Rivera, D; Obon, C; Cano, F; Robledo, A. *Introducción al mundo de las plantas medicinales en Murcia*. Ayuntamiento de Murcia, 1994.

Sánchez, MA; Soriano, MC; Pérez, O; Correal, E. Determination of condensed tannins in perennial forage legumes and its effect on their palatability for sheep. In: *18th International Conference on Polyphenols*. Bourdeaux (France); 1996, Vol.1, 243-244.

Sandral, GA; Dear, BS; Viragona, JM; Swan, AD; Orchard, BA. Changes in soil water content under annual- and perennial-based pasture systems in the wheatbelt of southern New South Wales. *Australian Journal of Agricultural Research*, 2006, 57, 321-333.

Scarisbrick, JJ; Taylor, P; Holtick, U; Makar, Y; Douglas, K; Berlin, G; Juvonen, E; Marshall, S. U.K. consensus statement on the use of extracorporeal photopheresis for treatment of cutaneous T-cell lymphoma and chronic graft-versus-host disease. *British Journal of Dermatology*, 2008, 158 (4), 659-78.

Singh, RJ. Landmark research in grain legumes. In: Singh RJ and Jauhar PP editors. *Genetic Resources, Chromosome Engineering, and Crop Improvement Series: Grain legumes*, CRC Press, Inc., Boca Ratón, Fla., 2005, Vol 1 1-9.

Sing, RJ. Landmark research in grain legumes. In: Singh RJ and Jauhar PP editors. *Genetic Resources, Chromosome Engineering, and Crop Improvement Series: Oilseed crops*. CRC Press, Inc., Boca Ratón, Fla., 2007, Vol 4, 1-12.

Singh, RJ; Chung, GH; Nelson, RL. Landmark research in legumes. *Genome*, 2007, 50, 525-537

Stenberg, M; Gishri, N; Mabjeesh, SJ. Effects of grazing on *Bituminaria bituminosa* (L.) Stirton: a potential forage crop in Mediterranean grasslands. *Journal of Agronomy and Crop Science*, 2006, 192, 399-407.

Stern, RS. Psoralen and ultraviolet A light therapy for psoriasis. *The New England Journal of Medicine*, 2007, 357, 682-690.

Stevenson, PC; Simmonds, MSJ; Yule, MA; Veitch, NC; Kite, GC; Irwin, D; Legg, M. Insect antifeedant furanocoumarins from *Tetradium daniellii*. *Phytochemistry*, 2003, 63, 41-46.

Tava, A; Pecetti, L; Ricci, M; Pagnotta, MA; Russi, L. Volatile compounds from leaves and flowers of *Bituminaria bituminosa* (L.) Stirt. (Fabaceae) from Italy. *Flavour and Fragance Journal*, 2007, 22 (5), 363-370.

Tournefort, J.P. *Institutiones rei Herbariae.* T.I. Paris, 1700.

Vasconsuelo, A; Boland, R. Molecular aspects of the early stages of elicitacion of secondary metabolites in plants. *Plant Science*, 2007, 172, 861-875.

Ventura, MR; Bastianelli, D; Hassoun, P; Flores, MP; Bonnal, L; González-García, E. Nutrient and tannins content of browsing shrub legumes informally used for small ruminant feeding in Canary *Island*, Spain. In: Acar Z, López-Francos A, Porqueddu C. editors. *New approaches for grassland research in a context of climate and socio-economic changes.* Zaragoza: Centre International de Hautes Etudes Agronomiques Méditerranéennes (CIHEAM) / Ondokuz Mayis University. (Options Méditerranénnes, Series A: Mediterranean Seminars) 2012, 102, 391-394.

Ventura, MR; Castañón, JIR; Méndez, P. Effect of season on tedera (*Bituminaria bituminosa*) intake by goats. *Animal Feed Science and Technology*, 2009, 153, 314-319.

Ventura, MR; Castañón, JIR; Pieltain, MC; Flores, MP. Nutritive value of forage shrubs: *Bituminaria bituminosa*, *Rumex lunaria*, *Acacia salicina*, *Cassia surtii* and *Adenocarpus foliolosus*. *Small Ruminant Research*, 2004, 52, 13-18.

Ventura, MR; Méndez, P; Flores, MP, Rodríguez and Castañón, JIR. Energy and protein content of tedera (*Bituminaria bituminosa*). In: *Cahiers Options Mediterranéennes*, 2000, 45, 219-221.

Verde, A; Rivera, D; Fajardo, J; Obón, C; Cebrián, F. *Guía de las plantas medicinales de Castilla la Mancha*. Altaban Ediciones. Albacete, 2008.

Villar, L; Palacín, JMª; Calvo, C; Gómez García, D; Montserrat Martí, G. *Plantas Medicinales del Pirineo Aragonés y demás tierras oscenses*. CSIC/Dip. Huesca, 1987.

Walker, DJ; Bernal, MP; Correal, E. The influence of heavy metals and mineral nutrient supply on *Bituminaria bituminosa*. *Water, Air & Soil Pollution,* 2007, 184, 335-345.

Walker, DJ; Martínez-Fernández, D; Correal, E; Romero-Espinar, P; Del Río, JA. Accumulation of furanocoumarins by *Bituminaria bituminosa* in relation to plant development and environmental stress. *Plant Physiology and Biochemistry,* 2012, 54, 133-139.

Walker, DJ; Moñino, I; Correal, E. Genome size in *Bituminaria bituminosa* (L.) C.H. Stirton (*Fabaceae*) populations: separation of "true" differences from environmental effects on DNA determination. *Environmental and Experimental Botany,* 2006, 55, 258-265.

Walker, DJ; Romero, P; Correal, E. Cold tolerance, water relations and accumulation of osmolytes in *Bituminaria bituminosa. Biologia Plantarum,* 2010, 54 (2), 293-298.

Wölfel, DJ. *Monumenta Linguae Canariae. Die kanarischen sprachdenkmäler. Eine Studie zur Vor- und Frühgeschichte Weißafrikas*. Akademische Druck-u. Verlagsanstalt. Graz (Austria); 1965.

Xin, Z; Browse, J. Cold comfort farm: the acclimation of plants to freezing temperatures. *Plant, Cell and Environment*, 2000, 23, 893-902.

In: Legumes
Editors: Hiroto Satou and Ren Nakamura

ISBN: 978-1-62808-280-7

Chapter 2

NUTRITIONAL CHARACTERIZATION OF WILD LEGUMES (*LATHYRUS* AND *VICIA* GENERA)

***Elena Pastor-Cavada*[1,*], *Julio E. Pastor*[2], *Rocío Juan*[2] *and Javier Vioque*[1]**
[1]Instituto de la Grasa (C.S.I.C.), Sevilla, Spain
[2]Departamento de Biología Vegetal y Ecología, Universidad de Sevilla, Sevilla, Spain

ABSTRACT

In several regions of the world, legumes play an important role in human and animal nutrition as they are a cheap source of protein and other nutrients. In recent years have been shown that the consumption of proteins, certain lipids, polyphenols, or bioactive peptides contained in legumes have a beneficial effect on health.

In addition, since the "green revolution," a large amount of the world's phytodiversity has been lost because local crops have been substituted by other genetically uniform crops, which have higher yields. However, the diversification of cultivars is necessary to adapt cultivars to soil and water conditions, or to limit phytosanitary risks. Hence, knowledge of local plants could provide useful data for the possible development of future crops.

Vicia and *Lathyrus* are major genera of the tribe *Fabeae* and have wild species in Southern Spain. A total of 46 different wild species have been studied. To our knowledge, this is the first time that a deep nutritional characterization of wild populations of *Lathyrus* and *Vicia* species has been made.

In this work the fatty acid composition of the seed oil, the polyphenol concentration and the antioxidant activity of the polyphenols extracted from the seeds, the amino acid composition and nutritional quality of seed proteins have been studied in 15 Lathyrus species and in 28 *Vicia* species (31 taxa).

The main seed oil fatty acids ranged from $C_{14:0}$ to $C_{20:0}$. All studied species showed higher amounts of total unsaturated fatty acids than saturated ones. Among unsaturated fatty acids, the most abundant were linoleic, oleic and linolenic acids. Among studied species, the ω6/ω3 ratio was variable, ranging from 1.7 in *V. articulata* to 17.3 in *V. faba*, there being nine species in which the ω6/ω3 ratio was below 5.

* Corresponding author: Tel. 954611550, Fax. 954616790, E-mail: epastor@ig.csic.es; epastor@cica.es.

Polyphenolic contents ranged from 1.9 mg/g meal in *V. narbonensis* to 29.2 mg/g meal in *L. sphaericus*. The highest specific polyphenols antioxidant activity (referred to as catechin equivalents in the extracts) was observed in the extracts from *Lathyrus annuus*, *Vicia parviflora* and *Vicia tenuifolia*. On the contrary, the highest total antioxidant activity (referred as to extract volume) was observed in *L. aphaca* y *V. sativa*. Results show that studied *Lathyrus* and *Vicia* species are rich in phenolic compounds with higher antioxidant activity than phenolics of widely consumed legumes such as soy, chickpea or lupin.

Protein contents in studied species ranged from 17.7% in *Lathyrus sativus* to 32% in *V. pyrenaica*. Among essential amino acids the most abundant were Leu, Lys, Phe, Thr and Val. Also, all species contained Lys above FAO recommendations. On the contrary, all species were limiting in Trp ranging from 0.5% to 1.0% content and were also limiting in sulphur containing amino acids, Met and Cys. In vitro protein digestibility (IVPD) in studied legumes ranged from 76.0% in *L. pratensis* to 86.5% in *L. annuus*. Nutritional parameters, such as protein efficiency ratio, corresponded to high quality proteins. Protein digestibility corrected amino acid score was similar to the observed in other commercial legumes such as lentil or peanut.

In conclusion, results confirm the interest of studying wild populations of *Lathyrus* and *Vicia* species as source of seeds with good nutritional characteristics and contribute to revalorize the cultivation of these legumes.

1. INTRODUCTION

Since the origin of life on Earth, the evolutive process has given rise to an enormous diversity of species which, by means of selection processes, have adapted to the different conditions on the globe. This genetic variability is essential for the balance of the ecosystem and constitutes the so-called “germoplasm” of the planet. Within this collection of species, the “plant genetic resources” consist of the genetic diversity of the vegetal world. Those resources constitute a World Heritage of an immeasurable value and its gradual loss actually means a severe threat for the balance of the ecosystems, the agricultural development, and the food security in the entire world. The use and exploitation of natural resources and in particular of the biodiversity is an important factor in the development of the countries.

A considerable proportion of the worldwide plant diversity is being lost gradually as the farmers disregard local varieties and adopt genetically-uniform and high-performance varieties. A key point to preserve the biodiversity is promoting the diversification of the primary sector. To do so, it is necessary to increase the sources and the knowledge about the species and varieties, as well as define the adequate methodologies to support a sustainable use by gathering the scientific knowledge and the data which are obtained by local communities (ethnobotanists) always in cooperation with other countries.

Throughout the years, human activities have proved to influence the decrease of species considerably not only in terms of size, but also in terms of genetic variability in wild populations; moreover, they have also caused an irreversible loss of habitats and ecosystems. It is a world-wide problem which is part of the so-called “biodiversity crisis”.

This issue has been addressed in the Convention on Biological Diversity [1] and the *International Treaty on Plant Genetic Resources for Food and Agriculture* [2]. Furthermore, the General Meeting of the United Nations proclaimed the year 2010 as “International Year of the Biological Diversity" in order to draw international attraction to the issue of the ongoing loss of biodiversity.

The leguminous plants, one the most-commonly used species in the diets, are not an exception for this problem, since the cultivation of some common legumes has also decreased over time.

Legumes, such as soybean (*Glycine max*), chickpea (*Cicer arietinum*) or green pea (*Pisum sativum*), are considered as highly nourishable and recommendable food considering their composition in proteins, carbohydrates, lipids, minerals, and vitamins [3].

In Southern Spain, the genus *Lathyrus* and *Vicia* (Tribu *Fabeae*, *Fabaceae Family*) represent a perfect model to perform a bioprospection of the proteins, fatty acids, and polyphenols of the seeds, not only by the number of cultivated species at present or in the past, but also by the number of wild taxa in that region; they may also be interesting as a means to obtain new species which might be useful in the diets and as a complementary, or even substitutive, genetic reservoir of the taxa which were traditionally cultivated.

1.1. Fatty Acids

The fat content of most legume seeds is fairly low, between 1-2% [4, 5, 6]. Nevertheless, there are some species like the peanut and the soybean in which the lipid content can be 23 % [7, 8].

The main fatty acids in legume seed lipids are palmitic, oleic, linoleic and linolenic acids [6, 9, 10]. According to Salunkhe et al. [11], the quantity of these fatty acids may vary depending on the species and the geographical conditions. In general, the most abundant fatty acids in leguminous plants in which seeds have the highest fat contents, such as the soybean and the peanut, are the linoleic and oleic acids. However, the dominant fatty acids in the leguminous plants with less fat seeds are the linoleic and linolenic acids [11].

Despite the fat content is not high in any of the leguminous plants, their fatty acid composition becomes more important as a crop, since they tend to have low concentrations of saturated fatty acids and high concentrations of unsaturated fatty acids [12].

The unsaturated fatty acids, in particular the poly-unsaturated fatty acids, such as the ω3, have had a greater nutritional relevance in the last two decades due to their beneficial effects on health (reduction of heart attack risk, improvement of arthritis, lowering of blood pressure, clinical benefits in subjects with personality disorders and antisocial behavior, etc.), as reported in different medical researches [13, 14, 15]. Some authors like Evans et al. [16] and Duranti [17], have highlighted the capability of legumes to reduce the glycemic index, cholesterolemia, prevent diabetes mellitus, and colon cancer, which is due not only to the protecting role of the dietetic fiber, but also to the favorable content in smaller components, like the fatty acids in the legumes [18].

With respect to the genus *Lathyrus*, the *L. maritimus*, *L. cicera* and, above all, *L. sativus* have been generally covered by other authors in the past [8, 9, 19, 20, 21, 22, 23, 24]. Other species of the genus *Lathyrus* include *L. aphaca* [25], *L. annuus*, *L. hirsutus*, *L. pratensis*, *L. setifolius* y *L. sphaericus* [26, 27]. In the case of the genus *Vicia,* the most-commonly used species in human diet and fodder is *V. faba* [28]. This species is the most common for authors [29] in order to study the composition of its seeds, although alike the rest of legumes the fatty content is low. Data about other species like *V. sativa*, *V. peregrina* or *V. hybrida* [12] are also available. Considering the fact that the composition in fatty acids of the lipid fraction of leguminous plants is rich in unsaturated fats, that they are beneficial for health and that there are not comprehensive researches on the composition of fatty acids of wild species of the genus *Lathyrus* and *Vicia* present in Southern Spain, it is interesting to widen the knowledge about these genera which may have not been traditionally cultivated or have been occasionally cultivated.

1.2. Polyphenols

The polyphenols are usually associated to the healthy effects of eating fruits and vegetables, and this positive effect is related to its antioxidant activity [30]. Thus, studies have pointed out that antioxidant-rich foods are inversely related to the occurrence of diseases like cancer, Alzheimer, and Parkinson. Moreover, conclusions generally indicate that there is a correlation between the consumption of phenol-rich foods and the prevention of the abovementioned diseases [31, 32, 33, 34, 35]. On the other hand, some polyphenols have also proved to have an anti-proliferative activity and apoptosis induction [36].

Even though phenols are a minor component in legumes, some authors, like Tharanathan & Mahadevamma [37], declare that the different bioactive compounds that they contain produce beneficial effects when eaten on a regular basis, since they can act on diseases like diabetes mellitus, colon cancer, and coronary disorders.

In the genus *Lathyrus*, the phenol composition has only been studied in a few species, like *L. sativus* [3, 38], or *L. maritimus* [39, 40]. Some authors have studied the polyphenol content in the genus *Vicia* and the antioxidant activity of *V. faba* [41] and *V. sativa* [42]. Ademabowo et al. [43] carried out an epidemiological study in which among several fruits and vegetables, only the intake of beans and lentils seem to have a direct effect on the lower incidence of breast cancer.

Considering the potential interest of the genera *Lathyrus* and *Vicia* as sources of functional compounds, such as polyphenols with an antioxidant activity, it would be interesting to expand these studies to a greater number of species.

1.3. Proteins

Legumes are considered as a group of plants with a high nutritional value, and therefore they constitute a very important component for the human diet in general and for a vegetarian diet in particular [44, 45]. The increasing popularity of vegetarian, ecological and healthy food is associated to the demand increase for fresh vegetal products with a high protein-content. The primary source for vegetal proteins is the seeds of legumes all around the world [46, 47]. Apart from the species like soybean, broad bean, green pea or lentil, the studies should pay special attention to the species almorta (*Lathyrus sativus*) which can grow in weather and ground conditions that are unfavorable for other crops [48].

In general, legume proteins have a similar amino acid content. They usually have low concentrations of sulphur-containing amino acids (methionine and cysteine), according to the studies on *Vicia faba, Lens culinaris, Cicer arietinum, Phaseolus vulgaris and Lathyrus sativus* [49, 50]. Furthermore, they have proteins with a rich content of aspartic acid and glutamic acid which give the acid character to legume proteins [51] and are very rich in lysine [52, 53]; therefore if legumes are combined with cereals (in which proteins are poor in lysine and rich in sulphur-containing amino acids) [54], the protein provision and the biological quality of the proteins in the resulting mix are fairly appropriate for human needs [55].

According to the bibliography available about the genus *Lathyrus*, the main species in which the protein composition in amino acids has been analyzed include *L. cicera*, *L. maritimus* and *L. sativus* [9, 20, 21, 50, 56, 57]. With respect to the genus *Vicia*, information about the amino acid content is available for some species like *V. ervilia*, *V. sativa* and above all *V. faba* [58, 59, 60, 61, 62, 63]. Considering that the quantities of essential amino acids were satisfactory in the species previously studied such as *L. sativus* [20], except for the content of sulfur amino acids which made the protein quality of these seeds similar to soybeans, analyzing the amino acid content of the proteins in other species of these genera represented in Southern Spain would be helpful in order to find out which species could be used as alternative crops. Also, according to Ferguson & Robertson [64], the use of wild species contributes to expand the genetic variability of cultivated species, so it would be interesting to apply them in the species covered in this study, as Rozan et al. [65] have already done with wild lentils.

As stated above, the need to preserve biodiversity, the nutritional and healthy factors all contribute to increase the value of marginal crops of legumes, and require the characterization of wild species.

This study includes 15 and 31 taxa of the genera *Lathyrus* and *Vicia* respectively, which can be divided into two groups depending if they have ever been cultivated or not

- Taxa which have ever been cultivated: *L. annuus*, *L. cicera*, *L. hirsutus*, *L. ochrus*, *L. sativus*, *L. tingitanus*, *V. angustifolia*, *V. articulata*, *V. benghalensis*, *V. dasycarpa*, *V. ervilia*, *V. faba*, *V. hirsuta*, *V. monantha* subsp. *calcarata*, *V. narbonensis*, *and V. sativa* subsp. *sativa*.
- Taxa which have never been cultivated: *L. amphicarpos*, *L. angulatus*, *L. aphaca*, *L. clymenum*, *L. filiformis*, *L. latifolius*, *L. pratensis*, *L. setifolius*, *L. sphaericus*, *V. altissima*, *V. cordata*, *V. disperma*, *V. eriocarpa*, *V. glauca* subsp. *giennensis*, *V. hybrida*, *V. incana*, *V. lathyroides*, *V. lutea* subsp. *cavanillesii*, *V. lutea* subsp. *lutea* var. *hirta*, *V. lutea* subsp. *lutea* var. *lutea*, *V. lutea* subsp. *vestita*, *V. monardi*, *V. onobrychioides*, *V. parviflora*, *V. peregrina*, *V. pseudocracca*, *V. pubescens*, *V. pyrenaica*, *V. tenuifolia*, *and V. vicioides*.

2. Fatty Acid Analysis

Considering that several authors [4, 5, 6, 9, 21] have previously verified that, in general, the fat richness contained in leguminous seeds is not higher than 2%, this study has calculated this specific value in some taxa (the fat content in *L. annuus* is 1.93%, in *L. cicera* is 0.64%, in *L. clymenum* is 1.99%, in *V. lutea* subsp *lutea* var. *hirta* is 1.97%, in *V. lutea* subsp. *vestita* is 1.59%, and in *V. monantha* subsp. *calcarata* is 1.41%) in order to check that the abovementioned ratio is maintained in this family. In the following section, the composition of the lipid fraction is examined in all the taxa studied in the genera *Lathyrus* and *Vicia* from Southern Spain.

2.1. Genus *Lathyrus*

Content of satured and unsatured fatty acids in the fifteen studied *Lathyrus* species is shown in Tables 1a and 1b.

All studied *Lathyrus* species had higher contents of total unsatured fatty acids (TUFAs) than satured ones (TSFAs) (Tables 1a and 1b).

As specified in those tables, the total content of both saturated and unsaturated fatty acids is fairly consistent in all species, since the first group ranges from 23.6 to 31.2%, and the second group ranges from 68.8 and 76.4%, not detecting significant differences between saturated and unsaturated fatty acids. These results are in accordance with tose obtained by other authors in species of this genus [8, 9, 19, 20, 21, 22, 24-27, 66], and in other legumes [12, 67, 68].

Table 1a. Saturated fatty acid composition (%) of the seed oil of studied *Lathyrus* species (mean ± s.d.) Different small letters indicate significant differences between values in the same column (Scheffe's test)

Species	n	$C_{14:0}$	$C_{15:0}$	$C_{16:0}$***	$C_{18:0}$**	$C_{20:0}$	TSFAs
L. amphicarpos	3	4.6±1.3	0.6±0.2	12.7±1.6^{ab}	8.9±0.1^{d}	0.9±0.1	27.8±0.2
L. angulatus	4	5.0±1.4	0.8±0.3	14.9±0.7^{ab}	5.4±0.4^{abcd}	0.9±0.1	27.0±1.8
L. annuus	6	4.7±1.2	1.2±0.3	15.5±0.8^{ab}	4.7±0.4^{ab}	0.6±0.1	26.7±1.6
L. aphaca	7	4.2±0.8	1.6±0.3	16.8±0.7^{ab}	5.0±0.6^{ab}	1.1±0.1	28.7±0.6
L. cicera	7	2.0±1.3	0.8±0.4	17.7±0.8^{b}	6.3±0.8^{bcd}	0.9±0.1	27.8±2.2
L. clymenum	7	2.6±0.7	0.5±0.2	16.3±0.9^{ab}	5.2±1.4^{abc}	0.8±0.2	25.5±1.9
L. filiformis	1	1.3	0.3	17.5^{ab}	4.8^{abcd}	1.0	25.0
L. hirsutus	3	4.8±2.1	1.5±1.1	16.0±0.9^{ab}	4.6±0.6^{abc}	0.9±0.0	27.7±3.6
L. latifolius	5	2.8±0.9	0.6±0.3	17.5±1.8^{ab}	3.5±0.5^{a}	0.7±0.2	25.1±2.9
L. ochrus	4	2.0±1.2	0.4±0.2	15.1±2.7^{ab}	6.8±0.3^{bcd}	0.6±0.2	24.9±3.2
L. pratensis	4	3.7±1.3	1.3±0.7	15.3±1.0^{ab}	4.6±0.9^{ab}	1.0±0.0	25.7±2.4
L. sativus	2	1.5±1.8	0.3±0.0	11.6±0.4^{a}	9.1±0.4^{d}	1.0±0.2	23.6±0.9
L. setifolius	3	2.4±0.4	0.7±2.0	14.4±0.6^{ab}	6.9±0.8^{bcd}	0.8±0.1	25.2±1.0
L. sphaericus	5	5.2±1.4	1.7±0.9	15.8±1.0^{ab}	7.3±1.3^{bcd}	1.2±0.2	31.2±2.6
L. tingitanus	7	3.3±1.0	0.5±0.2	13.7±0.8^{a}	7.5±0.3^{cd}	1.1±0.3	26.1±1.5

n = number of populations studied. **P<0.01, ***P<0.001. TSFAs = % total saturated fatty acids.

Table 1b. Unsaturated fatty acid composition (%) of the seed oil of studied *Lathyrus* species (mean ± s.d.) Different small letters indicate significant differences between values in the same column (Scheffe's test)

Species	n	$C_{16:1}$	$C_{18:1}$**	$C_{18:2}$*	$C_{18:3}$***	TUFAs
L. amphicarpos	3	0.5±0.1	12.5±4.0^{ab}	52.4±4.0^{ab}	6.8±0.9^{ab}	72.2±0.2
L. angulatus	4	0.7±0.2	10.7±1.5^{a}	50.5±2.9^{ab}	11.1±2.3^{ab}	73.0±1.8
L. annuus	6	0.3±0.2	19.5±2.3^{ab}	42.5±1.9^{a}	11.0±0.7^{b}	73.3±1.6
L. aphaca	7	0.3±0.2	5.4±1.2^{a}	56.9±1.2^{b}	8.6±0.6^{ab}	71.3±0.6
L. cicera	7	0.4±0.2	12.1±2.0^{a}	53.4±2.3^{ab}	6.2±0.6^{ab}	72.2±2.2
L. clymenum	7	0.4±0.1	6.2±1.5^{a}	60.1±1.7^{b}	7.8±0.9^{ab}	74.5±1.9
L. filiformis	1	1.0	21.9^{ab}	40.1^{ab}	12.0^{ab}	75.0
L. hirsutus	3	0.5±0.2	16.6±6.9^{ab}	48.1±5.5^{ab}	7.1±0.3^{ab}	72.3±3.6
L. latifolius	5	1.0±0.7	28.8±10.3^{b}	40.2±12.3^{a}	4.8±1.4^{a}	74.9±2.9
L. ochrus	4	0.2±0.1	13.8±6.2^{ab}	55.4±7.6^{ab}	5.6±1.2^{ab}	75.1±3.2
L. pratensis	4	0.5±0.2	12.8±4.3^{ab}	53.4±1.4^{ab}	7.6±2.3^{ab}	74.3±2.4
L. sativus	2	0.2±0.1	17.6±1.2^{ab}	54.6±2.3^{ab}	4.0±0.3^{ab}	76.4±0.9
L. setifolius	3	0.3±0.1	13.1±0.8^{ab}	53.9±1.9^{ab}	7.4±1.9^{ab}	74.8±1.0
L. sphaericus	5	0.3±0.1	7.5±1.4^{a}	50.0±3.1^{ab}	11.0±1.4^{b}	68.8±2.6
L. tingitanus	7	0.3±0.1	17.6±7.3^{ab}	49.0±5.4^{ab}	6.9±2.1^{ab}	73.9±1.5

n = number of populations studied. *$P<0.05$, **$P<0.01$, ***$P<0.001$. TUFAs = % total unsaturated fatty acids.

Among saturated fatty acids, palmitic acid ($C_{16:0}$) was the most abundant in all studied species with contents ranging from 11.6% in *L. sativus* to 17.7% in *L. cicera* (Table 1a). Significant differences ($p<0.001$) in palmitic acid contents were observed among *L. cicera* and two species of *Lathyrus* with lesser amounts of this acid (*L. sativus* and *L. tingitanus*). Another saturated fatty acid relatively abundant in legumes is stearic acid ($C_{18:0}$). Among studied species, the highest content of this fatty acid was observed in *L. sativus* with 9.1%, which shows significant differences ($p<0.01$) with respect to six of the remaining species (Table 1a). A negative correlation ($r^2 = 0.59$) between the $C_{16:0}$ and $C_{18:0}$ fatty acids contents was observed. Therefore, species with a low $C_{18:0}$ content tend to have a higher amount of $C_{16:0}$ (Table 1a). This is probably because $C_{16:0}$ is the biochemical substrate for the biosynthesis of $C_{18:0}$. *L. sativus* showed the highest $C_{18:0}$ content and the lowest $C_{16:0}$ content (Table 1a). $C_{18:0}$ is less hypercholesterolemic than $C_{16:0}$. Consumption of foods rich in saturated fatty acids from 12 to 16 carbon lengths is positively associated with low-density lipoprotein and cardiovascular disorders, whereas $C_{18:0}$ is considered neutral in this respect [69, 70, 71, 72]. Therefore, species with a low ratio $C_{14:0} + C_{15:0} + C_{16:0}/C_{18:0}$ would be more appropriate for human nutrition. The lowest ratio was observed in some species, which have been grown traditionally (*L. sativus* and *L. ochrus*), and in two of the nine non cultivated *Lathyrus* species (*L. amphicarpos* and *L. setifolius*). The highest ratio was observed in *L. latifolius* (Table 2).

Table 2. Different parameters based on the fatty acid composition of the seed oil of studied *Lathyrus* species. Data (TUFAs, TSFAs, PUFAs y MUFAs) are expressed as the average ± standard deviation. Different small letters indicate significant differences between values in the same column (Scheffe's test)

Species	n	$C_{14:0}+C_{15:0}+C_{16:0}/C_{18:0}$	TUFAs	TSFAs	TUFAS/ PUFAs	PUFAs**	MUFAs**	PUFAs/ MUFAs	O/L	ω6/ω3
L. amphicarpos	3	2.0	72.2 ± 0.2	27.8 ± 0.2	2.6	59.2 ± 4.2^{ab}	13.0 ± 4.0^{ab}	4.5	0.2	7.7
L. angulatus	4	3.8	73.0 ± 1.8	27.0 ± 1.8	2.7	61.6 ± 2.4^{ab}	11.4 ± 1.6^{a}	5.4	0.2	4.6
L. annuus	6	4.6	73.3 ± 1.6	26.7 ± 1.6	2.7	53.6 ± 1.8^{ab}	19.7 ± 2.3^{ab}	2.7	0.5	3.9
L. aphaca	7	4.5	71.3 ± 0.6	28.7 ± 0.6	2.5	65.5 ± 1.1^{b}	5.8 ± 1.1^{a}	11.3	0.1	6.6
L. cicera	7	3.3	72.2 ± 2.2	27.8 ± 2.2	2.6	59.7 ± 2.6^{ab}	12.5 ± 1.9^{a}	4.8	0.2	8.5
L. clymenum	7	3.7	74.5 ± 1.9	25.5 ± 1.9	2.9	67.9 ± 2.0^{b}	6.6 ± 1.4^{a}	10.2	0.1	7.7
L. filiformis	1	4.0	75.0	25.0	3.0	52.1^{ab}	22.9^{ab}	2.3	0.5	3.3
L. hirsutus	3	4.9	72.3 ± 3.6	27.7 ± 3.6	2.6	55.2 ± 5.6^{ab}	17.1 ± 7.2^{ab}	3.2	0.3	6.8
L. latifolius	5	6.0	74.9 ± 2.9	25.1 ± 2.9	3.0	45.1 ± 13.6^{a}	29.8 ± 10.9^{b}	1.5	0.7	8.3
L. ochrus	4	2.6	75.1 ± 3.2	24.9 ± 3.2	3.0	61.0 ± 8.6^{ab}	14.1 ± 6.2^{ab}	4.3	0.2	9.9
L. pratensis	4	4.4	74.3 ± 2.4	25.7 ± 2.4	2.9	61.0 ± 2.4^{ab}	13.3 ± 4.4^{ab}	4.6	0.2	6.7
L. sativus	2	1.5	76.4 ± 0.9	23.6 ± 0.9	3.2	58.5 ± 1.9^{ab}	17.9 ± 1.1^{ab}	3.3	0.3	13.8
L. setifolius	3	2.5	74.8 ± 1.0	25.2 ± 1.0	3.0	61.4 ± 1.5^{ab}	13.4 ± 1.0^{ab}	4.6	0.2	7.2
L. sphaericus	5	3.1	68.8 ± 2.6	31.2 ± 2.6	2.2	61.0 ± 2.2^{ab}	7.8 ± 1.4^{a}	7.8	0.2	4.6
L. tingitanus	7	2.3	73.9 ± 1.5	26.1 ± 1.5	2.8	56.0 ± 6.9^{ab}	17.9 ± 7.2^{ab}	3.1	0.4	7.1

n = number of populations studied. **$P<0.01$.

TUFAs total unsaturated fatty acids.

TSFAs total saturated fatty acids.

TUFAs/TSFAs total unsaturated fatty acids/saturated fatty acids ratio.

PUFAs total polyunsaturated fatty acids.

MUFAs monounsaturated fatty acids.

PUFAs/MUFAs polyunsaturated fatty acids/monounsaturated fatty acids ratio.

O/L oleic acid/linoleic acid ratio.

ω6/ω3 linoleic acid/linolenic acid ratio.

Among the four species previously mentioned with the lowest ratio $C_{14:0}$+$C_{15:0}$+$C_{16:0}$/$C_{18:0}$, *L. sativus* is the most remarkable by having the lowest content of total saturated fatty acids (23.6%). In the other three species, only *L. amphicarpos* shows a higher total saturated fatty acids (27.8%), although there are not significant differences.

Others saturated fatty acids, such as $C_{14:0}$, $C_{15:0}$ and $C_{20:0}$, were found in lesser concentrations and with similar values in all species examined.

Among unsaturated fatty acids, the most abundant in decreasing order were linoleic acid ($C_{18:2}$), oleic acid ($C_{18:1}$) and linolenic acid ($C_{18:3}$) (Table 1b), except for *Lathyrus angulatus*, *L. aphaca*, *L. clymenum*, *L. sphaericus* in which $C_{18:3}$ values were higher than $C_{18:1}$ contents. Nevertheless, within the 15 species examined, 6 species contain a remarkably higher content of oleic acid compared to the palmitic acid (*L. annuus*, *L. filiformis*, *L. hirsutus*, *L. latifolius*, *L. sativus,* and *L. tingitanus*). Regarding $C_{18:1}$, the highest percentages were observed in *L. latifolius* with 28.8%. which shows significant differences ($p<0.01$) with respect to *L. angulatus*, *L. aphaca*, *L. cicera*, *L. clymenum* y *L. sphaericus* (Tabla 1b). In most of the species studied, the quantity of $C_{18:1}$ was lower than $C_{18:2}$, and therefore the $C_{18:1}$/$C_{18:2}$ ratio is low (Table 2), as with other legumes [26, 66, 73]. A negative correlation (r2 = 0.61) between these two fatty acids was observed, which supports previous findings on *Lathyrus* [74], probably because $C_{18:1}$ is the substrate for the biosynthesis of $C_{18:2}$.

The $C_{18:3}$ contents ranged from 4% in *L. sativus* to 12% in *L. filiformis*. Finally, palmitoleic acid ($C_{16:1}$) is present in lesser amounts, with mean values below 1%.

Polyunsaturated fatty acids such as $C_{18:2}$ and $C_{18:3}$ not only contribute to a healthy diet but are also essential fatty acids for humans. Considering the negative effects on human health of saturated fatty acids, the FAO [75] recommend that the relation TUFAs/TSFAs should be over 0.84%. This ratio ranged from 2.2% in *L. sphaericus* to 3.2% *L. sativus* (Table 2), which are above the recommendations established by the FAO.

In addition to the total amounts of unsaturated fatty acids it is also important to establish the relative proportion of each fatty class. Thus, a diet rich exclusively in polyunsaturated fatty acids (PUFA) is not recommended if the proportion between ω6 and ω3 fatty acids is not balanced [76, 77, 78, 79]. $C_{18:2}$ and $C_{18:3}$ are the precursors of eicosanoids (prostaglandins, thromboxanes, and leukotrienes) with opposite effects in the human body. If derivatives of the linoleic acid are produced in small and controlled quantities, they can be used as structural components of plasmatic membranes [80]. Eicosanoids derived from $C_{18:2}$, such as arachidonic acid, increase the risk of cardiovascular diseases [81]. In contrast, eicosapentaenoic and docosahexaenoic acids derived from $C_{18:3}$ have opposite effects. Thus, the hypolipidemic, antithrombotic, and antiinflammatory effects of ω3 fatty acids have been extensively reported [82]. In this regard, research suggests that a high ω6/ω3 ratio may contribute to an increase in cardiovascular diseases [83]. According to the FAO [84], this ratio should be between 5/1 and 10/1, although some authors suggest that a 4/1 proportion is better, since an increase in $C_{18:3}$ improves the health of people suffering from asthma or arthritis [79, 85, 86]. The optimal proportion may also vary in function of the severity of the illness and the genetic predisposition [79].

Among the studied species, the ω6/ω3 ratio ranged from 3.3 in *L. filiformis* to 13.8 in *L. sativus* (Table 2). Four species (*L.* angulatus, *L. annuus*, *L. filiformis* y *L. sphaericus*) had an ω6/ω3 ratio below 5 and therefore a more balanced composition of polyunsaturated fatty acids. Thus, although *Lathyrus sativus* showed a good composition with respect to the TSFA contents and $C_{14:0}$+$C_{15:0}$+$C_{16:0}$/$C_{18:0}$ proportion, it had the worst ω6/ω3 ratio.

The results obtained for *L. sativus*, supports the results reported by Kuo et al. [20], which indicate a 56% of linoleic acid, 14.8% of palmitic acid, 16.7% of oleic acid, and 2.2% of linolenic acid, and the values reported by Grela & Günter [22] which are similar (57.1% of linoleic acid, 11.7% of palmitic acid, 16.2% of oleic acid, and 7.5% of linolenic acid).

Therefore, some of the species *Lathyrus* studied show a good composition of fatty acids in accordance to FAO requirements. This composition matches the vegetal nature of the samples which contain a high level of unsaturated fatty acids and a good ω6/ω3 ratio, which make them highly recommendable for human diet.

2.2. Genus *Vicia*

As in the previous genus, the main seed flour fatty acids in the *Vicia* species from Southern Spain ranged from myristic acid ($C_{14:0}$) to arachidic acid ($C_{20:0}$) with a predominance of unsaturated fatty acids of the series C_{18} (Tables 3a y 3b)

Table 3a. Saturated fatty acid composition (%) of the seed oil of studied *Vicia* taxa (mean ± s.d.) Different small letters indicate significant differences between values in the same column (Scheffe's test)

Taxa	n	$C_{14:0}$	$C_{15:0}$	$C_{16:0}$*	$C_{18:0}$***	$C_{20:0}$	TSFAs
V. altissima	1	0.2	0.3	19.3[ab]	3.1[ab]	0.8	23.8
V. angustifolia	5	1.9±0.9	0.4±0.1	17.6±0.6[ab]	3.3±0.3[a]	1.2±0.2	24.4±1.1
V. articulata	3	2.9±1.3	0.6±0.2	17.4±1.3[ab]	2.4±0.2[a]	1.1±0.1	24.4±2.2
V. benghalensis	4	0.3±0.0	0.2±0.0	17.6±1.8[ab]	2.7±0.2[a]	1.5±0.1	22.3±1.9
V. cordata	3	1.5±0.1	0.2±0.0	17.2±0.8[ab]	3.5±0.4[ab]	1.1±0.1	23.5±0.9
V. dasycarpa	3	1.5±1.1	0.4±0.3	19.0±1.2[ab]	3.2±0.3[ab]	0.9±0.1	25.1±1.1
V. disperma	6	1.2±0.8	0.4±0.1	17.5±0.7[ab]	3.5±0.6[a]	0.8±0.4	23.4±1.0
V. eriocarpa	4	1.4±0.4	0.3±0.1	17.6±1.5[ab]	3.0±0.1[a]	0.7±0.5	23.1±2.1
V. ervilia	3	3.1±0.9	0.8±0.2	17.2±0.8[ab]	2.4±0.3[a]	1.0±0.2	24.6±1.7
V. faba	1	0.2	0.2	17.5[ab]	3.1[ab]	1.5	22.6
V. glauca ssp. *giennensis*	1	0.3	0.3	17.4[ab]	2.9[ab]	1.1	21.9
V. hirsuta	3	2.7±1.4	0.6±0.2	20.4±0.6[b]	2.8±0.4[a]	1.4±0.2	27.9±2.4
V. hybrida	4	2.0±1.2	0.6±0.2	12.5±0.9[a]	3.7±0.3[ab]	0.6±0.0	19.3±1.9
V. incana	2	1.0±0.8	0.4±0.0	21.3±4.2[b]	5.1±2.6[ab]	1.2±0.7	29.0±6.6
V. lathyroides	3	0.4±0.1	0.5±0.1	19.3±0.6[ab]	4.4±0.3[ab]	1.1±0.1	25.7±1.2
V. lutea							
ssp. *cavanillesii*	1	1.7	0.3	18.2[ab]	3.7[ab]	0.9	24.8
ssp. *lutea* var. *hirta*	5	1.4±0.2	0.6±0.4	15.7±0.7[ab]	4.5±0.3[ab]	0.8±0.1	23.0±1.0
ssp. *lutea* var. *lutea*	3	1.1±0.7	1.0±1.2	15.1±0.9[ab]	4.0±0.7[ab]	0.7±0.1	21.9±2.7
ssp. *vestita*	2	1.8±1.5	0.9±0.9	14.6±1.2[ab]	3.5±0.1[ab]	0.8±0.0	21.5±1.2
V. monantha ssp. *calcarata*	3	0.9±0.7	0.2±0.1	18.7±1.1[ab]	1.9±0.2[a]	0.6±0.0	22.4±2.0
V. monardi	3	0.7±0.3	0.2±0.0	17.0±0.9[ab]	2.2±0.3[a]	1.0±0.2	21.1±1.3
V. narbonensis	3	1.9±1.2	0.3±0.3	14.3±0.7[ab]	3.1±0.8[ab]	1.1±0.3	20.7±1.0
V. onobrychioides	2	2.1±1.7	0.6±0.4	14.9±0.3[ab]	3.5±0.0[ab]	0.7±0.0	21.7±1.8
V. parviflora	5	2.9±0.7	0.4±0.2	15.6±0.4[ab]	5.9±0.6[ab]	1.0±0.2	25.7±0.8
V. peregrina	3	1.9±0.7	0.4±0.0	14.3±3.3[ab]	4.1±1.9[ab]	0.7±0.1	21.4±0.9

Taxa	n	$C_{14:0}$	$C_{15:0}$	$C_{16:0}$*	$C_{18:0}$***	$C_{20:0}$	TSFAs
V. pseudocracca	5	1.0±1.2	0.4±0.2	17.5±2.5^{ab}	3.7±0.8^{ab}	1.5±0.2	24.1±2.2
V. pubescens	3	3.1±2.1	0.8±0.2	14.6±0.4^{ab}	7.5±0.2^{b}	1.3±0.0	27.3±2.5
V. pyrenaica	1	0.3	0.4	17.7^{ab}	3.9^{ab}	0.9	23.1
V. sativa ssp. *sativa*	5	3.0±1.0	0.7±0.4	16.5±0.5^{ab}	3.7±0.3^{ab}	1.1±0.2	25.0±2.0
V. tenuifolia	4	3.2±1.4	0.6±0.4	17.5±1.2^{ab}	2.6±0.5^{a}	0.9±0.1	24.8±2.3
V. vicioides	3	0.4±0.1	0.2±0.1	21.0±2.9^{b}	3.2±0.5^{ab}	1.1±0.3	26.0±3.9

n = number of populations studied. *$P<0.05$, ***$P<0.001$. TSFAs = % total saturated fatty acids

Table 3b. Unsaturated fatty acid composition (%) of the seed oil of studied *Vicia* taxa (mean ± s.d.) Different small letters indicate significant differences between values in the same column (Scheffe's test)

Taxa	n	$C_{16:1}$	$C_{18:1}$**	$C_{18:2}$***	$C_{18:3}$***	TUFAs
V. altissima	1	0.3	15.9^{abc}	54.2^{abcd}	5.9^{abcd}	76.2
V. angustifolia	5	0.2±0.2	9.2±1.8^{a}	55.5±2.9^{bcd}	10.7±1.6^{abcd}	75.6±1.1
V. articulata	3	0.6±0.4	29.7±3.4^{bc}	28.7±1.9^{a}	16.6±2.9^{d}	75.6±2.2
V. benghalensis	4	0.1±0.1	20.3±7.3^{abc}	51.3±5.2^{bcd}	6.0±1.0^{ab}	77.7±2.0
V. cordata	3	0.1±0.1	12.9±0.6^{ab}	55.3±1.2^{bcd}	8.2±0.9^{abcd}	76.5±0.9
V. dasycarpa	3	0.1±0.1	18.8±0.2^{abc}	47.5±2.1^{abcd}	8.6±0.8^{abcd}	74.9±1.1
V. disperma	6	0.2±0.3	18.9±2.2^{abc}	52.4±1.3^{bcd}	5.1±0.4^{ab}	76.6±1.1
V. eriocarpa	4	0.4±0.3	18.3±4.7^{abc}	50.1±3.1^{abcd}	8.2±1.1^{abcd}	76.9±2.1
V. ervilia	3	0.9±0.1	32.5±0.9^{c}	36.8±2.4^{ab}	5.3±0.2^{ab}	75.4±1.7
V. faba	1	0.2	27.8^{abc}	46.8^{abcd}	2.7^{abcd}	77.4
V. glauca ssp. *giennensis*	1	0.3	18.5^{abc}	53.5^{abcd}	5.7^{abcd}	78.1
V. hirsuta	3	0.3±0.1	11.6±1.1^{ab}	48.9±3.3^{abcd}	11.3±0.8^{abcd}	72.1±2.5
V. hybrida	4	0.2±0.1	10.0±1.1^{a}	66.3±2.5^{d}	4.1±0.3^{ab}	80.7±1.9
V. incana	2	0.8±0.3	9.8±2.8^{ab}	50.5±7.9^{abcd}	9.9±3.8^{abcd}	71.0±6.6
V. lathyroides	3	0.7±0.1	11.9±1.5^{ab}	48.3±2.0^{abcd}	13.4±0.5^{bcd}	74.3±1.2
V. lutea						
ssp. *cavanillesii*	1	0.6	13.6^{abc}	54.9^{abcd}	6.1^{abcd}	75.2
ssp. *lutea* var. *hirta*	5	0.6±0.2	12.0±1.7^{a}	55.2±1.0^{bcd}	9.2±1.2^{abcd}	77.0±1.0
ssp. *lutea* var. *lutea*	3	0.5±0.4	21.4±5.0^{abc}	47.8±3.8^{abcd}	8.4±0.9^{abcd}	78.1±2.7
ssp. *vestita*	2	0.8±0.2	22.9±2.1^{abc}	50.1±0.3^{abcd}	4.6±0.3^{abc}	78.5±1.2
V. monantha ssp. *calcarata*	3	0.4±0.1	15.5±2.4^{abc}	57.7±2.0^{bcd}	4.1±0.5^{ab}	77.6±2.0
V. monardi	3	0.2±0.2	9.1±2.9^{a}	63.3±2.8^{cd}	6.3±1.3^{abc}	78.9±1.3
V. narbonensis	3	0.5±0.3	30.1±4.7^{bc}	45.5±4.6^{abc}	3.2±0.2^{a}	79.3±1.0
V. onobrychioides	2	0.5±0.3	21.2±4.3^{abc}	51.3±2.8^{abcd}	5.2±0.1^{abc}	78.3±1.8
V. parviflora	5	0.5±0.0	12±3.5^{a}	46.9±2.3^{abc}	14.8±1.8^{cd}	74.3±0.8
V. peregrina	3	0.2±0.1	17.5±2.1^{abc}	55.8±2.1^{bcd}	5.1±1.2^{ab}	78.6±0.9
V. pseudocracca	5	0.3±0.3	13.1±2.7^{ab}	54.6±4.4^{bcd}	7.9±1.5^{abcd}	75.9±2.2
V. pubescens	3	0.6±0.2	13.0±2.4^{ab}	42.5±2.9^{abc}	16.6±1.6^{d}	72.7±2.5
V. pyrenaica	1	0.4	7.2^{abc}	61.0^{bcd}	8.2^{abcd}	76.9
V. sativa ssp. *sativa*	5	0.8±0.3	13.3±1.5^{ab}	53.6±2.6^{bcd}	7.3±1.5^{abc}	75.0±2.0
V. tenuifolia	4	0.3±0.3	17.5±1.8^{abc}	50.2±3.8^{abcd}	7.2±1.8^{abc}	75.2±2.3
V. vicioides	3	0.4±0.1	14.6±4.3^{abc}	49.6±3.0^{abcd}	9.5±2.6^{abcd}	74.0±3.9

n = number of populations sudied. **$P<0.01$, ***$P<0.001$. TUFAs = % total unsaturated fatty acids.

All species studied had higher contents of total unsaturated fatty acids (TUFAs) than saturated ones (TSFAs) (Tables 3a y 3b). These results are in accordance with those obtained by other authors in other legumes [26, 66, 68, 87]. The content of TSFAs ranged between 19.3% in *V. hybrida* and 29.0% in *V. incana* (Table 3a), not observing significant differences between these two species. No significant difference was observed in the content of TUFAs either, which ranged from 71.0% in *V. incana* to 80.7% in *V. hybrida* (Table 3b).

Among saturated fatty acids, palmitic acid ($C_{16:0}$) was the most abundant with amounts ranging from 12.5% in *V. hybrida* to 21.3% in *V. incana*. $C_{16:0}$ contents in *V. hybrida* were significantly different only from the three species with the highest $C_{16:0}$ contents (*V. hirsuta*, *V. incana* and *V. vicioides*). The second most abundant saturated fatty acid was stearic acid $C_{18:0}$. *V. pubescens* had the highest content (7.5%) of this fatty acid and was significantly different from the amounts observed in the ten species with $C_{18:0}$ lowest percentages (Table 3a). The lowest ratio $C_{14:0}+C_{15:0}+C_{16:0}/C_{18:0}$ was observed in *V. pubescens* and the highest in *V. monantha* subsp. *calcarata* (Table 4). Most cultivated species, *V. faba* and *V. sativa* subsp. *sativa*, had intermediate ratios with 5.7 and 5.4, respectively. In general, $C_{14:0}+C_{15:0}+C_{16:0}/C_{18:0}$ ratio in the *Vicia* studied were higher than those observed in the related genus *Lathyrus* [88].

Other saturated fatty acids present in minor amounts were $C_{14:0}$, $C_{15:0}$ and $C_{20:0}$. However, no significant differences in their contents between the species studied were observed.

Among unsaturated fatty acids, the most abundant in decreasing order were $C_{18:2}$, $C_{18:1}$ and $C_{18:3}$ (Table 3b) except for *V. angustifolia*, *V. lathyroides*, *V. parviflora*, *V. pubescens*, *V. pyrenaica* in which $C_{18:3}$ values were higher than $C_{18:1}$ contents, and *V. hirusta* and *V. incana* where both fatty acids showed similar contents. The $C_{18:2}$ contents ranged from 28.7% in *V. articulata* to 66.3% in *V. hybrida*. The $C_{18:2}$ content in the latter species was significantly different ($P<0.001$) from those observed in *V. articulata*, *V. ervilia*, *V. narbonensis, V. parviflora* and *V. pubescens.*

The highest percentage of $C_{18:1}$ was observed in *V. ervilia* with 32.5%. Twelve of the studied species had $C_{18:1}$ significantly contents lower ($P<0.01$) than those reported for *V. ervilia*. In all studied species, with the exception of *V. articulata*, the quantity of $C_{18:1}$ was lower than $C_{18:2}$, and therefore the $C_{18:1}/C_{18:2}$ ratio was below one (Table 3b), as with other legumes [26]. The $C_{18:3}$ contents ranged from 2.7% in *V. faba* to 16.6% in *V. articulata* and *V. pubescens*. $C_{18:3}$ content in both species was significantly different ($P<0.001$) with respect to the 12 *Vicia* species with lowest contents of this fatty acid (Table 3b)

Among the species studied, the ω6/ω3 ratio ranged from 1.7% in *V. articulata* to 17.3 in *V. faba* (Table 4). The ω6/ω3 ratios between 5 and 10 were observed in 19 of the 31 taxa studied, and therefore indicating a more balanced distribution of polyunsaturated fatty acids.

Five of these taxa have a coefficient value which is lower than 5, and the species *V. articulata*, *V. lathyroides*, *V. parviflora* and *V. pubescens* have a remarkable value that is lower than 4, being a very balanced composition into poli-unsaturated fatty acids according to some authors, such as Simopoulos [79]. Among these species, only *V. articulata* has been cultivated in the past. It is noticeable that these species report better values for this coefficient than *V. faba*, which is the most-commonly used species in the diet.

Table 4. Different parameters based on the fatty acid composition of the seed oil of studied *Vicia* taxa Data (TUFAs, TSFAs, PUFAs y MUFAs) are expressed as the average ± standard deviation. Different small letters indicate significant differences between values in the same column (Scheffe's test)

Taxa	n	$C_{14:0}+C_{15:0}+C_{16:0}/C_{18:0}$	TUFAs	TSFAs	TUFAs/TSFAs	PUFAS***	MUFAs***	PUFAs/MUFAs	O/L	ω6/ω3
V. altissima	1	6.6	76.2	23.8	3.2	60.1^{abcd}	16.1^{abcd}	3.7	0.3	9.2
V. angustifolia	5	5.9	75.6 ± 1.1	24.4 ± 1.1	3.1	66.2 ± 2.0^{d}	9.4 ± 1.9^{d}	7.1	0.2	5.2
V. articulata	3	8.5	75.6 ± 2.2	24.4 ± 2.2	3.1	45.3 ± 2.1^{abc}	30.3 ± 3.0^{ab}	1.5	1.0	1.7
V. benghalensis	4	6.5	77.7 ± 2.0	22.3 ± 1.9	3.5	57.3 ± 5.9^{abcd}	20.4 ± 7.3^{abcd}	2.8	0.4	8.6
V. cordata	3	5.4	76.5 ± 0.9	23.5 ± 0.9	3.3	63.5 ± 0.5^{cd}	13.0 ± 0.6^{bcd}	4.9	0.2	6.7
V. dasycarpa	3	6.5	74.9 ± 1.1	25.1 ± 1.1	3.0	56.1 ± 1.3^{abcd}	18.8 ± 0.2^{abcd}	3.0	0.4	5.5
V. disperma	6	5.5	76.6 ± 1.1	23.4 ± 1.0	3.3	57.5 ± 1.4^{abcd}	19.1 ± 2.1^{abcd}	3.0	0.4	10.3
V. eriocarpa	4	6.4	76.9 ± 2.1	23.1 ± 2.1	3.3	58.2 ± 4.1^{abcd}	18.7 ± 4.9^{abcd}	3.1	0.4	6.1
V. ervilia	3	8.8	75.4 ± 1.7	24.6 ± 1.7	3.1	42.1 ± 2.6^{a}	33.3 ± 1.0^{a}	1.3	0.9	6.9
V. faba	1	5.7	77.4	22.6	3.4	49.4^{abcd}	20.8^{abcd}	1.8	0.6	17.3
V. glauca ssp. *giennensis*	1	6.2	78.1	21.9	3.6	59.2^{abcd}	18.9^{abcd}	3.1	0.3	9.4
V. hirsuta	3	8.3	72.1 ± 2.5	27.9 ± 2.4	2.6	60.2 ± 3.0^{d}	11.9 ± 1.1^{abcd}	5.1	0.2	4.3
V. hybrida	4	4.1	80.7 ± 1.9	19.3 ± 1.9	4.2	70.4 ± 2.5^{d}	10.3 ± 1.0^{d}	6.8	0.2	16.2
V. incana	2	4.4	71.0 ± 6.6	29.0 ± 6.6	2.5	60.5 ± 4.1^{d}	10.5 ± 2.5^{abcd}	5.7	0.2	5.1
V. lathyroides	3	4.5	74.3 ± 1.2	25.7 ± 1.2	2.9	61.7 ± 2.1^{d}	12.6 ± 1.5^{abcd}	4.9	0.2	3.6
V. lutea										
ssp. *cavanillesii*	1	5.5	75.2	24.8	3.1	61.0^{d}	14.2^{abcd}	4.3	0.2	9.0
ssp. *lutea* var. *hirta*	5	3.9	77.0 ± 1.0	23.0 ± 1.0	3.4	64.4 ± 1.3^{d}	12.6 ± 1.6^{cd}	5.1	0.2	6.0
ssp. *lutea* var. *lutea*	3	4.3	78.1 ± 2.7	21.9 ± 2.7	3.6	56.1 ± 3.6^{abcd}	21.9 ± 4.6^{abcd}	2.6	0.4	5.7
ssp. *vestita*	2	4.9	78.5 ± 1.2	21.5 ± 1.2	3.6	54.7 ± 0.5^{abcd}	23.8 ± 1.8^{abcd}	2.3	0.5	10.9
V. monantha ssp. *calcarata*	3	10.2	77.6 ± 2.0	22.4 ± 2.0	3.5	61.8 ± 2.1^{bcd}	15.8 ± 2.3^{abcd}	3.9	0.3	14.1
V. monardi	3	7.9	78.9 ± 1.3	21.1 ± 1.3	3.7	69.6 ± 3.1^{d}	9.3 ± 3.0^{d}	7.5	0.1	10.0
V. narbonensis	3	5.3	79.3 ± 1.0	20.7 ± 1.0	3.8	48.7 ± 4.7^{ab}	30.6 ± 4.8^{abc}	1.6	0.7	14.2
V. onobrychioides	2	5.1	78.3 ± 1.8	21.7 ± 1.8	3.6	56.6 ± 2.8^{abcd}	21.7 ± 4.6^{abcd}	2.6	0.4	9.9
V. parviflora	5	3.2	74.3 ± 0.8	25.7 ± 0.8	2.9	61.8 ± 3.4^{d}	12.5 ± 3.4^{bcd}	4.9	0.3	3.2

Table 4. (Continued)

Taxa	n	$C_{14:0}+C_{15:0}+C_{16:0}/C_{18:0}$	TUFAs	TSFAs	TUFAs/TSFAs	PUFAS***	MUFAs***	PUFAs/MUFAs	O/L	ω6/ω3
V. peregrina	3	4.0	78.6 ± 0.9	21.4 ± 0.9	3.7	60.9 ± 1.2^{abcd}	17.7 ± 2.1^{abcd}	3.4	0.3	10.9
V. pseudocracca	5	5.1	75.9 ± 2.2	24.1 ± 2.2	3.1	62.5 ± 3.1^{d}	13.4 ± 2.6^{bcd}	4.7	0.2	6.9
V. pubescens	3	2.5	72.7 ± 2.5	27.3 ± 2.5	2.7	59.1 ± 3.5^{bcd}	13.6 ± 2.3^{abcd}	4.3	0.3	2.6
V. pyrenaica	1	4.7	76.9	23.1	3.3	69.2^{bcd}	7.7^{abcd}	9.0	0.1	7.4
V. sativa ssp. *sativa*	5	5.4	75.0 ± 2.0	25.0 ± 2.0	3.0	60.9 ± 3.4^{d}	14.1 ± 1.5^{bcd}	4.3	0.2	7.3
V. tenuifolia	4	8.0	75.2 ± 2.3	24.8 ± 2.3	3.0	57.4 ± 2.9^{abcd}	17.8 ± 2.1^{abcd}	3.2	0.3	7.0
V. vicioides	3	6.7	74.0 ± 3.9	26.0 ± 3.9	2.9	59.0 ± 2.2^{bcd}	15.0 ± 4.1^{abcd}	3.9	0.3	5.2

n = number of populations studied. ***$P<0.01$. TUFAs total unsaturated fatty acids., TSFAs total saturated fatty acids, TUFAs/TSFAs total unsaturated fatty acids/saturated fatty acids ratio, PUFAs polyunsaturated fatty acids, MUFAs monounsaturated fatty acids, PUFAs/MUFAs polyunsaturated fatty acids/monounsaturated fatty acids ratio, O/L oleic acid/linoleic acid ratio, ω6/ω3 linoleic acid/linolenic acid ratio.

The fat composition of the diet rather than the amount of fat intake is more important in determining the risk factors for heart disease since the dietary fat is a major factor in determining the long-term fatty acid composition of adipose tissue [83]. Hence the good balance in the TSFA/TUFA and ω6/ω3 fatty acids ratios of the *Vicia* species studied may help us correct this current balance in the western diet. It has been observed that the processing and cooking of legumes do not affect the fatty acid composition, while antinutritional compounds present in legumes are removed or degraded [89, 90], thus making cooked legumes a possible source of important fatty acids.

3. Polyphenol Analysis

3.1. Genus *Lathyrus*

A. Polyphenol contents in Lathyrus seeds

Polyphenolic contents in studied *Lathyrus* species ranged between 3.8 mg/g seed flour in *L. setifolius* and 29.2 mg/g seed flour in *L. sphaericus* (Table 5).

Highest polyphenolic contents were observed in non-cultivated taxa such as *L. sphaericus* (29.2 mg/g), *L. angulatus* (27.5 mg/g), *L. aphaca* (23.9 mg/g), *L. filiformis* (20.8mg/g) and cultivated *L. tingitanus* (25.5mg/g), while lowest amounts were observed in cultivated *L. sativus* (4.3mg/g), *L. annuus* (4.6 mg/g) and non-cultivated *L. setifolius* (3.8 mg/g).

Table 5. Polyphenolic contents (means of three independent deterninations ± standard deviation) of *Lathyrus* seed flour (mg of catechin-equivalent polyphenolics/g flour) and seed size (mm2). Different small letters indicate significant differences between values in the same column (Tukey's test, P<0.001)

Species	Polyphenolic contents	Seed size
L. amphicarpos	12.4 ± 0.60^{bc}	74.7
L. angulatus	27.5 ± 0.51^{f}	17.3
L. annuus	4.6 ± 0.05^{a}	90.8
L. aphaca	23.9 ± 1.62^{ef}	23.4
L. cicera	6.0 ± 0.85^{a}	78.5
L. clymenum	6.2 ± 0.14^{a}	103.4
L. filiformis	20.8 ± 0.90^{de}	31.9
L. hirsutus	8.6 ± 0.25^{ab}	41.3
L. latifolius	5.9 ± 0.07^{a}	82.6
L. ochrus	6.1 ± 0.02^{a}	90.8
L. pratensis	17.5 ± 2.26^{cd}	33.6
L. sativus	4.3 ± 0.11^{a}	117.7
L. setifolius	3.8 ± 0.05^{a}	76.9
L. sphaericus	29.2 ± 1.55^{f}	27.9
L. tingitanus	25.5 ± 1.76^{ef}	116.7

Species that are marginally cultivated or were cultivated in the past are underlined.

Considering these data, the species that have been used in crops, in the past or at present (*L. annuus*, *L. sativus*), generally have lower polyphenol contents than the species which have never been cultivated at all, such as *L. sphaericus* or *L. angulatus* (Figure 1). This fact can be related to the selection made to grow plants with low content of polyphenols in the past, as they were traditionally considered as anti-nutritional substances [91, 92, 93].

Nevertheless, this perspective is now changing, since polyphenols are regarded as beneficial substances for health due to their anticarcinogenic and antioxidant properties [94, 95, 96, 97, 98]. As a result, the species which can be interesting from this point of view are those with high quantities of polyphenols or the ones which have polyphenols with a solid specific activity, even though their content in polyphenols is not remarkably high.

It is generally recognized that polyphenolic compounds are more abundant in the hull than in the seed cotyledons [99, 100]. Therefore within the same taxonomic group, in this case the genus *Lathyrus*, it is expected that smaller seeds, with a higher proportion of hull, will have higher polyphenolic contents. In studied *Lathyrus* species it was observed a negative correlation between polyphenolic contents and seed size (Figure 1).

Thus, smaller seeds contained more polyphenolics in the extracts than bigger ones due to their higher proportion of hulls. Four of the six cultivated *Lathyrus* species (white circles in Figure 1) with low polyphenolic contents and high seed size place in the down right corner. On the contrary, in the upper left corner are wild species with low seed size and high polyphenolic contents. An exception was *L. tingitanus*, a cultivated species, which with a high seed size possessed also high polyphenol contents. When *L. tingitanus* was not taken into account, the negative correlation between seed size and polyphenol contents increased from 0.37 to 0.59. This correlation was more evident when cultivated and non-cultivated species were considered independently. Thus, cultivated (excluding *L. tingitanus*) and non cultivated *Lathyrus* showed negative correlation indexes of 0.63 and 0.8 respectively [101].

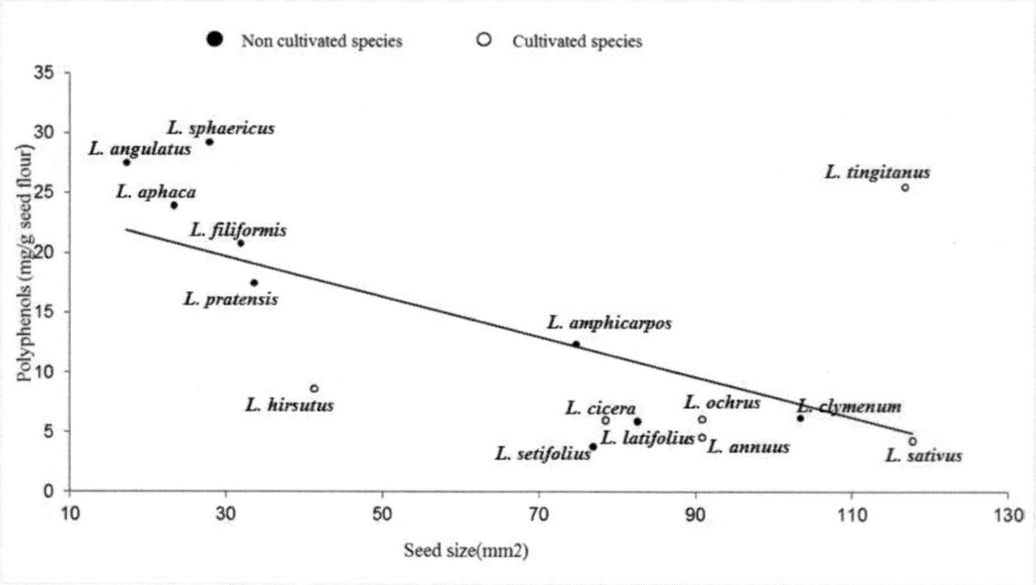

Figure 1. Correlation between polyphenolic contents (mg/g seed flour) and seed size in cultivated (white circles) and non-cultivated (black circles) *Lathyrus* species.

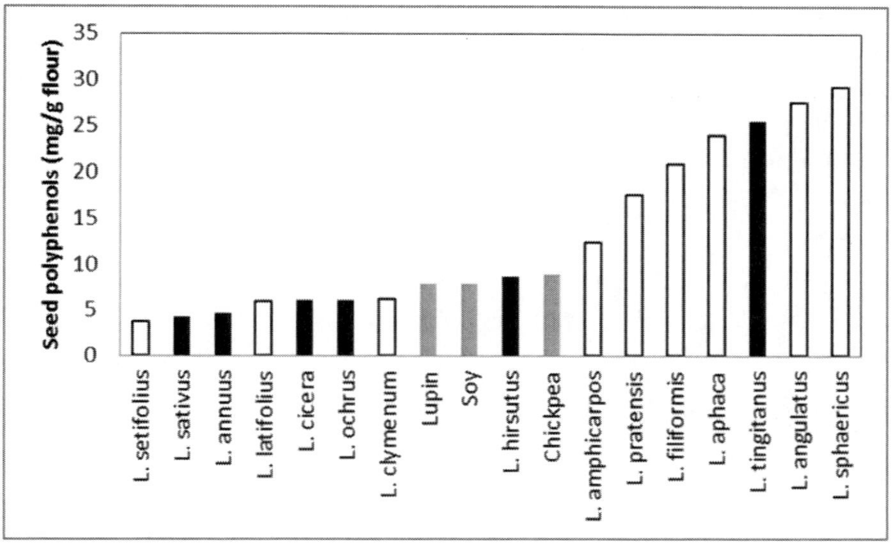

Figure 2. Seed polyphenolic contents in wild (white bars) and cultivated (black bars) *Lathyrus* species and some commercial legumes (grey bars).

In previous studies based in seeds of leguminous plants, low phenol contents have generally been obtained. Thus, Sosulski & Dabrowski [102] indicate that the total content of free phenols in the flour of some legumes like green peas, lentils, broad beans, lupins, and chickpeas ranged between 0.018 mg/g and 1.63 mg/g. Later, Chavan et al. [39] obtained 2.55 mg/g for *Pisum sativum*, 2.32 mg/g for *L. sativus* from Canada, and 1.99 mg/g for *L. sativus* from India. Cardador-Martínez et al. [103] obtained 2.09 mg of phenols/g in flour for the species *Phaseolus vulgaris*. Recently, Piecyk et al. [104] obtained values of 1.05 mg/g, 0.93 mg/g, and 7.82 mg/g for *Phaseolus vulgaris*, *Pisum sativum*, and *Lens culinaris*, respectively. Comparatively, the results obtained in this study are considerably higher than the results for other legumes, especially for *L. amphicarpos*, *L. pratensis*, *L. filiformis*, *L. aphaca*, *L. tingitanus*, *L. angulatus*, and *L. sphaericus* (Table 5, Figure 2). Having considered all the benefits attributable to these secondary metabolites, the use of the genus *Lathyrus* in human and animal diet should be taken into consideration.

B. Antioxidant activity of Lathyrus polyphenolic compounds

The antioxidant activity of the polyphenols extracted from *Lathyrus* seeds has been studied in two different experiments. In the first experiment, a phenol extract equivalent to 2 μg of catechine was obtained, whereas in the second experiment the antioxidant activity of the polyphenols was compared to a fixed quantity of flour (extract of 5 μL for each taxon). Four different formulas were used to calculate and express the antioxidant activity, as described by Oomah & Mazza [105] and Velioglu et al. [106].

In the first experiment, extracted *L. annuus* and *L. sativus* polyphenolics provided highest antioxidant activity (Table 6). Both species show equal AOX, DR and ORR values, while *L. annuus* shows slightly better AA and AAC values. *L. annuus* and *L. sativus* have been or are cultivated. Also, with high antioxidant activity per mg phenolics are other cultivated *Lathyrus* species such as *L. cicera* or *L. ochrus*. Among non-cultivated, *L. amphicarpos* and *L. setifolius* showed the best results. On the other hand *Lathyrus* species with lowest antioxidant activity are non-cultivated *L. filiformis* and *L. angulatus*. In general it was observed that seed polyphenolics of species with higher amounts of them possessed lower specific activity. These corresponded to wild taxa, while polyphenolics with highest specific activity are observed in cultivated taxa. Thus a negative correlation (r^2 = 0.61) between seed polyphenol contents and specific activity was observed (Figure 3).

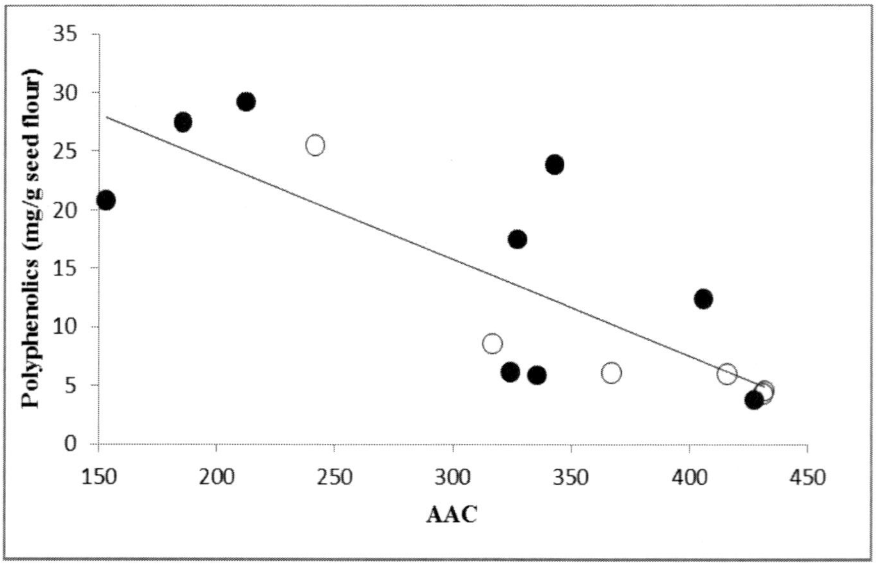

Figure 3. Correlation between polyphenolic contents (mg/g seed flour) and antioxidant activity coefficient in studied cultivated (white circles) and non-cultivated (black circles) *Lathyrus* species.

Table 6. Specific antioxidant activity of polyphenolics from *Lathyrus* seed flour. 2 μg of catechin-equivalent phenolics were assayed

Species	AOX[a]	DR[b]	AA[c]	ORR[d]	AAC[e]
L. amphicarpos	0.0030	0.007	56.83	0.431	405.94
L. angulatus	0.0045	0.011	34.24	0.658	185.53
L. annuus	0.0026	0.007	57.83	0.422	431.91
L. aphaca	0.0032	0.008	51.20	0.488	342.86
L. cicera	0.0027	0.007	56.26	0.437	415.96
L. clymenum	0.0034	0.009	48.76	0.512	324.30
L. filiformis	0.0049	0.012	28.47	0.715	152.88
L. hirsutus	0.0035	0.009	48.26	0.517	316.51
L. latifolius	0.0033	0.009	49.07	0.509	335.44
L. ochrus	0.0030	0.008	52.71	0.473	366.98
L. pratensis	0.0033	0.009	49.56	0.504	327.27
L. sativus	0.0026	0.007	57.78	0.422	431.17
L. setifolius	0.0027	0.007	56.66	0.433	427.46
L. sphaericus	0.0043	0.011	35.85	0.641	212.25
L. tingitanus	0.0040	0.010	39.96	0.600	241.56

[a] Antioxidant value. [b] Degradation rate. [c] Antioxidant activity. [d] Oxidation rate ratio. [e] Antioxidant activity coefficient.

Species that are marginally cultivated or were cultivated in the past are underlined.

In a second experiment, antioxidant activity of extracted polyphenolics was compared referred to a fixed amount of flour. Thus, 5 μL of extract were assayed for all taxa independently of phenolic concentration in each species. With this experiment the antioxidative activity of *Lathyrus* seeds per gram of flour was determined. If all polyphenolics extracted from the different *Lathyrus* species studied possessed the same specific antioxidant activity it would be expected that species containing more polyphenolics per gram of flour would yield higher antioxidant activity. However, we have already shown that specific activity of polyphenolics varies between *Lathyrus* species studied. Thus, *L. aphaca* yielded the best results for all values of antioxidant activity but, although with high phenolic contents, 23.9 mg/g seed flour, was not the one with highest concentration (Table 7). Thus, richest phenolic species such as *L. sphaericus*, *L. angulatus* or *L. tingitanus* showed only intermediate antioxidant values. This is in accordance with the lower specific activity of their seed polyphenolics. Thus *L. aphaca*, in addition to containing high amounts of polyphenolics in their seeds, these polyphenolics possessed a high specific antioxidative activity.

Antioxidant activity of *L. annuus* polyphenolic extracts was compared with that of commercial legumes chickpea, lupin and soy and with synthetic (BHT) and natural (catechin) antioxidants (Table 8). Antioxidant activity of *L. annuus* polyphenolics is twice that observed in chickpea, lupin or soy and only a few higher than 15 μg of catechin and 0.64 μg BHT.

In a similar way polyphenolic extracts from *L. aphaca*, a non-cultivated taxon, were characterized for their antioxidant activity (Table 9). 5 μl of *L. aphaca* methanolic extracts showed considerably higher antioxidant activity than same amounts of methanolic extracts from chickpea, lupin and soy. Antioxidant activity of 5 μL of *L. aphaca* population methanolic extracts was only a few higher than to 1.28 μg of BHT and 15 μg of catechin (Table 9).

Table 7. Antioxidant activity of polyphenols extracted from *Lathyrus* species per gram of flour extracted

Species	AOX[a]	DR[b]	AA[c]	ORR[d]	AAC[e]
L. amphicarpos	0.0025	0.0066	62.59	0.374	520.88
L. angulatus	0.0031	0.0081	54.40	0.456	430.39
<u>*L. annuus*</u>	0.0026	0.0073	58.85	0.412	483.76
L. aphaca	0.0020	0.0055	68.76	0.312	604.41
<u>*L. cicera*</u>	0.0026	0.0072	59.25	0.408	476.80
L. clymenum	0.0033	0.0086	51.53	0.485	409.51
L. filiformis	0.0036	0.0090	49.40	0.506	377.03
<u>*L. hirsutus*</u>	0.0031	0.0075	57.48	0.425	452.44
L. latifolius	0.0032	0.0083	52.78	0.472	423.43
<u>*L. ochrus*</u>	0.0028	0.0077	56.58	0.434	461.72
L. pratensis	0.0022	0.0060	65.97	0.340	586.45
<u>*L. sativus*</u>	0.0028	0.0075	57.75	0.423	465.20
L. setifolius	0.0030	0.0076	56.80	0.432	460.56
L. sphaericus	0.0032	0.0085	52.23	0.478	402.55
<u>*L. tingitanus*</u>	0.0028	0.0073	59.03	0.410	474.48

[a] Antioxidant value. [b] Degradation rate. [c] Antioxidant activity. [d] Oxidation rate ratio. [e] Antioxidant activity coefficient.

Species that are marginally cultivated or were cultivated in the past are underlined.

Table 8. Antioxidant activity of polyphenolics extracted from *L. annuus*, chickpea, lupin and soy. 2 µg of catechin-equivalent polyphenolics were assayed. BHT and catechin were assayed in the amounts per well indicated in the table. Results are the average of three independent determinations

	AOX[a]	DR[b]	AA[c]	ORR[d]	AAC[e]
L. annuus	0.0026	0.007	57.83	0.42	431.91
Chickpea	0.0057	0.013	20.79	0.79	130.95
Lupin	0.0055	0.0123	24.90	0.75	166.67
Soy	0.0052	0.0116	29.56	0.70	166.67
BHT (0.38 µg)	0.0044	0.0099	39.66	0.60	285.71
BHT (0.64 µg)	0.0039	0.0086	47.58	0.52	357.14
Catechin (10 µg)	0.0043	0.0097	41.02	0.59	297.62
Catechin (15 µg)	0.0038	0.0086	47.58	0.52	357.14

[a] Antioxidant value. [b] Degradation rate. [c] Antioxidant activity. [d] Oxidation rate ratio. [e] Antioxidant activity coefficient.

Table 9. Antioxidant activity per gram of seed meal of polyphenolics extracted from *L. aphaca*, chickpea, lupin and soy. BHT and catechin were assayed in the amounts per well indicated in the table. Results are the average of three independent determinations

	AOX[a]	DR[b]	AA[c]	ORR[d]	AAC[e]
L. aphaca	0.0020	0.0055	68.76	0.31	604.41
Chickpea	0.0067	0.0092	23.64	0.76	139.54
Lupin	0.0059	0.0081	32.77	0.67	209.30
Soy	0.0053	0.0073	39.46	0.61	267.44
BHT (1.3 µg)	0.0036	0.0049	59.46	0.41	453.19
Catechin (15 µg)	0.0036	0.0048	59.88	0.40	465.11

[a] Antioxidant value. [b] Degradation rate. [c] Antioxidant activity. [d] Oxidation rate ratio. [e] Antioxidant activity coefficient.

Among the studied *Lathyrus* species two of them yielded polyphenolics with highest antioxidant activity. Thus, cultivated *L. annuus* polyphenolics have the highest specific antioxidant activity, while non-cultivated *L. aphaca* have the seed flour with highest antioxidative activity.

Phenolic compounds are receiving increasing attention because of their health promoting effects, attributed to their antioxidant activity. Studied *Lathyrus* species contained polyphenolic compounds with a high antioxidative activity when compared with commercial legumes such as chickpea, lupin or soy. They may represent an interesting source of antioxidant polyphenolics that may favour the extension of their cultivation as new source of natural antioxidants in addition to containing high quality proteins for human or animal nutrition.

3.2. Genus *Vicia*

A. Polyphenol contents in Vicia seeds

The concentration of total polyphenols in the seeds corresponding to 28 *Vicia* species collected from Southern Spain was determined (Table 10).

Tabla 10. Polyphenolic contents (means of three independent deterninations ± standard deviation) of *Vicia* seed flour (mg of catechin-equivalent polyphenolics/g flour) and seed size (mm2). Different small letters indicate significant differences between values in the same column (Tukey's test, P<0.001)

Taxa	Polyphenolic contents	Seed size
V. altissima	4.4 ± 0.18^{abc}	44.0
V. angustifolia	11.4 ± 1.02^{c}	26.4
V. articulata	6.8 ± 0.05^{abc}	65.9
V. benghalensis	2.7 ± 0.07^{ab}	56.5
V. cordata	9.8 ± 0.74^{bc}	45.8
V. dasycarpa	2.1 ± 0.07^{ab}	50.2
V. disperma	2.8 ± 0.23^{ab}	39.6
V. eriocarpa	3.6 ± 0.25^{ab}	44.0
V. ervilia	7.9 ± 0.97^{abc}	53.4
V. faba	4.6 ± 0.18^{abc}	219.8
V. glauca ssp. *giennensis*	6.4 ± 0.09^{abc}	50.2
V. hirsuta	7.4 ± 0.67^{abc}	27.0
V. hybrida	7.3 ± 0.11^{abc}	40.8
V. incana	4.6 ± 0.44^{abc}	34.5
V. lathyroides	4.5 ± 0.41^{abc}	22.0
V. lutea		
ssp. *cavanillesii*	8.5 ± 0.41^{abc}	47.1
ssp. *lutea* var. *hirta*	2.1 ± 0.18^{ab}	62.8
ssp. *lutea* var. *lutea*	5.3 ± 0.58^{abc}	62.8
ssp. *vestita*	5.5 ± 0.00^{abc}	72.2
V. monantha ssp. *calcarata*	3.8 ± 0.76^{abc}	56.5
V. monardi	5.7 ± 0.21^{abc}	72.8
V. narbonensis	1.9 ± 0.09^{a}	94.2
V. onobrychioides	8.7 ± 0.25^{abc}	53.4
V. parviflora	3.2 ± 0.07^{ab}	18.8
V. peregrina	2.8 ± 0.14^{ab}	64.7
V. pseudocracca	2.2 ± 0.07^{ab}	39.6
V. pubescens	6.2 ± 0.32^{abc}	19.5
V. pyrenaica	11.6 ± 0.16^{c}	44.0
V. sativa ssp. *sativa*	21.3 ± 2.40^{d}	56.5
V. tenuifolia	2.8 ± 0.11^{ab}	37.7
V. vicioides	7.7 ± 0.32^{abc}	53.4

Taxa that are marginally cultivated or were cultivated in the past are underlined

Polyphenol concentrations observed in cultivated *Vicia* species ranged from 1.9 mg/g ground seed in *V. narbonensis* to 21.3 mg/g in *V. sativa* subsp. *sativa*. Polyphenols in non-cultivated *Vicia* species ranged from 2.1 mg/g ground seed in *V. lutea* subsp. *lutea* var. *hirta* to 11.6 mg/g in *V. pyrenaica*. However, when comparing the surface of the seeds in the 31 taxa studied of the species *Vicia* with the content in polyphenols, no correlation was observed (r^2= 0.0081), as opposed to the results previously obtained for the genus *Lathyrus* (r^2=0.37).

The polyphenolic contents of some commercial legumes were analyzed to compare with the observed in *Vicia* species. Thus, polyphenolic contents in lupin, soy and chickpea (grey bars) were in the range observed in *Vicia* species with less polyphenols than 9 mg/g (Figure 4). In commercial legumes polyphenolic contents ranged between 7.8 mg/g meal in lupin and soy and 8.9 mg/g meal in chickpea. Higher polyphenols contents than chickpea were observed in four *Vicia* species, non–cultivated *V. cordata* and *V. pyrenaica,* and cultivated *V. angustifolia* and *V. sativa* subsp. *sativa*.

There are not many data about the polyphenol content in the species of the genus *Vicia*. Amarowicz et al. [41] obtained total polyphenol values of 55.9 mg/g for flour of *V. faba*, which are higher than the results obtained in this study (Table 10). In general terms, the average polyphenol content in the species of the genus *Lathyrus* from Southern Spain was higher as compared to the values obtained for the species of the genus *Vicia* [107].

B. Antioxidant activity of Vicia polyphenolic compounds

Antioxidant activity of *Vicia* polyphenols was determined by measuring their inhibitory effect on the oxidative degradation of b-carotene in the presence of linoleic acid.

In a first series of experiments, this activity was referred to as polyphenol content (specific activity) by taking for the assays the volume of extract corresponding to 2 mg catechin equivalents (Table 11). Specific activity in *V. disperma*, *V. lathyroides*, *V. parviflora* and *V. tenuifolia* seeds were the highest among the seeds from non-cultivated *Vicia*, while *V. dasycarpa* yielded seeds with the highest specific activity among cultivated taxa (Table 11). The lowest antioxidant activity among non-cultivated and cultivated taxa was found in *V. monardi* and *V. faba*, respectively.

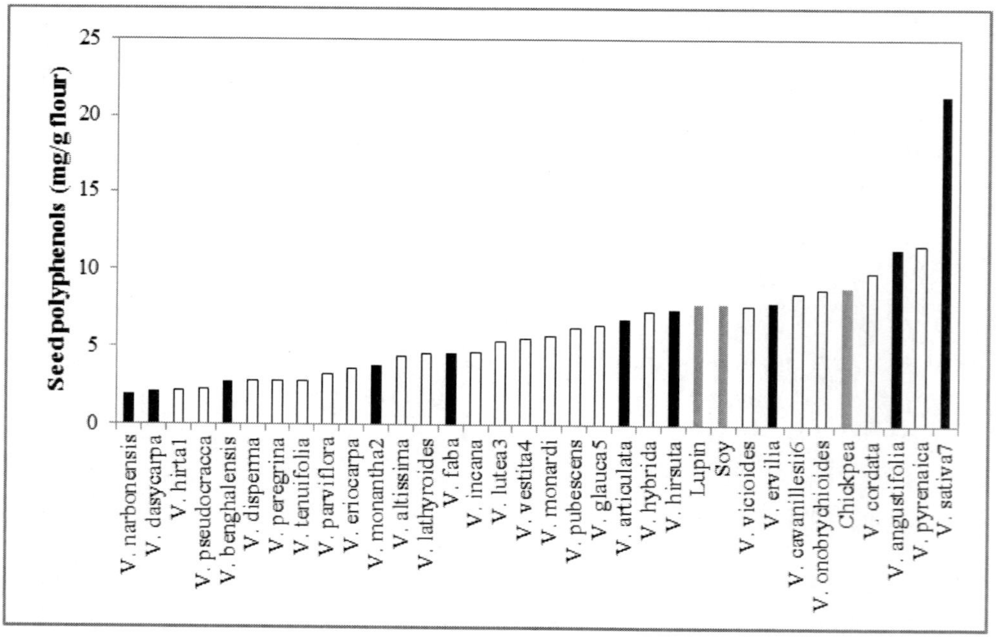

Figure 4. Seed polyphenolic contents in wild (white bars) and cultivated (black bars) Vicia taxa and some commercial legumes (grey bars). 1, *V. lutea subsp. lutea var. hirta;* 2, *V. monantha subsp. calcarata;* 3, *V. lutea subsp. lutea var. lutea;* 4, *V. lutea subsp. vestita;* 5, *V. glauca subsp. giennensis;* 6, *V. lutea subsp. cavanillesii;* 7, *V. sativa subsp. sativa.*

Table 11. Specific antioxidant activity of polyphenolics from *Vicia* seed flour. 2 μg of catechin-equivalent phenolics were assayed

Taxa	AOX[a]	DR[b]	AA[c]	ORR[d]	AAC[e]
V. altissima	0.0047	0.0103	56.40	0.44	392.27
V. angustifolia	0.0052	0.0117	50.72	0.49	332.60
V. articulata	0.0047	0.0113	52.02	0.48	346.96
V. benghalensis	0.0036	0.0083	65.17	0.35	482.87
V. cordata	0.0048	0.0113	52.13	0.48	348.07
V. dasycarpa	0.0034	0.0074	68.94	0.31	519.34
V. disperma	0.0033	0.0077	67.51	0.33	509.39
V. eriocarpa	0.0042	0.0096	59.63	0.40	427.62
V. ervilia	0.0042	0.0093	60.60	0.40	434.25
V. faba	0.0062	0.0141	40.62	0.59	244.20
V. glauca ssp. *giennensis*	0.0045	0.0101	57.46	0.43	398.90
V. hirsuta	0.0044	0.0105	55.65	0.44	382.32
V. hybrida	0.0056	0.0127	46.48	0.54	300.55
V. incana	0.0042	0.0094	60.23	0.40	426.52
V. lathyroides	0.0033	0.0077	67.69	0.32	522.65
V. lutea					
ssp. *cavanillesii*	0.0037	0.0081	65.78	0.34	485.08
ssp. *lutea* var. *hirta*	0.0045	0.0102	56.82	0.43	398.90
ssp. *lutea* var. *lutea*	0.0045	0.0108	54.65	0.45	372.38
ssp. *vestita*	0.0048	0.0108	54.59	0.45	371.27
V. monantha ssp. *calcarata*	0.0045	0.0095	60.07	0.40	429.83
V. monardi	0.0056	0.0128	45.86	0.54	289.50
V. narbonensis	0.0048	0.0112	52.67	0.47	346.96
V. onobrychioides	0.0049	0.0112	52.66	0.47	348.07
V. parviflora	0.0032	0.0074	68.76	0.31	527.07
V. peregrina	0.0040	0.0090	62.04	0.38	460.77
V. pseudocracca	0.0045	0.0103	56.59	0.43	390.06
V. pubescens	0.0036	0.0087	63.19	0.37	461.88
V. pyrenaica	0.0049	0.0114	51.78	0.48	344.75
V. sativa ssp. *sativa*	0.0046	0.0111	53.17	0.47	358.01
V. tenuifolia	0.0034	0.0077	67.65	0.32	527.07
V. vicioides	0.0052	0.0116	50.96	0.49	338.12

[a] Antioxidant value. [b] Degradation rate. [c] Antioxidant activity. [d] Oxidation rate ratio. [e] Antioxidant activity coefficient.

Taxa that are marginally cultivated or were cultivated in the past are underlined

In a second assay (Table 12), 5 μL polyphenolic extract was assayed in order to compare antioxidant activity in different taxa as referred to total extract (total activity), and not to polyphenol-equivalents in the extracts (specific activity) as shown in Table 11. Thus, correlation between polyphenol concentration and antioxidative activity was inconclusive. For instance, the extract from *V. sativa* subsp. *sativa*, which had the highest polyphenol concentration, also showed the highest antioxidative activity, but other taxa with high specific activity and average (e.g., *V. lutea* subsp. *cavanillesii* and *V. ervilia*) or even low (e.g., *V. tenuifolia*) polyphenol concentration also showed high total antioxidative activity (Table 12).

On the other hand, extracts with the lowest polyphenol concentrations also showed lower antioxidant activity, e.g., *V. narbonensis*, *V. lutea* subsp. *lutea* var. *hirta* and *V. dasycarpa*. The *V. sativa* subsp. *sativa* extract showed the highest antioxidant activity and polyphenol concentration of all the cultivated *Vicia*, as opposed to extracts from *V. faba* and *V. narbonensis* which showed the lowest antioxidant activity probably due to low specific activity and low polyphenol concentration, respectively. Extracts from *V. lutea* subsp. *cavanillesii* and *V. lutea* subsp. *lutea* var. *hirta* showed the highest and lowest antioxidant activity of all the non-cultivated taxa (Table 12).

Table 12. Antioxidant activity of polyphenols extracted from *Vicia* taxa per gram of flour extracted

Taxa	AOX[a]	DR[b]	AA[c]	ORR[d]	AAC[e]
V. altissima	0.0049	0.0128	42.15	0.58	273.03
V. angustifolia	0.0042	0.0102	54.03	0.46	377.36
V. articulata	0.0044	0.0118	46.43	0.54	309.66
V. benghalensis	0.0046	0.0122	44.98	0.55	294.12
V. cordata	0.0039	0.0098	55.76	0.44	395.12
V. dasycarpa	0.0052	0.0133	39.90	0.6	250.83
V. disperma	0.0046	0.0121	45.44	0.55	304.11
V. eriocarpa	0.0056	0.0148	32.96	0.67	196.45
V. ervilia	0.0036	0.0095	57.23	0.43	410.66
V. faba	0.0062	0.0157	28.77	0.71	165.37
V. glauca* ssp. *giennensis	0.0041	0.0107	51.48	0.49	357.38
V. hirsuta	0.0045	0.0111	49.66	0.50	338.51
V. hybrida	0.0053	0.0134	39.55	0.61	247.50
V. incana	0.0046	0.0119	46.12	0.54	305.22
V. lathyroides	0.0038	0.0098	55.71	0.44	399.56
V. lutea					
ssp. *cavanillesii*	0.0027	0.0073	67.07	0.33	524.97
ssp. *lutea* var. *hirta*	0.0057	0.0154	30.47	0.70	176.47
ssp. *lutea* var. *lutea*	0.005	0.0132	40.07	0.60	255.27
ssp. *vestita*	0.0047	0.0123	44.17	0.56	293.01
V. monantha* ssp. *calcarata	0.0048	0.0128	42.31	0.58	270.81
V. monardi	0.0051	0.0137	37.88	0.62	234.18
V. narbonensis	0.0057	0.0156	29.63	0.70	173.14
V. onobrychioides	0.0041	0.0106	52.21	0.48	368.48
V. parviflora	0.0039	0.0105	52.51	0.48	368.48
V. peregrina	0.0052	0.0134	39.38	0.61	248.61
V. pseudocracca	0.0056	0.0145	34.22	0.66	208.66
V. pubescens	0.0037	0.0097	56.20	0.44	399.56

Taxa	AOX[a]	DR[b]	AA[c]	ORR[d]	AAC[e]
V. pyrenaica	0.0039	0.0100	54.54	0.46	388.46
V. sativa* ssp. *sativa	0.0023	0.0059	73.21	0.27	597.11
V. tenuifolia	0.004	0.0098	55.82	0.44	400.66
V. vicioides	0.0046	0.0118	46.68	0.53	311.86

[a] Antioxidant value. [b] Degradation rate. [c] Antioxidant activity. [d] Oxidation rate ratio. [e] Antioxidant activity coefficient.

Taxa that are marginally cultivated or were cultivated in the past are underlined.

Table 13. Specific antioxidant activity of polyphenolics extracted from *V. parviflora, V. tenuifolia*, chickpea, lupin and soy. 2 μg of catechin-equivalent polyphenolics were assayed. BHT and catechin were assayed in the amounts per well indicated in the table. Results are the average of three independent determinations

	AOX[a]	DR[b]	AA[c]	ORR[d]	AAC[e]
V. parviflora	0.0032	0.0074	68.76	0.31	527.07
V. tenuifolia	0.0034	0.0077	67.65	0.32	527.07
Chickpea	0.0062	0.0137	24.06	0.759	152.78
Lupin	0.0062	0.0140	22.11	0.779	138.89
Soy	0.0056	0.0127	29.67	0.703	194.44
BHT (0.64 μg)	0.004	0.0089	50.70	0.493	375.00
BHT (1.28 μg)	0.0031	0.0070	61.20	0.388	500.00
BHT (1.92 μg)	0.0027	0.0059	66.98	0.330	555.56
Catechin (5 μg)	0.0047	0.0104	42.44	0.576	291.67
Catechin (10 μg)	0.0039	0.0085	52.99	0.470	388.89
Catechin (15 μg)	0.0034	0.0076	57.66	0.423	444.44

[a] Antioxidant value. [b] Degradation rate. [c] Antioxidant activity. [d] Oxidation rate ratio. [e] Antioxidant activity coefficient.

Table 14. Antioxidant activity per gram of seed meal of polyphenolics extracted from *Vicia sativa* subsp. *sativa*, chickpea, lupin and soy. BHT and catechin were assayed in the amounts per well indicated in the table. Results are the average of three independent determinations

	AOX[a]	DR[b]	AA[c]	ORR[d]	AAC[e]
V. sativa subsp. *sativa*	0.0023	0.0059	73.21	0.27	597.11
Chickpea	0.0054	0.0124	24.02	0.760	138.17
Lupin	0.0053	0.0121	25.22	0.748	151.05
Soy	0.0046	0.0105	35.51	0.645	227.17
BHT (0.64 μg)	0.004	0.0089	50.70	0.493	375.00
BHT (1.28 μg)	0.0031	0.0070	61.20	0.388	500.00
BHT (1.92 μg)	0.0027	0.0059	78.00	0.330	605.56
Catechin (5 μg)	0.0047	0.0104	42.44	0.576	291.67
Catechin (10 μg)	0.0039	0.0085	52.99	0.470	388.89
Catechin (15 μg)	0.0034	0.0076	57.66	0.423	444.44

[a] Antioxidant value. [b] Degradation rate. [c] Antioxidant activity. [d] Oxidation rate ratio. [e] Antioxidant activity coefficient.

Tables 13 and 14 showed that the *Vicia* extracts had the highest antioxidant activity relative to the extracts of the corresponding grain legumes chickpea, lupin and soy, and the synthetic antioxidant BHT and the natural antioxidant catechin. Activity was referred to catechin equivalents in the extracts (specific activity, Table 13) or to extract volume (total activity, Table 14). Both specific (*V. parviflora* and *V. tenuifolia*) and total activity (*V. sativa* subsp. *sativa*) in these extracts at least doubled the activity in the extracts corresponding to the grain legumes, and were similar to the activity of 1.92 µg of BHT and 15 µg of catechin.

The antioxidant activity in most of the extracts corresponding to the *Vicia* species was higher than in the extracts of corresponding commercial legumes as chickpea, lupin and soy. Therefore, *Vicia* may represent a useful source of polyphenols with high antioxidant activity.

This study highlights the potential of *Vicia* species, marginally cultivated nowadays, as a source of polyphenols as functional components for foods.

4. Protein Analysis

4.1. Genus *Lathyrus*

A. Protein contents and amino acid analysis

Protein content in studied *Lathyrus* species ranged from 17.7% in *L. sativus* to 25,6% in *L. tingitanus* (Table 15). Only *L. sativus* showed significant differences ($P<0.05$) with *L. tingitanus* protein contents [108]. Protein contents observed in studied *Lathyrus* are in the range reported (21-25%) for legumes in general [87, 109, 110].

Wild *L. sativus* populations in Southern Spain have shown an average richness which is a little lower than the results obtained by Chavan *et al.* [21] for the populations in Canada and India of the same species (23.6% and 21.3%, respectively). Nevertheless, Monsoor & Yusuf [111] obtained a higher protein richness (29.9%) with *L. sativus* seeds from Bangladesh. The differences observed in the same species could be caused by environmental conditions, the plot characteristics, and the method used for the analysis, as reported by Hanbury et al. [9].

Considering that some of the studies performed in different legumes reported an increase of the amino acid levels in the seeds, which was proportional to maturity [50, 112, 113], in the present work the amino acid analysis of legumes has been done with their greatest physiological maturity which takes place, according the referred bibliography related to *Lathyrus* (*L. sativus*), when the seeds are richer in nitrogen components, including amino acids [89].

The nutritional quality of food proteins depends mainly of its amino acid composition. Amino acid composition of seed proteins of *Lathyrus* species is shown in Tables 16a and 16b. Among essential amino acids the most abundant in decreasing order were Leucine, Lysine, Phenylalanine, Threonine and Valine . Leucine contents were above FAO recommendations in all taxa [114]. *L. clymenum* contained the highest Leucine percentages (9.8%) which were significantly different from most species ($P<0.01$). Also, all species contained Lysine above FAO recommendations [114] because legumes have high contents in this amino acid [55]. *L. aphaca* showed Lysine contents (8.5%) significantly higher ($P<0.001$) than those observed in *L. amphicarpos*, *L. angulatus*, *L. annuus*, *L. cicera*, *L. hirsutus*, *L. latifolius* y *L. tingitanus* (Tabla 16a). All species showed high contents in aromatic amino acids (Phenylalanine and

Tyrosine) with values above FAO recommendations [114]. Similarly, Histidine, Isoleucine and Valine percentages are in all studied species above FAO recommendations [114]. The only limitation found in the amino acid composition of *Lathyrus* is a constant feature in other leguminous plants. In other words, they do not reach the minimally required levels neither in sulphur-containing amino acids (methionine and cysteine), nor in tryptophan [49, 115, 116].

However, it should be remarked that populations in Cadiz of *L. ochrus* (*L. ochrus*[2], in Table 16a) actually reach the recommended value, that is 2.5, and that *L. latifolius* is fairly close to this value (2.2%). On the contrary, the populations from Jaen and Seville of *L. ochrus* (*L. ochrus*[1], in Table 16a) show a lower content of these amino acids. Such wide differences observed between the populations of this species (1.3% and 0.3%, respectively for methionine) are not observed in any other species. These variations might be due to the plot characteristics and environmental conditions, since seeds were harvested when they were completely matured in all cases. Also, previous studies in *Teucrium*, Juan *et al.* [117] indicated that annual species show higher sulphur amino acids contents, although if *L. ochrus* populations from Cadiz were annual or biannual could not be confirmed. Finally, all species were limiting in Tryptophan ranging from 0.5% to 0.8% content, below the FAO recommendation of 1% [114].

With respect to non-essential amino acids, the most abundant in decreasing order were Glutamic, Aspartic, Arginine and Serine (Table 16b). The content of aspartic acid has been similar in all species (10.6-12.8%), although the glutamic proportion ($P<0.05$) had significant differences between *L. annuus,* with the highest value (21.5%), and *L. aphaca*, *L. filiformis*, *L. latifolius* and *L. sphaericus* with the lowest values (Table 16b). The richness in these two amino acids is similar in other leguminous plants [51].With respect to arginine, significant differences ($P<0.05$) have been observed between *L. tingitanus*, which has the highest value (12.9%), and *L. clymenum*.

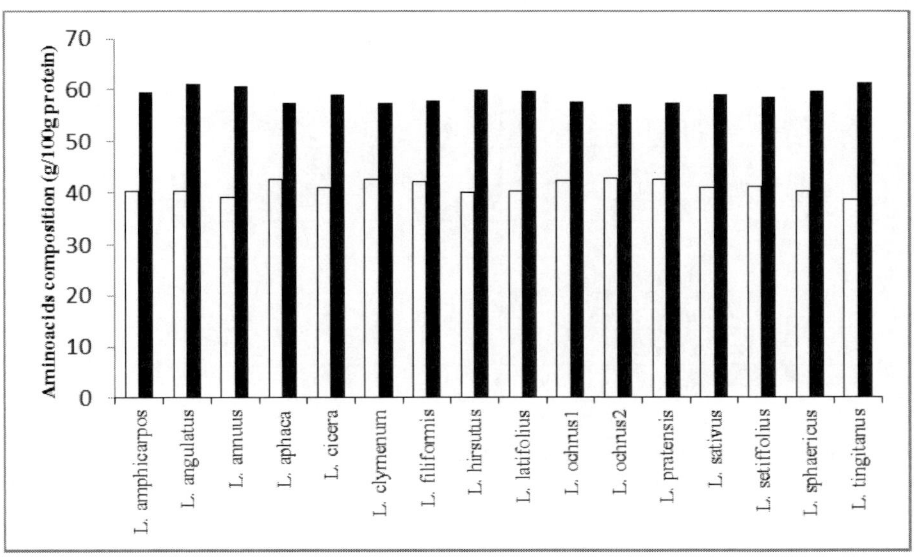

Figure 5. Total essential (white bars) and non-essential (black bars) amino acids composition (g/100g protein) of *Lathyrus* species studied. *L. ochrus*$_1$ (populations from Seville and Jaén). *L. ochrus*$_2$ (populations from Cádiz).

In general, amino acid composition is similar to the reported for other *Lathyrus*, such as *L. sativus* [20, 21, 50, 118], and *L. maritimus* [21].

Figure 5 illustrates the total quantity of essential and non-essential amino acids for each species examined in this study. Despite non-essential amino acid proportion is higher than essential amino acid proportion in all species, its detailed composition shows that the species of this genus may have more balanced values (*L. annuus*, *L. clymenum*, *L. tingitanus*, *L. cicera*, etc) than other commercial legumes, such as lupins, haricots, broad beans, lentils or kidney beans. For example, all species of the *Lathyrus* genus have a richer content in lysine compared to the abovementioned commercial legumes [49, 55, 119].

B. In vitro protein digestibility

Other factors than the amino acid composition may determine the nutritional quality of food proteins. To make use of amino acids in food proteins these must first be digested in order to release them. Thus, protein digestibility by digestive enzymes is also important determinant of the nutritional quality of a protein [116, 120]. From a nutritional point of view a food protein should have an equilibrated amino acid composition and a high protein digestibility.

Table 15. Seed protein content (%) in *Lathyrus* species studied. Data are means ± standard deviation of indicated number of populations studied. Superscripts letters indicate significant differences between values in the column (Scheffe`s test, P<0.05)

Species	n	Seed protein content (%)
L. amphicarpos	3	19.9 ± 2.6^{ab}
L. angulatus	4	23.8 ± 2.0^{ab}
L. annuus	5	21.0 ± 1.7^{ab}
L. aphaca	6	20.8 ± 0.5^{ab}
L. cicera	6	21.5 ± 1.0^{ab}
L. clymenum	7	21.6 ± 1.8^{ab}
L. filiformis	2	21.2 ± 1.1^{ab}
L. hirsutus	3	25.1 ± 2.9^{ab}
L. latifolius	4	24.4 ± 1.8^{ab}
L. ochrus	4	20.2 ± 1.1^{ab}
L. pratensis	4	23.2 ± 0.9^{ab}
L. sativus	2	17.7 ± 2.4^{a}
L. setifolius	4	22.7 ± 3.8^{ab}
L. sphaericus	5	23.5 ± 1.6^{ab}
L. tingitanus	7	25.6 ± 1.9^{b}

n = number of populations studied.

Table 16a. Seed protein essential amino acid composition of *Lathyrus* species studied. Data expressed as g/100 g protein are means ± standard deviation of indicated number of populations studied. Superscripts letters indicate significant differences between values of the same column (Scheffe`s Test)

Species	n	Cys	Phe	His*	Ile	Leu**	Lys***	Met	Tyr	Thr***	Trp	Val
L. amphicarpos	3	1.3±0.1	4.8±0.0	2.8±0.1[abc]	3.5±0.1	8.0±0.2[ab]	7.3±0.2[a]	0.6±0.3	2.6±0.0	4.5±0.2[a]	0.7±0.1	4.4±0.2
L. angulatus	4	0.8±0.1	5.2±0.2	2.5±0.1[abc]	3.7±0.3	8.6±0.3[abc]	7.2±0.2[a]	0.6±0.2	2.4±0.1	4.3±0.1[a]	0.5±0.1	4.3±0.2
L. annuus	5	1.3±0.2	4.6±0.2	2.6±0.2[abc]	3.5±0.3	7.9±0.1[ab]	7.0±0.1[a]	0.5±0.1	2.3±0.1	4.5±0.2[a]	0.8±0.1	4.3±0.2
L. aphaca	6	1.4±0.4	5.4±0.2	2.7±0.1[abc]	3.8±0.2	8.3±0.3[ab]	8.5±0.3[b]	0.4±0.3	2.4±0.2	4.5±0.3[a]	0.7±0.1	4.5±0.2
L. cicera	6	1.2±0.3	5.0±0.1	2.7±0.1[abc]	3.6±0.2	8.0±0.2[ab]	7.4±0.1[a]	0.7±0.2	2.8±0.4	4.4±0.1[a]	0.8±0.1	4.4±0.2
L. clymenum	7	0.9±0.1	5.1±0.1	2.5±0.2[ab]	3.8±0.3	9.8±0.6[c]	7.7±0.2[ab]	0.7±0.1	2.6±0.1	4.2±0.1[a]	0.7±0.0	4.6±0.2
L. filiformis	2	1.0±0.3	4.8±0.2	3.3±0.4[c]	3.9±0.1	7.8±0.25[ab]	7.5±0.1[ab]	0.8±0.0	2.3±0.2	5.0±0.3[ab]	0.6±0.1	5.1±0.2
L. hirsutus	3	1.4±0.2	4.6±0.1	2.5±0.1[abc]	3.2±0.1	7.8±0.1[ab]	7.3±0.3[a]	0.5±0.1	2.3±0.2	5.8±0.2[b]	0.7±0.0	4.0±0.1
L. latifolius	4	1.5±0.4	5.1±0.2	3.0±0.1[abc]	3.6±0.2	7.0±0.7[a]	7.2±0.2[a]	0.7±0.4	2.6±0.1	4.4±0.1[a]	0.6±0.0	4.4±0.2
L. ochrus[1]	2	0.7±0.1	5.4±0.1	2.5±0.0[abc]	4.0±0.2	9.2±0.1[bc]	8.0±0.3[ab]	0.3±0.0	2.5±0.2	4.3±0.1[a]	0.6±0.0	4.9±0.1
L. ochrus[2]	2	1.2±0.3	5.2±0.3	2.7±0.1[abc]	3.6±0.2	9.0±0.03[bc]	7.8±0.1[ab]	1.3±0.1	2.7±0.1	4.2±0.1[a]	0.7±0.0	4.5±0.2
L. pratensis	4	1.2±0.3	5.1±0.2	3.1±0.5[bc]	3.8±0.3	8.3±0.2[ab]	7.8±0.3[ab]	0.6±0.2	2.4±0.2	4.7±0.1[ab]	0.8±0.3	4.8±0.3
L. sativus	2	1.2±0.0	5.0±0.1	2.7±0.1[abc]	3.7±0.1	8.1±0.1[abc]	7.6±0.1[ab]	0.3±0.1	2.6±0.0	4.6±0.3[ab]	0.6±0.1	4.5±0.2
L. setifolius	4	0.9±0.0	5.1±0.1	2.5±0.2[abc]	3.6±0.2	8.4±0.1[abc]	7.9±0.2[ab]	0.7±0.2	2.5±0.1	4.5±0.2[a]	0.7±0.0	4.5±0.2
L. sphaericus	5	0.8±0.1	5.4±0.4	2.5±0.3[abc]	3.1±0.2	8.1±0.5[ab]	7.6±0.3[ab]	0.4±0.2	2.4±0.1	4.4±0.1[a]	0.7±0.1	4.3±0.4
L. tingitanus	7	0.8±0.2	4.8±0.5	2.4±0.1[a]	3.6±0.3	7.7±0.6[ab]	7.1±0.4[a]	0.4±0.1	2.6±0.5	4.2±0.4[a]	0.6±0.0	4.5±0.3
FAO[3]				1.9	2.8	6.6	5.8	2.5♠	6.3♠♠	3.4	1.1	3.5

n = number of populations studied. *$P<0.05$, **$P<0.01$, ***$P<0.001$. Cys, cisteine. Phe, phenylalanine. His, histidine. Ile, isoleucine. Leu, leucine. Lys, lisine. Met, methionine. Tyr, tyrosinae Thr, threonine. Trp, tryptophan. Val, valine. *L. ochrus*[1] (populations from de Sevilla and Jaén). *L. ochrus*[2] (populations from Cádiz). [3] Suggested pattern of amino acid requirements (FAO/WHO/UNU, 1985). ♠ = Met+Cys, ♠♠ = Phe+Tyr.

Table 16b. Seed protein non essential amino acid composition of *Lathyrus* species studied. Data expressed as g/100 g protein are means ± standard deviation of indicated number of populations studied. Superscripts letters indicate significant differences between values of the same column (Scheffe`s Test P<0.05)

Species	n	Ala	Arg*	Asp[1]	Glu[2]*	Gly	Pro	Ser
L. amphicarpos	3	4.7 ± 0.1	9.8 ± 0.4^{ab}	12.5 ± 0.7	19.7 ± 0.2^{ab}	4.5 ± 0.1	2.1 ± 0.5	6.2 ± 0.1
L. angulatus	4	4.7 ± 0.1	10.9 ± 2.9^{ab}	12.4 ± 0.7	19.1 ± 0.7^{ab}	5.0 ± 0.2	1.7 ± 0.7	6.1 ± 0.2
L. annuus	5	6.0 ± 1.0	8.9 ± 0.4^{ab}	11.8 ± 0.8	21.5 ± 0.5^{b}	4.4 ± 0.2	2.3 ± 0.8	5.8 ± 0.2
L. aphaca	6	4.8 ± 0.2	9.6 ± 1.2^{ab}	12.3 ± 1.0	17.9 ± 1.0^{a}	4.7 ± 0.8	1.9 ± 1.0	6.1 ± 0.3
L. cicera	6	4.8 ± 0.1	9.5 ± 0.3^{ab}	12.6 ± 0.7	19.7 ± 0.4^{ab}	4.5 ± 0.1	1.6 ± 0.9	6.2 ± 0.2
L. clymenum	7	4.9 ± 0.1	8.7 ± 0.3^{a}	12.2 ± 0.7	19.3 ± 0.5^{ab}	4.3 ± 0.2	1.5 ± 0.8	6.4 ± 0.2
L. filiformis	2	6.1 ± 0.2	9.3 ± 0.6^{ab}	12.3 ± 2.0	17.5 ± 0.1^{a}	5.0 ± 0.4	1.1 ± 0.2	6.3 ± 0.7
L. hirsutus	3	4.9 ± 0.2	10.3 ± 2.0^{ab}	12.3 ± 1.1	20.0 ± 0.3^{ab}	4.4 ± 0.1	1.7 ± 0.6	5.9 ± 0.2
L. latifolius	4	5.0 ± 0.2	12.2 ± 1.1^{ab}	10.6 ± 1.3	18.8 ± 1.0^{a}	4.9 ± 0.5	1.8 ± 0.4	6.4 ± 0.3
L. ochrus[1]	2	4.7 ± 0.0	8.9 ± 0.3^{ab}	12.7 ± 1.1	19.8 ± 0.5^{ab}	4.3 ± 0.1	0.5 ± 0.1	6.5 ± 0.0
L. ochrus[2]	2	4.9 ± 0.0	8.8 ± 0.2^{ab}	12.8 ± 0.8	19.3 ± 0.5^{ab}	4.4 ± 0.2	0.3 ± 0.0	6.5 ± 0.1
L. pratensis	4	5.1 ± 0.3	9.9 ± 1.0^{ab}	11.4 ± 1.3	19.0 ± 0.7^{ab}	4.9 ± 0.1	1.0 ± 0.7	6.0 ± 0.1
L. sativus	2	5.0 ± 0.1	10.8 ± 1.4^{ab}	12.1 ± 0.1	19.3 ± 0.1^{ab}	4.9 ± 0.1	0.2 ± 0.3	6.4 ± 0.0
L. setifolius	4	5.1 ± 0.2	8.6 ± 0.2^{ab}	12.6 ± 0.5	19.6 ± 0.4^{ab}	4.5 ± 0.2	1.8 ± 0.9	6.2 ± 0.2
L. sphaericus	5	4.4 ± 0.1	12.6 ± 1.7^{ab}	12.5 ± 1.6	18.3 ± 0.5^{a}	4.2 ± 0.3	1.6 ± 0.5	5.9 ± 0.4
L. tingitanus	7	5.3 ± 1.3	12.9 ± 2.5^{b}	11.1 ± 0.9	19.4 ± 1.2^{ab}	5.3 ± 1.4	1.6 ± 0.7	5.6 ± 0.2

n = number of populations studied. *L. ochrus*[1] (populations from de Sevilla and Jaén). *L. ochrus*[2] (populations from Cádiz). Ala, alanine. Arg, arginine. Asp[1], aspartic + asparaguine. Glu[2], glutamic + glutamine. Gly, glycine. Pro, proline. Ser, serine.

In vitro protein digestibility (IVPD) in studied *Lathyrus* ranged from 76.0% in *L. pratensis* to 86.5% in *L. annuus* (Table 17). There only were significant differences (P<0.01) between these species.

Table 17. *In vitro* protein digestibility in *Lathyrus* species studied. Data are means ± standard deviation of indicated number of populations studied. Superscripts letters indicate significant differences between values in the column (Scheffe`s test, P<0.01)

Species	n	IVPD (%)
L. amphicarpos	3	81.3 ± 0.6^{ab}
L. angulatus	3	80.3 ± 1.5^{ab}
L. annuus	3	86.5 ± 2.8^{b}
L. aphaca	3	79.0 ± 3.0^{ab}
L. cicera	3	84.0 ± 0.0^{ab}
L. clymenum	3	80.3 ± 1.5^{ab}
L. filiformis	2	76.6 ± 0.0^{ab}
L. hirsutus	3	80.6 ± 0.6^{ab}
L. latifolius	3	79.0 ± 1.7^{ab}
L. ochrus	4	79.2 ± 1.7^{ab}
L. pratensis	3	76.0 ± 1.0^{a}
L. sativus	2	80.2 ± 0.3^{ab}
L. setifolius	3	80.6 ± 0.8^{ab}
L. sphaericus	3	81.0 ± 0.0^{ab}
L. tingitanus	3	84.7 ± 2.9^{ab}

IVPD (%), *In vitro* protein digestibility n = number of .populations studied.

If protein digestibility of *Lathyrus* species is compared *in vitro* to the values of some legumes studied by other authors [49, 55, 111, 119, 121], it is observed that even though some values may be higher for *L. annuus* or higher for *L. pratensis* than the values obtained in this study, the results are fairly similar (Table 18). Nevertheless, the fact that the 15 species of *Lathyrus* have a higher digestibility than kidney bean [49] justifies considering some of the species of this genus as good candidates for future cultivation.

C. Nutritional Assessment of Proteins in Lathyrus Seeds

With the amino acid composition and protein digestibility several nutritional parameters in 15 *Lathyrus* species have been calculated (Table 19).

The highest proportion of essential amino acids with respect to the total (%EAA/TAA) was observed in *L. clymenum* (42.6%) which were not significantly different from the remaining studied species.

L. clymenum and *L. ochrus* from Cádiz province showed highest amino acid score but all studied species showed good values (Table 19).

The biological value (BV) measures the proportion of ingested proteins that are incorporated in the organism [122]. The biological values ranged between 24.5% in *L. hirsutus* and 55.4% in *L. ochrus* populations from Cádiz, but there were not significant differences between studied species.

All the parameters mentioned above are important, however, the most remarkable are the theoretical Protein Efficiency Rate (PER) [123] and above all the Protein Digestibility Corrected Amino Acid Score (PDCAAS) which is the most reliable assessment [124]. The protein efficiency ratio (PER) has been recognized as a valuable method for the calculation of protein quality. Theoretical PER values, that bear a good relationship with real PER values were calculated [123]. Low quality proteins have PER values below 1.5, and above 2 are considered high quality proteins [116].

Theoretical PER values (PER_1, PER_2 y PER_3) are similar in all studied species (Table 19). The best values were observed in *L. clymenum* with significant differences ($P < 0.01$) for the three PER values with *L. amphicarpos*, *L. annuus*, *L. aphaca*, *L. cicera*, *L. latifolius*, *L. sphaericus* y *L. tingitanus*. These values were higher than the reported for other Mediterranean legumes, such as *Lupinus* [125]. In general, the values obtained for the theoretical PER in this study are very high, in accordance to the results obtained by Alsmeyer et al. [123] with some foods that include legumes and by Wolzak et al. [126] for soybeans.

Table 18. *In vitro* protein digestibility (IVPD %) of various legumes according to some authors

	In vitro protein digestibility (%)				
	Carbonaro et al. **(1997)**	**Plahar *et al.* (1997)**	**Monnsoor & Yusuf (2002)**	**Lqari *et al.* (2002)**	**Kalpanadevi & Mohan (2013)**
Lupin	--	--	--	80.0	--
Kidney bean	73.76	--	--	--	--
Chickpea	78.28	--	89.01	--	--
Faba bean	83.07	--	--	--	--
Lathyrus sativus	--	--	93.39	--	--
Lentil	82.5	--	95.19	--	--
Vigna unguiculata	--	78.0	--	--	71.3

Table 19. Nutritional characteristics of studied *Lathyrus* seed proteins. Results are means ± standard deviation of number of populations indicated in Table 16b for each specie. Superscripts letters indicate significant differences between values in the same column (Scheffe`s test)

Species	% EAA/TAA	AAS	VB	PER_1**	PER_2**	PER_3**	PDCAAS*
L. amphicarpos	40.4 ± 0.7	119.3 ± 2.0	44.7 ± 5.1	2.88 ± 0.1^{ab}	2.96 ± 0.1^{ab}	2.87 ± 0.1^{a}	0.52 ± 0.1^{ab}
L. angulatus	40.1 ± 1.6	118.2 ± 4.8	37.5 ± 4.8	3.16 ± 0.1^{abc}	3.19 ± 0.1^{abc}	3.46 ± 0.2^{ab}	0.41 ± 0.0^{ab}
L. annuus	39.3 ± 1.3	115.8 ± 3.7	42.6 ± 9.4	2.80 ± 0.1^{ab}	2.86 ± 0.0^{ab}	2.88 ± 0.1^{a}	0.59 ± 0.1^{b}
L. aphaca	42.5 ± 1.4	125.6 ± 4.5	29.9 ± 9.7	3.01 ± 0.2^{ab}	3.07 ± 0.1^{ab}	3.09 ± 0.2^{a}	0.39 ± 0.0^{ab}
L. cicera	41.0 ± 0.7	121.0 ± 2.0	44.8 ± 7.1	2.88 ± 0.1^{ab}	2.87 ± 0.1^{ab}	2.64 ± 0.5^{a}	0.55 ± 0.0^{ab}
L. clymenum	42.6 ± 1.0	125.7 ± 2.8	42.0 ± 3.0	3.72 ± 0.2^{c}	3.71 ± 0.3^{c}	4.25 ± 0.5^{b}	0.49 ± 0.1^{ab}
L. filiformis	42.5 ± 2.1	125.0 ± 7.0	33.0 ± 2.8	2.81 ± 0.1^{ab}	2.82 ± 0.1^{ab}	3.13 ± 0.0^{ab}	0.45 ± 0.0^{ab}
L. hirsutus	40.4 ± 1.1	119.0 ± 3.4	24.5 ± 2.7	2.77 ± 0.1^{ab}	2.97 ± 0.0^{abc}	2.83 ± 0.1^{a}	0.53 ± 0.0^{ab}
L. latifolius	40.3 ± 0.9	118.6 ± 2.5	45.0 ± 9.1	2.42 ± 0.3^{a}	2.44 ± 0.3^{a}	2.15 ± 0.3^{a}	0.45 ± 0.0^{ab}
L. ochrus[1]	42.5 ± 0.9	125.3 ± 2.5	28.2 ± 0.3	3.50 ± 0.1^{bc}	3.45 ± 0.0^{bc}	3.70 ± 0.0^{ab}	0.34 ± 0.0^{a}
L. ochrus[2]	43.0 ± 1.3	127.0 ± 3.6	55.4 ± 3.8	3.40 ± 0.0^{bc}	3.35 ± 0.0^{abc}	3.82 ± 0.1^{ab}	0.54 ± 0.0^{ab}
L. pratensis	42.3 ± 2.0	124.9 ± 0.4	37.0 ± 4.7	3.04 ± 0.1^{abc}	3.03 ± 0.1^{abc}	3.31 ± 0.2^{ab}	0.49 ± 0.0^{ab}
L. sativus	41.2 ± 1.0	121.5 ± 2.8	33.2 ± 6.0	3.02 ± 0.0^{abc}	3.04 ± 0.1^{abc}	2.84 ± 0.1^{ab}	0.50 ± 0.0^{ab}
L. setifolius	41.5 ± 0.5	122.5 ± 1.4	36.9 ± 7.9	3.07 ± 0.1^{abc}	3.08 ± 0.0^{abc}	3.17 ± 0.0^{ab}	0.48 ± 0.0^{ab}
L. sphaericus	40.3 ± 2.2	116.2 ± 3.2	35.5 ± 9.9	2.92 ± 0.2^{ab}	2.97 ± 0.2^{ab}	2.84 ± 0.3^{a}	0.38 ± 0.0^{a}
L. tingitanus	39.5 ± 0.8	116.5 ± 2.5	33.2 ± 6.5	2.86 ± 0.1^{ab}	2.85 ± 0.1^{ab}	2.49 ± 0.5^{a}	0.43 ± 0.1^{ab}

% EAA/TAA = % Essential amino acids/total amino acids. AAS = Amino acids score. BV = Biological value. $PER_{1,2,3}$ = Protein efficiency ratio. PDCAAS = Protein digestibility corrected amino acid score. L. ochrus[1] (populations from Sevilla and Jaen). L. ochrus[2] (Populations from Cadiz). *$P<0.05$, **$P<0.01$

With respect to PDCAAS value, *L. annuus* (0.59) showed significant differences (P<0.05) with *L. sphaericus* (0.38) and populations of *L. ochrus* from outside Cádiz (0.34). Considering that a value close to 1 in this rate indicates that the quality of the proteins is the best or the most complex [127], it can be stated that the values obtained for some species (PDCAAS ≥ 0.5) are quite acceptable (Table 19). Although all species showed a lower PDCAAS value than soy, black bean and pea [128], some of the studied species (*L. amphicarpos*, *L. annuus*, *L. cicera*, *L. hirsutus*, *L. ochrus* from Cádiz, *L. pratensis*, *L. sativus* and *L. setifolius*) showed a PDCAAS value equal or higher than lentil or peanut [127].

4.2. Genus *Vicia*

A. Protein contents and amino acid analysis

Vicia seed protein contents ranged from 20.1% in *V. articulata* to 32% in *V. pyrenaica* whit absence of significant differences among the studied taxa (Table 20) [129]. Mostly, the protein contents of several of the studied *Vicia* were similar than those observed in other legumes [109, 110, 130, 131].

In the past, some species like *V. faba* and *V. ervilia* had shown a slightly higher protein content compared to the results obtained in the present study (Elsherbeeny & Robertson, [132] and Sadeghi et al., [62], respectively). However, after analyzing 31 taxa of the genus *Vicia*, it has been observed that in general they have a higher protein richness than other so commonly-used legumes, like chickpeas or lentils (Figure 6).

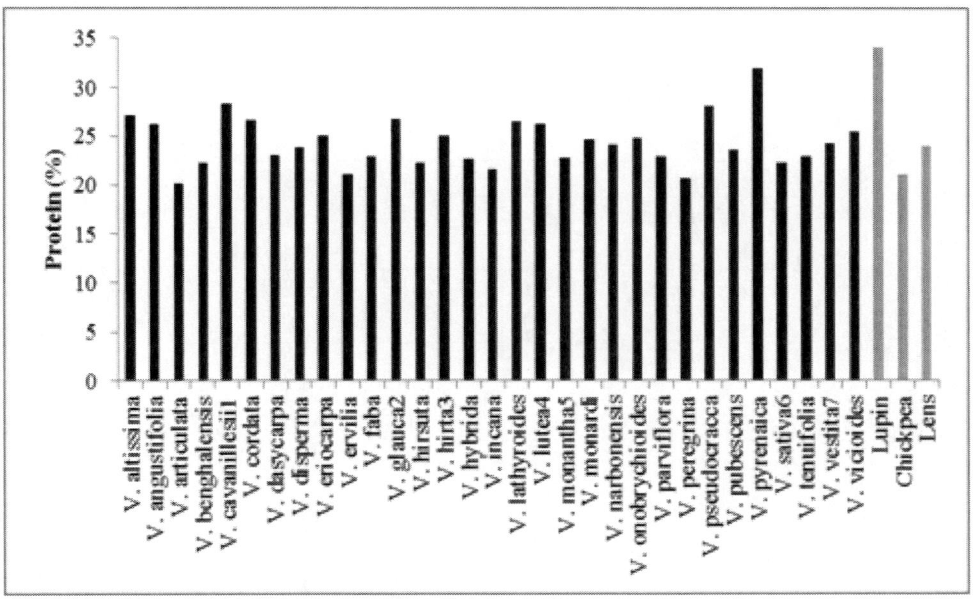

Figure 6. Protein content (%) of *Vicia* taxa studied (black bars) and some commercial legumes (grey bars). 1, *V. lutea* subsp. *cavanillesii;* 2, *V. glauca* subsp. *giennensis*; 3, *V. lutea* subsp. *lutea* var. *hirta*; 4, *V. lutea* subsp. *lutea* var. *lutea*; 5, *V. monantha* subsp. *calcarata*; 6, *V. sativa* subsp. *sativa*; 7, *V. lutea* subsp. *vestita.*

Table 20. Seed protein content (%) in *Vicia* taxa studied. Data are means ± standard deviation of indicated number of populations studied. Column without superscript letters indicates the absence of significant differences among taxa (Scheffe`s test)

Taxa	n	Seed protein content (%)
V. altissima	1	27.2
V. angustifolia	6	26.2 ± 3.1
V. articulata	3	20.1 ± 4.8
V. benghalensis	4	22.2 ± 3.4
V. cordata	3	26.6 ± 0.9
V. dasycarpa	3	23.0 ± 0.5
V. disperma	6	23.8 ± 1.3
V. eriocarpa	4	25.0 ± 1.6
V. ervilia	3	21.1 ± 2.6
V. faba	1	22.9
V. glauca ssp. *giennensis*	1	26.7
V. hirsuta	3	22.3 ± 0.9
V. hybrida	4	22.7 ± 1.5
V. incana	2	21.6 ± 2.9
V. lathyroides	3	26.5 ± 2.2
V. lutea		
ssp. *cavanillesii*	1	28.4
ssp. *lutea* var. *hirta*	8	25.1 ± 1.8
ssp. *lutea* var. *lutea*	3	26.2 ± 3.1
ssp. *vestita*	3	24.3 ± 2.2
V. monantha ssp. *calcarata*	3	22.8 ± 2.6
V. monardi	3	24.7 ± 1.9
V. narbonensis	4	24.1 ± 1.7
V. onobrychioides	3	24.8 ± 3.1
V. parviflora	6	22.9 ± 0.9
V. peregrina	3	20.6 ± 2.5
V. pseudocracca	8	28.1 ± 2.4
V. pubescens	3	23.6 ± 1.2
V. pyrenaica	1	32.0
V. sativa ssp. *sativa*	5	22.2 ± 3.0
V. tenuifolia	4	22.9 ± 0.6
V. vicioides	3	25.5 ± 0.8

n = number of populations studied

Table 21a. Seed protein essential amino acid composition of *Vicia* taxa studied. Data expressed as g/100 g protein are means ± standard deviation of indicated number of populations studied. Superscripts letters indicate significant differences between values of the same column (Scheffe`s Test)

Taxa	n	Cys***	His	Ile	Leu*	Lys***	Met	Phe*	Thr***	Trp	Tyr	Val
V. altissima	1	2.6^{bc}	3.0	4.1	7.9^{ab}	7.1^{abc}	0.6	5.0^{ab}	4.4^{ab}	0.7	2.4	5.0
V. angustifolia	6	1.2 ± 0.1^{abc}	2.9±0.2	3.8±0.3	7.9 ± 0.2^{ab}	6.8 ± 0.2^{abc}	0.7±0.1	4.6 ± 0.1^{a}	4.4 ± 0.3^{ab}	0.7±0.1	2.7±0.2	4.5±0.3
V. articulata	3	1.0 ± 0.3^{abc}	2.2±0.1	3.8±0.3	8.5 ± 0.2^{ab}	7.6 ± 0.2^{abc}	0.7±0.1	5.6 ± 0.1^{ab}	4.0 ± 0.0^{ab}	0.7±0.1	2.4±0.1	4.7±0.3
V. benghalensis	4	1.7 ± 0.3^{c}	2.5±0.2	3.6±0.1	7.4 ± 0.2^{ab}	6.8 ± 0.2^{abc}	0.6±0.2	4.7 ± 0.2^{ab}	4.0 ± 0.2^{a}	0.6±0.0	2.4±0.1	4.6±0.1
V. cordata	3	1.2 ± 0.1^{abc}	2.9±0.1	3.9±0.1	8.1 ± 0.2^{ab}	7.0 ± 0.1^{abc}	0.6±0.2	4.8 ± 0.1^{ab}	4.6 ± 0.3^{ab}	0.8±0.0	2.9±0.1	4.8±0.0
V. dasycarpa	3	1.0 ± 0.2^{abc}	2.9±0.4	3.9±0.3	7.9 ± 0.2^{ab}	7.0 ± 0.1^{abc}	0.7±0.6	5.0 ± 0.3^{ab}	4.6 ± 0.2^{ab}	0.7±0.1	2.6±0.2	4.8±0.4
V. disperma	6	0.8 ± 0.1^{abc}	2.6±0.2	3.7±0.2	7.8 ± 0.4^{ab}	7.1 ± 0.4^{abc}	0.5±0.2	5.3 ± 0.3^{ab}	5.5 ± 0.3^{ab}	0.7±0.0	2.5±0.2	4.6±0.4
V. eriocarpa	4	1.0 ± 0.3^{abc}	2.4±0.2	3.8±0.2	7.4 ± 0.1^{ab}	6.4 ± 0.1^{a}	0.5±0.2	4.8 ± 0.1^{ab}	4.6 ± 0.1^{ab}	0.8±0.1	2.6±0.2	4.5±0.1
V. ervilia	3	0.8 ± 0.1^{abc}	2.7±0.0	3.7±0.0	8.2 ± 0.2^{ab}	8.0 ± 0.1^{c}	0.6±0.1	5.3 ± 0.0^{ab}	4.3 ± 0.1^{ab}	0.8±0.1	2.4±0.0	4.6±0.2
V. faba	1	0.8^{abc}	2.4	3.4	8.2^{ab}	7.6^{abc}	0.2	4.9^{ab}	5.0^{ab}	1.0	2.3	4.3
V. glauca ssp. *giennensis*	1	1.3^{abc}	3.0	4.0	8.0^{ab}	6.9^{abc}	1.0	5.0^{ab}	5.1^{ab}	0.6	27	5.8
V. hirsuta	3	1.3 ± 0.0^{abc}	3.1±0.1	3.4±0.0	8.3 ± 0.0^{ab}	7.7 ± 0.1^{abc}	0.9±0.2	5.7 ± 0.2^{ab}	4.3 ± 0.0^{ab}	0.7±0.0	2.3±0.1	4.3±0.1
V. hybrida	4	0.8 ± 0.2^{abc}	2.8±0.3	4.3±0.0	8.0 ± 0.4^{ab}	7.7 ± 0.0^{abc}	0.5±0.2	5.9 ± 0.1^{b}	5.4 ± 0.3^{ab}	0.7±0.1	2.9±0.1	4.9±0.1
V. incana	2	1.0 ± 0.1^{abc}	2.9±0.3	3.7±0.3	7.7 ± 0.2^{ab}	7.1 ± 0.1^{abc}	1.0±0.1	4.8 ± 0.0^{ab}	5.4 ± 0.0^{ab}	0.7±0.0	2.5±0.1	5.3±0.7
V. lathyroides	3	1.5 ± 0.1^{abc}	2.7±0.0	4.0±0.1	8.1 ± 0.1^{ab}	7.2 ± 0.1^{abc}	0.7±0.0	5.0 ± 0.1^{ab}	4.5 ± 0.1^{ab}	0.6±0.0	2.5±0.1	5.1±0.0
V. lutea												
ssp. *cavanillesii*	1	1.1^{abc}	3.2	4.3	8.0^{ab}	7.7^{abc}	0.4	5.3^{ab}	4.8^{ab}	0.5	2.3	5.0
ssp. *lutea* var. *hirta*	8	1.1 ± 0.1^{abc}	2.9±0.2	4.3±0.1	8.0 ± 0.3^{ab}	7.4 ± 0.1^{abc}	0.4±0.1	5.3 ± 0.2^{ab}	4.5 ± 0.4^{ab}	0.6±0.1	2.5±0.2	4.8±0.2
ssp. *lutea* var. *lutea*	3	1.1 ± 0.1^{abc}	3.1±0.2	4.2±0.3	7.9 ± 0.1^{ab}	7.3 ± 0.0^{abc}	0.4±0.1	5.2 ± 0.2^{ab}	4.2 ± 0.0^{ab}	0.5±0.1	2.5±0.2	5.0±0.5
ssp. *vestita*	3	1.2 ± 0.1^{abc}	3.1±0.2	4.3±0.0	8.1 ± 0.1^{ab}	7.5 ± 0.1^{abc}	0.6±0.1	5.4 ± 0.0^{ab}	4.6 ± 0.4^{ab}	0.8±0.0	2.6±0.3	5.1±0.3
V. monantha ssp. *calcarata*	3	1.4 ± 0.1^{abc}	3.1±0.3	3.7±0.1	7.8 ± 0.3^{ab}	6.7 ± 0.3^{abc}	0.8±0.2	4.9 ± 0.3^{ab}	4.5 ± 0.2^{ab}	0.7±1.0	2.7±0.1	5.2±0.9
V. monardi	3	1.1 ± 0.1^{abc}	2.9±0.4	3.6±0.4	7.7 ± 0.3^{ab}	6.7 ± 0.2^{abc}	0.7±0.2	4.9 ± 0.3^{ab}	6.1 ± 0.3^{b}	0.7±0.0	2.7±0.1	4.6±0.7
V. narbonensis	4	1.7 ± 0.3^{c}	2.9±0.6	3.7±0.3	7.8 ± 0.3^{ab}	7.2 ± 0.1^{abc}	0.4±0.2	4.8 ± 0.4^{ab}	4.4 ± 0.3^{ab}	0.7±0.1	2.2±0.1	4.4±0.2

Table 21a. (Continued)

Taxa	n	Cys***	His	Ile	Leu*	Lys***	Met	Phe*	Thr***	Trp	Tyr	Val
V. onobrychioides	2	1.5±0.1[abc]	2.7±0.0	4.1±0.1	8.1±0.2[ab]	7.2±0.1[abc]	0.3±0.0	5.2±0.1[ab]	5.5±0.9[ab]	0.6±0.0	2.5±0.0	4.9±0.1
V. parviflora	6	0.7±0.1[ab]	3.0±0.1	3.7±0.2	8.2±0.0[ab]	7.6±0.1[abc]	0.6±0.1	5.3±0.1[ab]	4.3±0.1[ab]	0.6±0.1	2.3±0.0	4.6±0.1
V. peregrina	3	1.4±0.1[abc]	2.8±0.2	3.9±0.0	7.8±0.2[ab]	7.2±0.3[abc]	0.5±0.1	5.2±0.2[ab]	5.4±0.8[ab]	0.8±0.1	2.5±0.2	4.9±0.0
V. pseudocracca	8	0.9±0.2[abc]	3.0±1.0	4.1±0.1	7.3±0.2[a]	6.8±0.2[abc]	0.4±0.1	5.1±0.3[ab]	4.5±0.4[ab]	0.7±0.0	2.4±0.2	4.8±0.1
V. pubescens	3	0.5±0.2[a]	2.5±0.7	3.7±0.3	8.9±0.2[b]	7.9±0.2[bc]	0.5±0.2	5.7±0.2[ab]	4.6±0.4[ab]	0.6±1.0	2.2±0.3	4.5±0.2
V. pyrenaica	1	1.9[abc]	2.7	3.9	7.6[ab]	6.9[abc]	0.2	4.5[ab]	5.3[ab]	0.5	2.5	5.1
V. sativa ssp. sativa	5	0.9±0.1[abc]	2.8±0.2	3.7±0.2	8.0±0.2[ab]	7.1±0.3[abc]	0.5±0.2	4.7±0.1[a]	4.5±0.2[ab]	0.7±0.1	2.5±0.0	4.6±0.3
V. tenuifolia	4	1.0±0.1[abc]	2.8±0.3	3.6±0.4	7.4±1.1[ab]	6.7±0.5[abc]	0.7±0.5	4.6±0.6[a]	5.7±0.3[ab]	0.7±0.1	2.4±0.3	4.6±0.5
V. vicioides	3	0.9±0.1[abc]	2.8±0.4	4.2±0.2	7.4±0.2[ab]	6.4±0.1[ab]	0.9±0.2	5.5±0.3[ab]	6.2±0.1[b]	0.6±1.0	2.5±0.0	5.8±0.6
FAO[1]			1.9	2.8	6.6	5.8	2.5♠♠		3.4	1.1	6.3♠	3.5

n = number of populations studied. *P<0.05, ***P<0.001. Cys, cisteine. Phe, phenylalanine. His, histidine. Ile, isoleucine. Leu, leucine. Lys, lisine. Met, methionine. Tyr, tyrosinae Thr, threonine. Trp, tryptophan. Val, valine.

[1] Suggested pattern of amino acid requirements (FAO/WHO/UNU, 1985) . ♠♠ = Met+Cys, ♠ = Phe+Tyr.

Table 21b. Seed protein non essential amino acid composition of *Vicia* taxa studied. Data expressed as g/100 g protein are means ± standard deviation of indicated number of populations studied. Superscripts letters indicate significant differences between values of the same column (Scheffe`s Test)

Taxa	n	Ala	Arg	Asp[1]*	Gly**	Glu[2]***	Pro	Ser
V. altissima	1	4.3	10.0	11.4^{ab}	3.9^{ab}	20.2^{ab}	2.3	6.0
V. angustifolia	6	4.7 ± 0.1	9.9 ± 1.1	13.3 ± 1.0^{b}	4.8 ± 0.2^{b}	19.4 ± 1.3^{b}	2.2 ± 0.2	5.8 ± 0.1
V. articulata	3	4.6 ± 0.1	9.9 ± 1.3	12.9 ± 0.2^{ab}	4.4 ± 0.2^{ab}	18.5 ± 0.2^{ab}	2.5 ± 0.3	6.1 ± 0.3
V. benghalensis	4	4.4 ± 0.1	15.3 ± 1.6	11.1 ± 0.5^{ab}	4.1 ± 0.1^{ab}	18.3 ± 0.2^{ab}	2.5 ± 0.3	5.4 ± 0.2
V. cordata	3	4.8 ± 0.1	9.6 ± 0.4	12.4 ± 1.0^{ab}	4.9 ± 0.2^{b}	18.8 ± 0.7^{ab}	2.3 ± 0.1	5.8 ± 0.3
V. dasycarpa	3	4.8 ± 0.2	11.7 ± 3.0	11.7 ± 0.6^{ab}	4.6 ± 0.1^{ab}	18.0 ± 0.9^{ab}	2.5 ± 0.1	5.9 ± 0.2
V. disperma	6	4.5 ± 0.3	14.1 ± 5.3	11.2 ± 0.8^{ab}	4.1 ± 0.2^{ab}	16.4 ± 1.0^{ab}	2.4 ± 0.3	6.0 ± 0.4
V. eriocarpa	4	4.5 ± 0.1	16.7 ± 0.9	11.5 ± 0.4^{ab}	4.2 ± 0.1^{ab}	16.5 ± 0.6^{ab}	2.3 ± 0.3	5.6 ± 0.2
V. ervilia	3	5.0 ± 0.2	8.2 ± 0.2	12.6 ± 0.8^{ab}	4.7 ± 0.1^{ab}	19.6 ± 0.1^{ab}	2.3 ± 0.3	6.3 ± 0.2
V. faba	1	5.0	10.1	12.7^{ab}	4.7^{ab}	18.5^{ab}	3.4	6.0
V. glauca ssp. *giennensis*	1	5.0	11.4	9.5^{ab}	4.7^{ab}	18.7^{ab}	1.9	5.7
V. hirsuta	3	5.0 ± 0.0	8.2 ± 0.1	12.5 ± 0.6^{ab}	4.5 ± 0.1^{ab}	18.7 ± 0.6^{ab}	2.7 ± 0.1	6.7 ± 0.0
V. hybrida	4	4.6 ± 0.2	13.3 ± 1.5	11.1 ± 0.8^{ab}	3.9 ± 0.2^{ab}	14.8 ± 1.0^{a}	2.7 ± 0.3	5.6 ± 0.3
V. incana	2	5.2 ± 0.4	11.3 ± 1.7	11.6 ± 0.7^{ab}	4.7 ± 0.2^{ab}	17.2 ± 1.3^{ab}	2.4 ± 0.3	5.7 ± 0.4
V. lathyroides	3	4.6 ± 0.0	9.7 ± 0.2	12.3 ± 0.2^{ab}	4.2 ± 0.2^{ab}	19.6 ± 0.4^{ab}	2.0 ± 0.2	5.9 ± 0.1
V. lutea								
ssp. *cavanillesii*	1	4.6	11.8	11.6^{ab}	4.5^{ab}	16.5^{ab}	2.3	6.4
ssp. *lutea* var. *hirta*	8	4.3 ± 0.1	11.9 ± 0.6	11.4 ± 0.2^{ab}	4.2 ± 0.1^{ab}	17.8 ± 0.6^{ab}	2.1 ± 0.2	6.5 ± 0.4
ssp. *lutea* var. *lutea*	3	4.4 ± 0.1	12.8 ± 0.9	10.6 ± 1.0^{ab}	4.2 ± 0.1^{ab}	18.5 ± 0.8^{ab}	1.9 ± 0.4	6.1 ± 0.0
ssp. *vestita*	3	4.4 ± 0.1	10.5 ± 0.9	11.4 ± 0.5^{ab}	3.9 ± 0.0^{ab}	18.1 ± 0.9^{ab}	2.2 ± 0.3	6.2 ± 0.2
V. monantha ssp. *calcarata*	3	4.6 ± 0.2	11.2 ± 1.2	11.9 ± 0.8^{ab}	4.6 ± 0.2^{ab}	18.4 ± 0.5^{ab}	2.1 ± 0.3	5.9 ± 0.4
V. monardi	3	4.8 ± 0.2	12.5 ± 3.4	11.3 ± 0.9^{ab}	4.5 ± 0.2^{ab}	17.3 ± 0.4^{ab}	2.2 ± 0.4	5.8 ± 0.4
V. narbonensis	4	4.4 ± 0.1	11.6 ± 2.0	11.8 ± 0.4^{ab}	4.3 ± 0.1^{ab}	19.2 ± 0.7^{ab}	2.3 ± 0.5	6.3 ± 0.3
V. onobrychioides	2	4.6 ± 0.1	9.9 ± 0.3	10.8 ± 0.7^{ab}	4.3 ± 0.1^{ab}	19.6 ± 0.1^{ab}	2.2 ± 0.2	6.2 ± 0.2
V. parviflora	6	4.8 ± 0.1	11.1 ± 0.7	12.2 ± 0.7^{ab}	4.6 ± 0.1^{b}	17.7 ± 0.7^{ab}	2.6 ± 0.2	6.1 ± 0.1
V. peregrina	3	4.6 ± 0.1	8.9 ± 2.2	12.3 ± 0.9^{ab}	4.4 ± 0.0^{ab}	17.9 ± 1.3^{ab}	3.2 ± 0.6	6.6 ± 0.8
V. pseudocracca	8	4.3 ± 0.2	18.1 ± 1.5	10.4 ± 0.4^{a}	3.8 ± 0.2^{a}	15.4 ± 0.6^{a}	2.0 ± 0.4	5.9 ± 0.4
V. pubescens	3	4.8 ± 0.1	9.7 ± 0.1	12.3 ± 0.7^{ab}	4.8 ± 0.2^{ab}	18.4 ± 0.8^{ab}	2.1 ± 0.4	6.5 ± 0.3
V. pyrenaica	1	4.8	9.7	11.3^{ab}	5.2^{ab}	20.7^{ab}	1.5	5.8
V. sativa ssp. *sativa*	5	4.6 ± 0.1	9.5 ± 0.6	12.9 ± 0.7^{ab}	4.7 ± 0.1^{b}	20.1 ± 0.7^{b}	2.4 ± 0.4	5.9 ± 0.2
V. tenuifolia	4	5.0 ± 0.7	15.6 ± 6.3	10.8 ± 0.7^{ab}	4.3 ± 0.4^{ab}	16.4 ± 1.6^{ab}	2.6 ± 0.4	5.5 ± 0.7
V. vicioides	3	5.2 ± 1.3	14.6 ± 1.3	11.2 ± 1.3^{ab}	4.0 ± 0.3^{ab}	14.0 ± 0.2^{a}	2.1 ± 0.1	6.0 ± 0.5

n = number of populations studied. *$P < 0.05$ **$P < 0.01$ ***$P < 0.001$. Ala, alanine. Arg, arginine. Asp[1], aspartic + asparaguine. Glu[2], glutamic + glutamine. Gly, glycine. Pro, proline. Ser, serine.

Amino acid composition of *Vicia* is shown in Tables 21a and 21b. *Vicia* species show an amino acid pattern characteristic of legumes [49]. According to FAO recommendations [114], all taxa examined are deficient in sulfur amino acids (Methionine and Cysteine) and Tryptophan, except *V. altissima*, which was only deficient in Tryptophan.

In studied *Vicia* taxa, the most abundant essential amino acids in decreasing order were Leucine, Lysine, Phenylalanine, Threonine, and Valine. Leucine values in all taxa are above FAO recommendations [114], with the highest contents observed in *V. pubescens*, which were significantly different (P < 0.05) from those observed in *V. pseudocracca* (Table 21a). Lysine, was not limiting in any studied *Vicia*, with the highest percentages observed in *V. ervilia*, which were significantly different (P < 0.001) from those observed in *V. eriocarpa* and *V. vicioides*. Contents in aromatic amino acids, Phenylalanine and Tyrosine, are also above FAO recommendations [114]. These contents were similar in all studied taxa, and only in *V. hybrida* Phenylalanine percentages were significantly different (P < 0.05) from those observed in *V. angustifolia*, *V. sativa* subsp. *sativa*, and *V. tenuifolia*. Histidine, Isoleucine and Valine contents are similar in all studied species with values in all cases above FAO recommendations [114].

The only limitation observed in the amino acid composition of *Vicia* is a constant feature in other leguminous plants, as previously pointed out for the genus *Lathyrus*. This means that in general they do not reach the minimally required levels neither in sulphur-containing amino acids (cysteine and methionine), nor in tryptophan [49, 115, 116]. Nevertheless, among the studied species, *V. altissima* appears as the one with the most balanced amino acid composition, because it was only deficient in Tryptophan. Among the remaining taxa, *V. benghalensis* and *V. glauca* subsp. *giennensis* showed the best sulfur amino acids contents (2.3%). Respect to Tryptophan, *V. faba* showed the highest value (1%), which was not significantly different from those observed among the remaining taxa.

In relation to non-essential amino acids, the most abundant are the glutamic acid, aspartic acid, arginine, and serine in descending order (Table 21b). The arginine and serine content is similar in all the taxa analyzed; nevertheless, significant differences have been found in the proportion of glutamic acid and aspartic acid. In relation to glutamic acid, *V. vicioides* (14%), *V. hybrida* (14.8%), and *V. pseudocracca* (15.4%) have the lowest values and are considerably distinguished (P<0.001) from *V. angustifolia* (19.4%) and *V. sativa* subsp. *sativa* (20.1%). In the case of aspartic acid, there is only a relevant difference (P<0.05) between *V. pseudocracca* (10.4%) and *V. angustifolia* (13%). Another non-essential amino acid which is found in smaller quantities but has greater differences among species is glycine. In this case there are significant differences (P<0.01) between *V. pseudocracca*, with the lowest value (3.8%) and *V. angustifolia*, *V. cordata*, *V. parviflora,* and *V. sativa* subsp. *sativa*. With respect to the other non-essential amino acids (alanine and proline), results are very similar among different taxa and no significant differences are observed between them.

Figure 7 shows the total quantity of essential and non-essential amino acids for each species examined. Very similar contents are found for both types of amino acids. Despite the proportion of non-essential amino acid is greater than the proportion of essential amino acids, its composition indicates that the taxa of this genus may have more balanced values than other commercial legumes, such as lupins or wild cowpeas, since all taxa of the *Vicia genus* are richer in lysine than the abovementioned commercial legumes [55, 119].

B. In vitro protein digestibility (IVPD)

In vitro methods of protein evaluation are useful for determining the nutritional value of new protein foods because they are faster than in vivo methods [111]. Hsu et al. [120] used a method that obtains in vitro digestibility value which has high correlation with in vivo digestibility observed in rats. IVPD in studied *Vicia* species ranged from 78% in *V. incana* to 86.3% in *V. hirsuta* (Table 22). Among the studied taxa, 15 of them showed an IVPD higher (Table 22) than *V. faba* (83.2%), which is the most cultivated species. Furthermore, most possessed an IVPD higher than those observed in other legumes such as bean (73.76%) [49]

or chickpea (76.2%) [130]. The fact that all taxa in the genus *Vicia* have a higher protein digestibility compared to commercial legumes means that some at least some taxa of this genus could be good candidates for cultivation in the future.

C. Nutritional Assessment of Proteins in Vicia Seeds

With the amino acid composition and protein digestibility several nutritional parameters in 31 *Vicia* taxa have been calculated (Table 23).

The highest proportion of essential amino acids with respect to the total (% EAA/TAA) was observed in *V. hybrida* (43.9%) and the lowest in *V. benghalensis* (38.8%) but there were not significant differences between species.

The chemical score is a simplified model for predicting dietary protein quality, but it does not take into account amino acid digestibility or availability [133]. The amino acid score was the highest in *V. hybrida* with an average value of 129.5 (Table 23). The lowest value was observed in *V. benghalensis* and *V. eriocarpa* with average values of 114.2.

The highest Biological Value (BV) was observed in *V. benghalensis* with values of 67, whereas the lowest was observed in *V. hybrida* with 18.7 ($P<0.01$). The best biological values observed in this study were similar to those reported in cereals such as wheat (61.6) or triticale (65.3) [116].

The theoretical PER is a good method for evaluating protein quality. Theoretical PER values were calculated according to the formulae proposed by Alsmeyer et al. [123], which have been shown to bear a good relationship with real PER values. The best PER values were observed in *V. pubescens* with PER_1, PER_2, and PER_3 of 3.27, 3.29, and 3.77, respectively. The lowest values were calculated for *V. tenuifolia* with PER_1 of 2.51 and PER_2 of 2.58, and *V. eriocarpa* with PER_3 of 2.20. Similar values to those reported here for *Vicia* have been described for other legumes [123]. In this way, Sujak et al. [134] used PER_2 to evaluate protein quality of *Lupinus angustifolius* (2.36) *and Lupinus albus* (2.87).

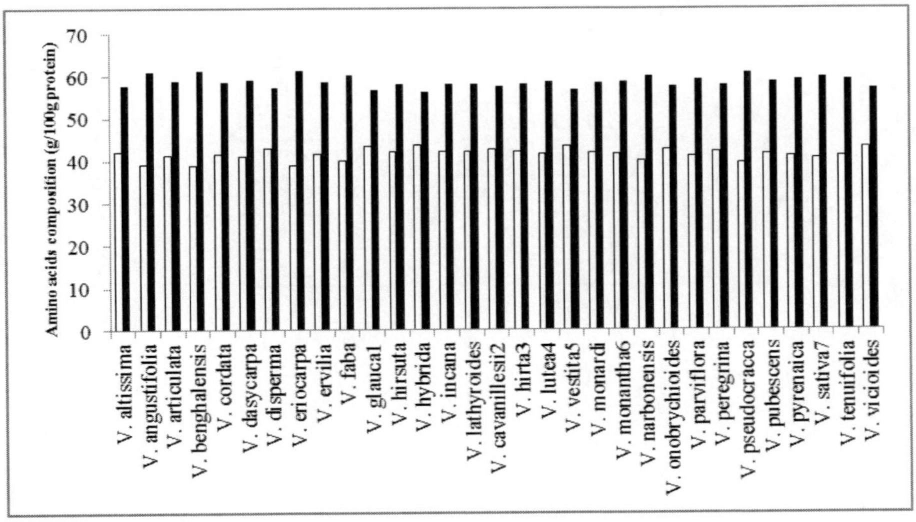

Figure 7. Total essential (white bars) and non-essential (black bars) amino acids composition (g/100g protein) of Vicia taxa studied. 1, *V. glauca subsp. giennensis;* 2, *V. lutea subsp. cavanillesii;* 3, *V. lutea subsp. lutea var. hirta;* 4, *V. lutea subsp. lutea var. lutea;* 5, *V. lutea subsp. vestita;* 6, *V. monantha subsp. calcarata;* 7, *V. sativa subsp. sativa.*

Table 22. *In vitro* protein digestibility in *Vicia* taxa studied. Data are means ± standard deviation of indicated number of populations studied. Column without superscript letters indicates the absence of significant differences among taxa (Scheffe`s test)

Taxa	n	IVPD (%)
V. altissima	1	82.0
V. angustifolia	3	83.0 ± 1.0
V. articulata	3	83.3 ± 1.1
V. benghalensis	4	83.5 ± 1.7
V. cordata	3	83.7 ± 1.1
V. dasycarpa	3	83.0 ± 1.0
V. disperma	5	81.2 ± 2.1
V. eriocarpa	4	82.0 ± 0.8
V. ervilia	3	85.7 ± 0.6
V. faba	1	83.2
V. glauca ssp. *giennensis*	1	83.0
V. hirsuta	3	86.3 ± 0.6
V. hybrida	3	82.0 ± 2.0
V. incana	2	78.0 ± 4.2
V. lathyroides	3	83.2 ± 0.7
V. lutea		
ssp. *cavanillesii*	1	81.0
ssp. *lutea* var.*hirta*	3	83.3 ± 1.5
ssp. *lutea* var. *lutea*	3	84.5 ± 1.3
ssp. *vestita*	3	84.9 ± 0.2
V. monantha ssp. *calcarata*	3	85.2 ± 0.7
V. monardi	3	80.2 ± 0.7
V. narbonensis	3	82.0 ± 1.0
V. onobrychioides	2	85.0 ± 1.4
V. parviflora	4	85.9 ± 0.8
V. peregrina	3	84.3 ± 2.3
V. pseudocracca	3	80.3 ± 1.1
V. pubescens	3	85.7 ± 2.3
V. pyrenaica	1	80.0
V. sativa ssp. *sativa*	4	84.5 ± 2.4
V. tenuifolia	4	81.7 ± 1.5
V. vicioides	3	79.8 ± 0.3

IVPD (%), *In vitro* protein digestibility. n = number of populations studied

Table 23. Nutritional characteristics of studied *Vicia* seed proteins. Data are means ± standard deviation of indicated number of populations studied. Superscripts letters indicate significant differences between values in the same column (Scheffe`s test, P<0.01)

Taxa	n	PDCAAS	% EAA/TAA	AAS	VB**	PER_1	PER_2	PER_3
V. altissima	1	0.52	42.1	124.2	53.0^{ab}	2.79	2.85	2.98
V. angustifolia	3	0.52 ± 0.04	39.0 ± 0.6	115.0 ± 1.8	53.5 ± 1.9^{ab}	2.81 ± 0.0	2.84 ± 0.1	2.80 ± 0.1
V. articulata	3	0.54 ± 0.10	41.2 ± 0.7	121.4 ± 2.1	51.2 ± 6.4^{ab}	3.06 ± 0.1	3.13 ± 0.1	3.33 ± 0.1
V. benghalensis	4	0.44 ± 0.02	38.8 ± 1.1	114.2 ± 2.8	67.0 ± 4.4^{b}	2.54 ± 0.1	2.62 ± 0.1	2.46 ± 0.2
V. cordata	3	0.54 ± 0.01	41.4 ± 0.2	122.4 ± 0.8	43.3 ± 13.0^{ab}	2.88 ± 0.1	2.95 ± 0.2	2.65 ± 0.2
V. dasycarpa	3	0.49 ± 0.10	40.9 ± 1.4	120.7 ± 4.1	35.9 ± 11.6^{ab}	2.78 ± 0.1	2.82 ± 0.1	2.73 ± 0.5
V. disperma	5	0.51 ± 0.03	42.0 ± 2.2	123.8 ± 6.3	20.1 ± 4.2^{ab}	2.84 ± 0.1	2.88 ± 0.1	2.78 ± 0.1
V. eriocarpa	4	0.47 ± 0.10	38.9 ± 0.5	114.2 ± 1.4	35.5 ± 9.2^{ab}	2.58 ± 0.1	2.61 ± 0.0	2.20 ± 0.1
V. ervilia	3	0.46 ± 0.02	41.4 ± 0.5	122.1 ± 1.6	35.4 ± 1.5^{ab}	2.95 ± 0.1	3.01 ± 0.1	3.12 ± 0.1
V. faba	1	0.32	39.9	117.7	21.5^{ab}	2.87	2.99	2.94
V. glauca ssp. *giennensis*	1	0.45	43.2	127.6	36.8^{ab}	2.85	2.86	2.94
V. hirsuta	3	0.51 ± 0.04	41.9 ± 0.1	123.5 ± 0.4	51.5 ± 5.1^{ab}	2.99 ± 0.0	3.07 ± 0.0	3.55 ± 0.1
V. hybrida	4	0.38 ± 0.05	43.9 ± 0.5	129.5 ± 1.7	18.7 ± 4.5^{a}	2.86 ± 0.2	2.87 ± 0.2	2.48 ± 0.3
V. incana	2	0.49 ± 0.01	42.0 ± 0.3	123.6 ± 0.6	29.9 ± 1.2^{ab}	2.71 ± 0.1	2.75 ± 0.1	2.84 ± 0.1
V. lathyroides	3	0.48 ± 0.02	42.0 ± 0.0	123.5 ± 0.5	48.7 ± 1.6^{ab}	2.93 ± 0.1	2.96 ± 0.0	3.03 ± 0.0
V. lutea								
ssp. *cavanillesii*	1	0.37	42.4	125.2	28.7^{ab}	2.85	2.92	3.10
ssp. *lutea* var.*hirta*	9	0.46 ± 0.04	41.9 ± 0.7	123.6 ± 2.1	33.2 ± 7.8^{ab}	2.86 ± 0.1	2.90 ± 0.1	2.85 ± 0.2
ssp. *lutea* var. *lutea*	3	0.40 ± 0.02	41.4 ± 1.5	122.1 ± 4.4	39.0 ± 4.4^{ab}	2.84 ± 0.0	2.88 ± 0.0	2.86 ± 0.1
ssp. *vestita*	3	0.55 ± 0.05	43.2 ± 0.6	127.2 ± 1.7	37.2 ± 8.9^{ab}	2.89 ± 0.0	2.91 ± 0.0	2.91 ± 0.1
V. monantha ssp. *calcarata*	3	0.54 ± 0.01	41.4 ± 0.2	122.1 ± 0.8	50.2 ± 10.0^{ab}	2.76 ± 0.1	2.78 ± 0.1	2.71 ± 0.1
V. monardi	3	0.51 ± 0.03	41.6 ± 2.3	123.0 ± 7.0	21.1 ± 2.0^{ab}	2.72 ± 0.1	2.74 ± 0.1	2.53 ± 0.3

Table 23. (Continued)

Taxa	n	PDCAAS	% EAA/TAA	AAS	VB**	PER_1	PER_2	PER_3
V. narbonensis	4	0.51 ± 0.10	40.2 ± 1.3	118.7 ± 3.8	51.7 ± 12.2^{ab}	2.77 ± 0.1	2.85 ± 0.1	2.96 ± 0.1
V. onobrychioides	3	0.46 ± 0.00	42.5 ± 0.2	125.2 ± 0.8	26.6 ± 12.0^{ab}	2.88 ± 0.1	2.93 ± 0.1	2.85 ± 0.1
V. parviflora	6	0.40 ± 0.05	40.7 ± 0.2	120.2 ± 1.3	33.9 ± 5.2^{ab}	2.91 ± 0.0	3.0 ± 0.0	3.30 ± 0.1
V. peregrina	3	0.53 ± 0.10	42.1 ± 1.5	124.1 ± 4.6	29.7 ± 11.4^{ab}	2.70 ± 0.1	2.80 ± 0.1	2.67 ± 0.2
V. pseudocracca	8	0.39 ± 0.05	40.0 ± 0.9	117.9 ± 2.8	34.4 ± 14.3^{ab}	2.54 ± 0.0	2.59 ± 0.1	2.46 ± 0.2
V. pubescens	3	0.34 ± 0.10	41.5 ± 1.3	122.3 ± 3.8	24.7 ± 10.1^{ab}	3.27 ± 0.1	3.29 ± 0.1	3.77 ± 0.1
V. pyrenaica	1	0.36	41.0	121.0	31.0^{ab}	2.72	2.72	2.41
V. sativa* ssp. *sativa	5	0.40 ± 0.07	39.9 ± 0.8	117.8 ± 2.2	35.4 ± 8.5^{ab}	2.86 ± 0.1	2.91 ± 0.1	2.84 ± 0.2
V. tenuifolia	4	0.43 ± 0.10	41.0 ± 3.3	117.8 ± 10.4	23.6 ± 6.8^{ab}	2.51 ± 0.5	2.58 ± 0.5	2.49 ± 0.5
V. vicioides	3	0.44 ± 0.00	43.0 ± 0.9	127.0 ± 2.6	18.8 ± 2.8^{ab}	2.56 ± 0.1	2.59 ± 0.1	2.51 ± 0.0

% EAA/TAA = % Essential amino acids/total amino acids. AAS = Amino acids score. BV = Biological value. $PER_{1, 2, 3}$ = Protein efficiency ratio. PDCAAS = Protein digestibility corrected amino acid score.

The PDCAAS is nowadays the most recommended theoretical parameter for evaluating the nutritional quality of food proteins. It is based on FAO recommendations of amino acids requirements and IVPD. It is superior to other methods for evaluating the protein quality of food for humans because it determines the quality of a protein based on the amino acid requirements of a 2- to 5-year-old child adjusted for digestibility. The highest PDCAAS value for a given protein is 1.0, this protein will provide 100% of the indispensable amino acids recommended by FAO/WHO [114], and values ≥ 0.5 are considered quite good. PDCAAS values in *Vicia* ranged from 0.32 in *V. faba* to 0.55 in *V. lutea* subsp. *vestita* (Table 23). How PDCAAS ≥0.5 are considered quite good, *V. altissima*, *V. angustifolia*, *V. articulata*, *V. cordata*, *V. disperma*, *V. hirsuta*, *V. lutea* subsp. *vestita*, *V. monantha* subsp. *calcarata*, *V. monardi*, *V. narbonensis*, y *V. peregrina* have a high nutritional quality proteins, showing similar PDCCAS values to those observed in other legumes, such as lentil (0.52) although lower than pea (0.69) [127].

CONCLUSION

The lipid content in the seed of the genera *Lathyrus* and *Vicia* is low, although it has a prevalence of unsaturated fatty acids which are very beneficial for health. In the two genera analyzed in the present study, the unsaturated/saturated ratio (TUFAs/TSFAs) is higher than the minimum recommendations established by the FAO [75]. The main unsaturated fatty acids in decreasing order were linoleic acid ($C_{18:2}$), oleic acid ($C_{18:1}$) and linolenic acid ($C_{18:3}$). Among the taxa studied, *Lathyrus filiformis*, *L. latifolius*, *Vicia articulata*, *V. ervilia*, *V. faba*, and *V. narbonensis*, are the ones which show significantly higher levels of oleic acid and lower levels of linoleic acid, improving the oleic/linoleic (O/L) and poly-unsaturated/mono-unsaturated (PUFAs/ MUFAs) ratio. On the other hand, even though most taxa show an adequate proportion of ω3 and ω6 fatty acids according to the FAO, the best contents are observed in *L. annuus*, *L. filiformis*, *V. articulata*, *V. lathyroides*, *V. parviflora*, and *V. pubescens*.

Moreover, several taxa examined could be alternative anti-oxidant sources, since 26 of them have a high total anti-oxidant activity, either because of its great content of polyphenols, like in the case of *Lathyrus angulatus*, *L. tingitanus* or *Vicia sativa* subsp. *sativa,* among others, or because they have polyphenols with a high antioxidant activity, like in the case of *Lathyrus annuus*, *L. clymenum*, *Vicia pubescens*, and many more. Thus, the *Lathyrus* and *Vicia* genera could be a useful source of polyphenols with a higher anti-oxidant activity than other commercial legumes, such as chickpea, lupin or soybean.

In relation to the protein study, the analysis of the amino acids show that the taxa observed actually reach the minimum values recommended by the FAO [114] in most essential amino acids. They have only insufficient sulphur-containing amino acids (cysteine + methionine) and tryptophan. However, in relation to sulphur-containing amino acids, it is remarkable that the populations from Cadiz of *L. ochrus* actually reach the recommended value of 2.5, and that *V. altissima* exceeds it (3.2%). Furthermore, even though no species has very high levels of tryptophan, all of them have a quantity of ≥0.5, which is the minimum required for an adult. With respect to protein digestibility *in vitro*, *L. annuus* stands out with 86.5%, being the highest level among the 46 taxa analyzed. Digestibility values observed in the wild taxa examined are similar to the results obtained from cultivated legumes, such as kidney beans, chickpeas or lentils. Finally, in this study most species in the two genera showed a PDCAAS score which is similar to cultivated legumes like kidney beans. In the taxa analyzed, *L. annuus* stands out with a value of 0.59 and *V. lutea* subsp. *vestita* with a value of 0.55.

The results obtained in this study demostrate the importance of promoting these wild species which can be useful and helpful in human diet, as well as serving as a complementary genetic reservoir of the cultivated taxa, thus supporting the preservation of the biodiversity.

REFERENCES

[1] Convenio sobre la diversidad biologica. (1992). *Conferencia de las Naciones Unidas sobre el Medio Ambiente y el Desarrollo*. Rio de Janeiro.

[2] Tratado Internacional sobre los Recursos Fitogenéticos para la Alimentación y la Agricultura. (2001). *Comisión de Recursos Genéticos para la Alimentación y la Agricultura (FAO).*

[3] Deshpande, S. S. & Campbell, C. G. (1992). Genotype variation in BOAA, condensed tannins, phenolics and enzyme-inhibitors of grass pea (Lathryus sativus). *Canadian Journal of Plant Sciences*, *72*, 1037-1047.

[4] Aykroyd, W. R. & Doughty, J. (1964). *Legumes in human nutrition*. Food and Agriculture Organisation of the United Nations, *19*, 138. Rome.

[5] Harborne, J. B., Boulter, D. & Turner, B. E. (1971). *Chemotaxonomy of the Leguminosae*. London. Academic Press.

[6] Mabaleha, M. B. & Yeboha, S. O. (2004). Characterization and compositional studies of the oils from some legume cultivars, Phaseolus vulgaris, grown in southern Africa. *Journal of American Oil Chemical Society*, *81*, 361-364.

[7] Howells, R. H., Brim, C. A. & Rinne, R. H. (1972). The plant geneticists contribution towards changing lipid and amino acid composition of soybeans. *Journal of American Oil Chemists' Society*, *49*, 30-32.

[8] Grela, E. R. & Günter, K. D. (1995). Fatty acid composition and tocopherol content of some legume seeds. *Animal Feed Science and Technology*, *52*, 325-331.

[9] Hanbury, C. D., White, C. L., Mullan, B. P. & Siddique, K. H. M. (2000). A review of the potential of Lathyrus sativus L. and L. cicera L. grain for use as animal feed. *Animal Feed Science and Technology*, *87*, 1-27.

[10] Niveditha, V. R., Sridhar, K. R. & Chatra, S. K. R. (2012). Fatty acid composition of cooked and fermented beans of the wild legumes. *International Food Research Journal*, *19(4)*, 1401-1407.

[11] Salunkhe, D. K., Sathe, S. K. & Reddy, N. R. (1983). Legume lipids. En: Arora, S. K. (ed.), *Chemistry and Biochemistry of Legumes*. Edward Arnold. London.

[12] Akpinar, N., Akpinar, M. A. & Türkoglu, S. (2001). Total lipid content and fatty acid composition of the seeds of some Vicia L. species. *Food Chemistry*, *74*, 449-453.

[13] Connor, W. E. (2000). Importance of n-3 fatty acids in health and disease. *American Journal of Clinical Nutrition*, *71*, 171S-175S.

[14] Gesch, C. B., Hammond, S. M., Hampson, S. E., Eves, A. & Crowder, M. J. (2002). Influence of supplementary vitamins, minerals and essential fatty acids on the antisocial behaviour of young adult prisoners. Randomised, placebo-controlled trial. *British Journal of Psychiatry, 181*, 22-28.

[15] Zanarini, M. C. & Frankenburg, F. R. (2003). Omega-3 fatty acid treatment of women with borderline personality disorder: a double-blind, placebo-controlled pilot study. *American Journal of Psychiatry, 160*, 167-169.

[16] Evans, A. J., Cheung, P. C. K. & Cheetham, N. W. H. (1993). The carbohydrate composition of cotyledons and hulls of cultivars of Lupinus angustifolius from western-Australia. *Journal of the Science of Food and Agriculture, 61*, 189-194.

[17] Duranti, M. (2006). Grain legume proteins and nutraceutical properties. *Fitoterapia, 77*, 67-82.

[18] Rochfort, S. & Panozzo, J. (2007). Phytochemical for Elath, the role of pulses. *Journal of Agricultural and Food Chemistry, 55*, 7981-7994.

[19] Senatore, F. & Basso, F. (1994). Fatty acid and sterol composition from seeds of some Lathyrus ecotypes. *Rivista Italiana Sostanza Grasse, 71*, 567-569.

[20] Kuo, Y. H., Bau, H. M., Quemener, B., Khan, J. K. & Lambein, F. (1995). Solid-state fermentation of Lathyrus sativus seeds using Aspergillus oryzae and Rhizopus oligosporus SP T-3 to eliminate the neurotoxin beta-odap without loss of nutritional value. *Journal of the Science of Food and Agriculture, 69*, 81-89.

[21] Chavan, U. D., Shahidi, F., Bal, A. K. & McKenzie, D. B. (1999). Physico-chemical properties and nutrient composition of beach pea (Lathyrus maritimus L.). *Food Chemistry, 66*, 43-50.

[22] Grela, E. R., Jensen, S. K. & Jakobsen, K. (1999). Fatty acid composition and content of tocopherols and carotenoids in raw and extruded grass pea (Lathyrus sativus L). *Journal of the Science of Food and Agriculture, 79*, 2075-2078.

[23] Chinnasamy, G., Davis, P. J. & Bal, A. K. (2003). Seasonal changes in oleosomic lipids and fatty acids of perennial root nodules of beach pea. *Journal of Plant Physiology, 160*, 355-365.

[24] Chinnasamy, G., Bal, A. K. & Mckenzie, D. B. (2005). Fatty acid composition of grass pea (Lathyrus sativus L.) seeds. *Lathyrus Lathyrism Newsletter, 4*, 2-4.

[25] Bagci, E., Genc, H. & Sahim, A. (2001). Fatty acid composition of four Lathyrus aphaca L. varieties, a chemotaxonomic approach. *Pakistan Journal of Biological Sciences, 4*, 872-874.

[26] Maestri, D. M., Fortunato, R. H., Guzmán, C.A., Torres, M. M. & Lamarque, A. L. (2002). Seed compositional Studies of some species of Papilionoideae (Leguminosae) native to Argentina. *Journal of the Science of Food and Agriculture, 82*, 248-251.

[27] Bagci, E. & Sahim, A. (2004). Fatty acid patterns of the seed oils of some Lathyrus species L. (Papilionideae) from Turkey, a chemotaxonomic approach. *Pakistan Journal of Botany, 36(2)*, 403-413.

[28] Madar, Z. & Stark, A. H. (2002). New legumes sources as therapeutic agents. *British Journal of Nutrition, 88(S3)*, S287-S292.

[29] Yoshida, H., Tomiyama, Y., Yoshida, N., Saiki, M. & Mizushina, Y. (2008). Lipid classes, fatty acid distributions and triacyglycerol molecular species of broad beans (Vicia faba). *Journal of American Oil Chemical Society, 85*, 535-541.

[30] Balasundram, N., Sundram, K. & Samman, S. (2006). Phenolic compounds in plants and agri-industrial by-products: Antioxidant activity, occurrence, and potential uses. *Food Chemistry*, *99*, 191-203.

[31] Azevedo, A., Gomes, J. C., Stringheta, P. C., Contijo, A. M. C., Padovani, C. R. & Riberio, L. R. Z. (2003). Black bean (Phaseolus vulgaris L.) as a protective agent against DNA damage in mice. *Food and Chemical Toxicology*, *41*, 1671-1676.

[32] Ramassamy, C. (2006). Emerging role of polyphenolic compounds in the treatment of neurodegenerative diseases: A review of their intracellular targets. *European Journal of Pharmacology*, *545*, 51-64.

[33] Boudet, A. M. (2007). Evolution and current status of reseach in phenolic compounds. *Phytochemistry*, *68*, 2722-2735.

[34] Seifried, H. E., Anderson, D. E., Fisher, E. I. & Milner, J. A. (2007). A review of the interaction among dietary antioxidants and reactive oxygen species. *Journal of Nutritional Biochemistry*, *18*, 567-579.

[35] Winham, D. M. & Hutchins, A. M. (2007). Baked bean consumption reduces serum cholesterol in hypercholesterolemic adults. *Nutrition Research*, *27*, 380-386.

[36] Pan, M.-H., Ghai, G. & Ho, C.-T. (2008). Food bioactives, apoptosis, and cancer. *Molecular Nutrition and Food Research*, *52*, 43-52.

[37] Tharanathan, R. N. & Mahadevamma, S. (2003). A review: Grain legumes a boon to human nutrition. *Trends in Food Science and Technology*, *14*, 507-518.

[38] Wang, X. F., Warkentin, T .D., Briggs, C. J., Oomah, B. D., Campbell, C. G. & Woods, S. (1998). Total phenolics and condensed tannins in field pea (Pisum sativum L.) and grass pea (Lathyrus sativus L.). *Euphytica*, *101*, 97-102.

[39] Chavan, U. D., Amarowicz, R., & Shahidi, F. (1999). Antioxidant activity of phenolic fractions of beach pea (Lathyrus maritimus L.). *Journal of Food Lipids*, *6*, 1-11.

[40] Shahidi, F., Chavan, U. D., Naczk, M. & Amarowicz, R. (2001). Nutrient distribution and phenolic antioxidants in air-classified fractions of beach pea (Lathyrus maritimus L.). *Journal of Agricultural and Food Chemistry*, *49*, 926-933.

[41] Amarowicz, R., Troszynska, A., Barylko-Pikielna, N. & Shahidi, F. (2004). Polyphenolics extracts from legume seeds: correlations between total antioxidant activity, total phenolics content, tannins content and astringency. *Journal of Food Lipids*, *11*, 278-286.

[42] Amarowicz, R. & Pegg R. B. (2008). Legumes as a source of natural antioxidants. *European Journal of Lipid Science and Technology*, *110*, 865-878.

[43] Ademabowo, C. A., Cho, E., Sampson, L., Katan, M. B., Spiegelman, D., Willett, W. C. & Holmes, M. D. (2005). Dietary flavonols and flavonol-rich foods intake and the risk of breast cancer. *International Journal of Cancer*, *114*, 628-633.

[44] Attia, R. S., Shehata A. M. E., Aman, M. E. & Hamza, M. A. (1994). Effect of cooking and decortication on the physical properties, the chemical composition and the nutritive value of chickpea (Cicer arietinum L.). *Food Chemistry*, *50*, 125-131.

[45] Siddhuraju, P., Vijayakumari, K. & Janarhanan, K. (1996). Chemical composition and protein quality of the little known legume, velvet bean (Mucura pruriens (l.) DC). *Journal of Agricultural and Food Chemistry*, *44*, 2636-2641.

[46] Morrow, B. (1991). The rebirth of legumes: legume production, consumption, and export are increasing as more people aware of legumes nutritional benefits. *Food Technology*, *45*(9), 96-121.

[47] Onweluzo, J. C., Obanu, Z. A. & Onuoha, K. C. (1994). Functional properties of some lesser known tropical legumes. *Journal of Food Science and Technology*, *31*, 302-306.

[48] Przybylska, J., Zimniak-Przybylska, Z. & Krajewski, P. (1999). Diversity of seed albumins in some Lathyrus species related to L. sativus L.: An electrophoretic study. *Genetic Resources and Crop Evolution*, *46*, 261-266.

[49] Carbonaro, M., Cappelloni, M., Nicoli, S., Lucarini, M. & Carnovale, E. (1997). Solubility-Digestibility relationship of legume proteins. *Journal of Agricultural and Food Chemistry*, *45*, 3387-3394.

[50] Korus, A., Lisiewska, Z. & Kmiecik, W. (2003). Content of amino acids in fresh and preserved physiologically immature grass pea (Lathyrus sativus L.) seeds. *European Food Research and Technology*, *217*, 148-153.

[51] Derbyshire, E., Wright, D. J. & Boulter, D. (1976). Legumin and vicilin, storage proteins of legume seeds. *Phytochemistry*, *15*, 3-24.

[52] Farzana, W. & Khalil, I. A. (1999). Protein quality of tropical food legumes. *Journal of Science and Technology*, *23*, 13-19.

[53] Wang, T. L., Domoney, C., Hedley, C. L., Casey, R. & Grusak, M. A. (2003). Can we improve the nutritional quality of legume seeds? *Plant Physiology*, *131*, 886-891.

[54] Amjad, I., Khalil, I.A., & Shah, H. (2003). Nutritional yield and amino acid profile of rice protein as influenced by nitrogen fertilizer. *Sarhad Journal of Agriculture*, *19*, 127-134.

[55] Plahar, W. A., Annan, N. T. & Nti, C. A. (1997). Cultivar and processing effects on the pasting characteristics, tannin content and protein quality and digestibility of cowpea (Vigna unguiculata). *Plants Foods for Human Nutrition*, *51*, 343-356.

[56] Fikre, A., Korbu, L., Kuo, Y. & Lambein, F. (2008). The contents of the neuro-excitatory amino acid B-ODAP (B-N-oxalyl-L-a,b-diaminopropionic acid), and other free and protein amino acids in the seeds of different genotypes of grass pea (Lathyrus sativus L.). *Food Chemistry*, *110*, 422-427.

[57] Mullan, B. P., Pluske, J. R., Trezona, M., Harris, D. J., Allen, J. G., Siddique, K. H. M., Hanbury, C. D., Van Barneveld, R. J. & Kim, J. C. (2009). Chemical composition and standardised ileal digestible amino acid contents of Lathyrus (Lathyrus cicera) as an ingredient in pig diets. *Animal Feed Science and Technology*, *150*, 139-143.

[58] Aguilera, J. F., Bustos, M. & Molina, E. (1992). The degradability of legume seed meals in the rumen: effect of heat treatment. *Animal Feed Science and Technology*, *36*, 101-112.

[59] Seabra, M., Carvalho, S., Freire, J., Ferreira, R., Mourato, M., Cunha, L., Cabral, F., Teixeira, A. & Aumaitre, A. (2001). Lupinus luteus, Vicia sativa and Lathyrus cicera as protein sources for piglets: ileal and total tract apparent digestibility of amino acids and antigenic effects. *Animal Feed Science and Technology*, *89*, 1-16.

[60] Palander, S., Laurinen, P., Perttila, S., Valaja, J. & Partanen, K. (2006). Protein and aminoacid digestibility and metabolizable energy value of pea (Pisum sativum), faba bean (Vicia faba) and lupin (Lupinus angustifolius) seeds for turkeys of different age. *Animal Feed Science and Technology*, *127*, 89-100.

[61] Lisiewska, Z., Kmiecik, W. & Slupski, J. (2007). Content of amino acids in raw and frozen broad beans (Vicia faba var. major) seeds at milk maturity stage, depending on the processing method. *Food Chemistry*, *105*, 1468-1473.

[62] Sadeghi, G. H., Pourreza, J., Samei, A. & Rahmani, H. (2009). Chemical composition and some anti-nutrient content of raw and processed bitter vetch (Vicia ervilia) seed for use as feeding stuff in poultry diet. *Tropical Animal Health and Production*, *41*, 85-93.

[63] Yáñez-Ruiz, D. R., Martín-García, A. I., Weisbjerg, M. R., Hvelplund, T. & Molina-Alcaide, E. (2009) A comparison of different legume seeds as protein supplement to optimise the use of low quality forages by ruminants. *Archives of Animal Nutrition*, *63*(1), 39-55.

[64] Ferguson, M. E. & Robertson, L. D. (1999). Morphological and phonological variation in the wild relatives of lentil. *Genetic Resources and Crop Evolution*, *46*, 3-12.

[65] Rozan, P., Kuo, Y. H. & Lambein, F. (2001). Amino acids in seeds and seedlings of the genus Lens. *Phytochemistry*, *58*, 281-289.

[66] Yoshida, H., Tomiyama, Y., Tanaka, M. & Mizushina, Y. (2007). Distribution of fatty acids in triacylglycerols and phosoholipids from peas (Pisum sativum L.). *Journal of the Science of Food and Agriculture*, *87*, 2709-2714.

[67] Hemavathy, J. & Prabhakar, J. V. (1989). Lipid composition of fenugreek (Trigonella foenumgraecum L.) seeds. *Food Chemistry*, *31*, 1-7.

[68] Ajayi, A., Oderinde, R. A., Kajogbola, D. O. & Uponi, J. I. (2006). Oil content and fatty acid composition of some underutilized legumes from Nigeria. *Food Chemistry*, *99*, 115-120.

[69] Emken, E. A. (1994). Metabolism of dietary stearic acid relative to other fatty acids in human subjects. *American Journal of Clinical Nutrition*, *60*, 1023S-1028S.

[70] Clevidence, B. A., Judd, J. T., Schaefer, E. J. Jenner, J. L., Lichtenstein, A. H., Muesing, R. A., Wittes, J. & Sunkin, M. E. (1997). Plasma lipoprotein (a) levels in men and women consuming diets enrichedin satured, cis-, or trans-monounsaturated fatty acids. *Arteriosclerosis, Thrombosis, and Vascular Biology*, *17*, 1657-1661.

[71] Schaefer, E. J. (1997). Effects of dietary fatty acids on lipoproteins and cardiovascular disease risk. *American Journal of Clinical Nutrition*, *65*(5), S1655-S1656.

[72] Baer, D. J., Judd, J. T., Kris-Etherton, P. M., Zhao, G. & Emken, E. A. (2003). Stearic Acid Absorption and its metabolizable energy value are minimally lower than other fatty acids in healthy men fed mixeddiets. *Journal of Nutrition.*, *133*, 4129-4134.

[73] Lamarque, A. L., Fortunato, R. H., Maestri, D. M. & Guzmán, C. A. (2000). Seed components and taxonomy of some Acacia species. *Biochemical Systematics and Ecology*, *28*, 53-60.

[74] Bagci, E. & Sahim, A. (2004). Fatty acid patterns of the seed oils of some Lathyrus species L. (Papilionideae) from Turkey, a chemotaxonomic approach. *Pakistan Journal of Botany*, *36*(2), 403-413.

[75] FAO/WHO (1994). Fats and oils in human nutrition. Report of a joint expert consultation. Food and Agriculture Organisation of the United Nations, Rome.

[76] Leaf, A. & Weber, P. C. (1987). A new era for science in nutrition. *American Journal of Clinical Nutrition*, *45*(S), 1048-1053.

[77] Simopoulos, A. P. (1991). Omega-3 fatty acids in health and disease and in growth and development. *American Journal of Clinical Nutrition*, *54*, 438-463.

[78] Simopoulos A. P. (2002). Omega-3 fatty acids in inflammation an autoimmune diseases. *Journal of the American College of Nutrition*, *21*, 495-505.

[79] Simopoulos, A. P. (2006). Evolutionary aspects of diets, the omega-6/omega-3 ratio and genetic variation: nutritional implications for chronic diseases. *Biomedicine and Pharmacotherapy*, *60*, 502-507.

[80] Vles, R. O. & Gottenbos, J. J. (1989). Nutritional characteristics and food uses of vegetable oils. En: Robblen, G., Downey, R.K. & Ashri, A. (eds.), *Oil crops of the world*. McGraw Hill. New York.

[81] Simopoulos, A. P. (1999). Essential fatty acids in health and chronic disease. *American Journal of Clinical Nutrition*, *70*, 560S-569S.

[82] Weber, P. C. & Leaf, A. (1991). Cardiovascular effects of omega-3 fatty acids. Atherosclerosis risk factor modification by omega-3 fatty acids. *World Review of Nutrition and Dietetics*, *66*, 218-232.

[83] Weber, P. C. (1989). *Are we what we eat?. Fatty acids in nutrition and in cell membranes: cell functions and disorders induced by dietary conditions*. Svanoy Foundation. Norway.

[84] FAO/WHO (1998). Preparation and use of food-based dietary guidelines. Report of join FAO/WHO consultation. *Technical report series nº 880*. Geneva.

[85] Lorgeril, M. de, Renaud, S., Mamelle, N., Salen, P., Martín, J. L. & Monjaud, I. (1994). Mediterranean a-linolenic acid-rich diet in secundary prevention of coronary heart disease. *Lancet*, *343*, 1454-1459.

[86] Weiss, L. A., Barret-Connor, E. & Von Muhlenm, D. (2005). Ratio of n-6 to n-3 essential fatty acid and bone mineral density in older adults: the Rancho San Bernardo study. *American Journal of Clinical Nutrition*, *81*, 934-938.

[87] Katoch, R. (2013). Nutritional potential of rice bean (Vigna Umbellata): an underutilized legume. *Journal of Food Science*, *78*(1), 8-16.

[88] Pastor-Cavada, E., Juan, R., Pastor, J. E., Alaiz, M. & Vioque, E. (2009). Chemical composition and nutritional characteristics of the seed oil of wild Lathyrus, Lens and Pisum species from southern Spain. *Journal American Oil Chemistry Society*, *86*, 329–335.

[89] Rotter, R. G., Marquardt, R. R. & Campbell, C. G. (1991). The nutritional value of low lathyrogenic Lathyrus (Lathyrus sativus) for growing chicks. *British Poultry Science*, *32*, 1055-1067.

[90] Pirman, T. & Stibilj, V. (2003). An influence of cooking on fatty acid composition in three varieties of common beans and in lentil. *Eur Food Res Technol*, *217*, 498–503.

[91] Elías, L. G., Fernández, D. G. & Brezan, R. (1979). Posible effects of seed coat polyphenolics on the nutricional quality of bean protein. *Journal of Food Science*, *44*, 524-531.

[92] Salunkhe, D. K., Kadam, S. S. & Chavan, J. K. (1985). *Postharvest Biotechnology of Food Legumes*. CRC Press. Florida.

[93] Gatel, F. & Grosjean, F. (1990). Composition and nutritive value of peas for pigs: a review of European results. *Livestock Production Science*, *26*, 155-175.

[94] Muir, A. D. (1995). Antioxidative activity of condensed tannins. In F. Shahidi (Ed.), *Natural antioxidants, chemistry, health effects and applications* (204-212). AOCS Press. Champaign.

[95] Katiyar, S. K. & Mukhtar, H. (1996). Tea in chemoprevention of cancer: epidemiological and experimental studies. *International Journal of Oncology*, *8*, 221-238.

[96] Chung, K. T., Wong, T. Y., Wei, C. I., Huang, Y. W. & Lin, Y. (1998). Tannins and human health: a review. *Critical Reviews in Food Science and Nutrition*, *38*, 421-464

[97] Gamez, E. J. C., Luyengi, L., Lee, S. K., Zhu, L., Zhou, B., Fong, H. H. S., Pezzuto, J. M. & Kinghorn, A. D. (1998). Antioxidant flavonoid glycosides from Daphniphyllum calycinum. *Journal of Natural Products*, *61*, 706-708

[98] Amarowicz, R. & Troszynska, A. (2003). Antioxidant activity of extract of pea and its fraction of low molecular phenolics and tannins. *Polish Journal of Food and Nutrition Sciences*, *53*, 10-15.

[99] Adebooye, O. C. & Singh, V. (2007). Effect of cooking on the profile of phenolics, tannins, phytate, amino acid, fatty acid and mineral nutrients of whole-grain and decorticated vegetable cowpea (Vigna unguiculata L.Walp). *Journal of Food Quality*, *30*, 1101-1120.

[100] Rocha-Guzmán, N. E., González-Laredo, R. F., Ibarra Pérez, F. J., Nava-Berúmen, C. A. & Gallegos-Infante, J. A. (2007). Effect of pressure cooking on the antioxidante activity of extracts from three common bean (Phaseolus vulgaris L.) cultivars. *Food Chemistry*, *100*, 31-35.

[101] Pastor-Cavada, E., Juan, R., Pastor, J. E., Alaiz, M. & Vioque, J. (2009). Antioxidant activity of seed polyphenols in fifteen wild Lathyrus species from South Spain. *LWT-Food Science and Technology*, *42*, 705-709.

[102] Sosulski, F. W. & Dabrowski, K. J. (1984). Composition of free hydrolysable phenolic acids in flours and hull of ten leguminous species. *Journal of Agricultural and Food Chemistry*, *32*, 131-133.

[103] Cardador-Martinez, A., Loarca-Piña, G. & Oomah, B. D. (2002). Antioxidant activity in common beans (Phaseolus vulgaris L.). *Journal of Agricultural and Food Chemistry*, *50*, 6975-6980.

[104] Piecyk, M., Wołosiak, R., Druzynska, B. & Worobiej, E. (2012). Chemical composition and starch digestibility in flours from Polish processed legume seeds. *Food Chemistry*, *135*, 1057-1064.

[105] Oomah, B. D. & Mazza, G. (1996). Flavonoids and antioxidative activities in buckwheat. *Journal of Agricultural and Food Chemistry*, *44*, 1746-1750.

[106] Velioglu, Y. S., Mazza, G., Gao, L. & Oomah, B. D. (1997). Antioxidant activity and total phenolics in selected fruits, vegetables and grain products. *Journal of Agricultural and Food Chemistry*, *45*, 304-309.

[107] Pastor-Cavada, E., Juan, R., Pastor, J. E., Alaiz, M., Girón-Calle, J. & Vioque, J. (2011). Antioxidative activity in the seeds of 28 Vicia species from southern Spain. *Journal of Food Biochemistry*, *35*, 1373-1380.

[108] Pastor-Cavada, E., Juan, R., Pastor, J. E., Alaiz, M. & Vioque, J. (2011). Nutritional characteristics of seed proteins in 15 Lathyrus apecies (Fabaceae) from southern Spain. *LWT-Food Science and Technology*, *44*, 1059-1064.

[109] Solanki, I. S., Kapoor, A. C. & Singh, U. (1999). Nutritional parameters and yield evaluation of newly developed genotypes of lentil (Lens culinaris Medik). *Plant Foods for Human Nutrition*, *54*, 79-87.

[110] Aparna, K., Khatoon, N. & Prakash, J. (2000). Cooking quality and in vitro digestibility of legumes cooked I different media. *Journal of Food Science and Technology*, *37*, 169-173.

[111] Monsoor, M. A. & Yusuf, H. K. M. (2002). In Vitro protein digestibility of lathyrus pea (Lathyrus sativus), lentil (Lens culinaria) and chickpea (Cicer arietinum). *International Journal of Food Science and Technology*, *37*, 97-99.

[112] Islam, M. N. & Lea, R. A. (1981). Effect of maturity on the amino acid composition and freezing quality of emerald soybean. *Journal of Food Science*, *46*, 658-665.

[113] Daveby, Y. D., Abrahamsson, M. & Aman, P. (1993). Changes in chemical composition during development of 3 different types pf peas. *Journal of the Science of Food and Agriculture*, *63*, 21-28.

[114] FAO/WHO/UNU (1985). Energy and protein requirements, report of the joint FAO/WHO/UNU expert consultation. *Technical report series* nº 724. Geneva.

[115] Sarwar, G. & Peace, R. W. (1986). Comparisons between true digestibility of total nitrogen and limiting amino acids I vegetable proteins. *Journal of Nutrition*, *116*, 1172-1184.

[116] Friedman, M. (1996). Nutritional value of proteins from different food sources. A review. *Journal of Agricultural and Food Chemistry*, *44*, 6-29.

[117] Juan, R., Pastor, J., Millán, F., Alaiz, M. & Vioque, J. (2004). Amino acids composition of Teucrium nutlet proteins and their systematic significance. *Annals of Botany*, *94*, 615-621.

[118] Lisiewska, Z., Kmiecik, W. & Korus, A. (2001). Content of nitrogen compounds in raw and preserved seeds of grass pea (Lathyrus sativus L.). *European Food Research and Technology*, *213*, 343-348.

[119] Lqari, H., Vioque, J., Pedroche, J. & Millán, F. (2002). Lupinus angustifolius protein isolates: chemical composition, functional properties and protein characterization. *Food Chemistry*, *76*, 349-356.

[120] Hsu, H. W., Vavak, D. L., Satterlee, L. D. & Miller, G. A. (1977). A multienzyme technique for estimating protein digestibility. *Journal of Food Science*, *42*(5), 1269-1273.

[121] Kalpanadevi, V. &. Mohan, V. R. (2013). Effect of processing on antinutrients and in vitro protein digestibility of the underutilized legume, Vigna unguiculata (L.) Walp subsp. unguiculata. *LWT - Food Science and Technology*, *51*, 455-461.

[122] Morup, I. K. & Olesen, E. S. (1976). New method for prediction of protein value from essential amino acid patterBfrien. *Nutritional Report International*, *13*, 355-365.

[123] Alsmeyer, R. H., Cunningham, A. E. & Happich, M. L. (1974). Equations predict PER from amino acid analysis. *Food Technology*, *28*, 34-38.

[124] Schaafsma, G. (2000). The protein digestibility-corrected amino acid score. *Journal of Nutrition*, *130*, 1865-1867.

[125] Pastor-Cavada, E., Juan, R., Pastor, J. E., Alaiz, M. & Vioque, J. (2009). Analytical nutritional characteristics of seeds proteins in six wild Lupinus species from suthern Spain. *Food Chemistry*, *117*, 466-469.

[126] Wolzak, A., Elias, L. G. & Bressani, R. (1981). Protein quality of vegetable proteins as determined by traditional biological methods and rapid chemical assays. *Journal of Agricultural and Food Chemistry*, *29*, 1063-1068.

[127] Henley, E. C. & Kuster J. M. (1994). Protein quality evaluation by protein digestibility-corrected amino acid scoring. *Food Technology*, *48*, 74-77.

[128] Sarwar, G. (1997). The protein digestibility-corrected amino acid score method overestimates quality of proteins containing antinutritional factors and of poorly

digestible proteins supplemented with limiting amino acids in rats. *Journal of Nutrition*, *127*, 758-764.

[129] Pastor-Cavada, E., Juan, R., Pastor, J. E., Alaiz, M. & Vioque, J. (2011). Nutritional characteristics of seed proteins in 28 Vicia apecies (Fabaceae) from southern Spain. *Journal Food Science*, *76*, 1118-1124.

[130] Sánchez-Vioque, R., Clemente, A., Vioque, J., Bautista, J. & Millán, F. (1999). Protein isolates from chispea (Cicer arietinum L.): chemical composition, functional properties and protein characterization. *Food Chemistry*, *64*, 237-243.

[131] Boye, J. A., Zare, F. B. & Pletch, A. C. (2010). Pulse proteins: processing, characterization, functional properties and applications in food and feed. *Food Research International*, *43*(2), 414-431.

[132] Elsherbeeny, M. H. & Robertson, L. D. (1992). Protein content variation in a pure line faba bean (Vicia-faba) collection. *Journal of Science of Food and Agriculture*, *58*, 193-196.

[133] Moughan, P.J. (2005). Dietary protein quality in humans – An overview. *Journal of AOAC International*, *88*, 874-876.

[134] Sujak, A., Kotlarz, A. & Strobel, W. (2006). Compositional and nutritional evaluation of several lupin seeds. *Food Chemistry*, *98*, 711-719.

In: Legumes
Editors: Hiroto Satou and Ren Nakamura

ISBN: 978-1-62808-280-7

Chapter 3

LEGUMES LEADING THE WAR AGAINST "DIABESITY-THE OBESITY-DIABETES EPIDEMIC": AN OVERVIEW

Rocio Campos-Vega[1,2], B. Dave Oomah[3], Guadalupe Loarca-Piña[1] and Haydé Azeneth Vergara-Castañeda[4*]

[1]Programa de Posgrado en Alimentos del Centro de la República (PROPAC), Research and Graduate Studies in Food Science,
School of Chemistry, Universidad Autónoma de Querétaro,
Querétaro, Qro, México
[2]Kellogg de México, Carretera al Campo Militar,
San Antonio de la Punta, Querétaro, Qro., México
[3]National Bioproducts and Bioprocesses Program,
Pacific Agri-Food Research Centre,
Agriculture and Agri-Food Canada, Summerland,
British Columbia, Canada
[4]Nucitec, S. A. de C. V., Departamento de Investigación,
Colonia Peñuelas, Queretaro, México

ABSTRACT

The rates of obesity and diabetes are worrying. The worldwide prevalence of type 2 diabetes mellitus (T2DM) is rising rapidly primarily due to the surge in obesity, combined with ageing populations and a trend towards urbanization. Overweight and obesity are key features of diabetes and prevention of excessive weight gain is a fiscally prudent global health priority. The new term, 'diabesity' has been coined to describe people burdened with obesity and type 2 diabetes. Individuals with diabesity are at increased risk of multiple comorbidities (including chronic and degenerative diseases) and represent a huge economic burden on health services around the world. Some

* Corresponding author: telephone (442) 240 9331 Ext 104. E-mail address: hvergara@nucitec.com; jaiakiryvergara@hotmail.com.

adjuvants can assist in weight loss and reduction in blood glucose. In this regard, there is strong evidence that a diet high in whole grains, like legumes, is associated with lower body mass index, waist circumference and reduced risk of overweight, obesity and various associated chronic and degenerative diseases. Therefore, diets high in whole grains and legumes can help reduce weight gain and elicit weight loss with energy controlled diets high in cereals and legumes. Legume consumption has been highlighted as the perfect way to improve and/or promote glycemic control, food intake and body weight in obese, glucose-intolerant and diabetic subjects. The mechanism of action that could explain those effects is linked to the concomitant presence of bioactive compounds like glycoproteins and fiber, among others. This chapter explores the role of legumes, particularly their bioactive compounds on the war against obesity and diabetes.

INTRODUCTION

Obesity is one of the strongest diabetes risk factors, and weight control is a key component of the diabetes agenda for prevention and treatment. In fact, diabetes and obesity are linked to such extent in terms of etiology and policies for treatment and prevention, that the term "diabesity" emphasizing the obesity-dependent nature of diabetes, is gaining popularity.

Although there are several pharmacological options to treat both conditions, adherence to these treatments and therefore their effectiveness is often not expected, and the economic cost and attrition rate are often so high that health professionals constantly seek other options to reduce the prevalence of diabesity in the world. The current trend is trying to counteract this epidemic through the consumption of natural foods because they contain bioactive compounds. Moreover, diet manipulation alone may reduce or eliminate the need for pharmacologic therapy of diabetes. In this context, legumes have gained great importance due to their reported effects on diabetes and obesity, among other pathologies, associated with their nutrient content, which includes fiber, protein, nutraceuticals, as well as being low in fat and glycemic index.

Research has focused on demonstrating the modulation of metabolism by legumes, adding value in order to have a natural and effective option to treat and/or prevent diabesity. This chapter compiles studies and scientific data that support the use of legumes in controlling diabesity, elucidate the molecular mechanisms involved and its effectiveness.

CURRENT TRENDS IN TREATMENT OF DIABESITY - LEGUMES AS POTENTIAL TREATMENT

The World Health Organization (WHO) considers obesity a global epidemic. The most recent WHO data indicate that 1.6 billion adults worldwide are overweight (BMI 25.0–29.9 kg/m^2) and 400 million are obese (BMI $\geq$ 30 kg/m^2). By 2015, 2.3 billion adults will be overweight and more than 700 million will be obese, according to WHO projections [1]. The risk is particularly high when excess adipose tissue is distributed intra-abdominally (as characterized by a high waist circumference). Obesity is particularly associated with an increased risk of developing non-insulin-dependent or type 2 diabetes [2]. Together, being overweight with abdominal fat distribution, and obesity probably account for about 80–90% of all type 2 diabetes (T2DM) and are important obstacles to its successful long-term diabetic management. Women with a BMI of 23–25 have a 4-fold higher risk of type 2 diabetes than those with a BMI < 20; those with a BMI 24–25 have a 5-fold increased risk, and a BMI > 35 increases the risk of developing type 2 diabetes 93-fold [3].

Given the strong association between excess weight and T2DM, the prevention of weight gain in people with diabetes, as well as in individuals with impaired glucose tolerance (IGT), should be a cardinal focus in treatment strategies [4].

The coexistence of T2DM and obesity presents a complex therapeutic challenge. T2DM confers an elevated risk of developing a broad range of complications, including macrovascular disorders (e.g. cardiovascular disease, stroke) and microvascular disease (e.g. retinopathy, neuropathy and nephropathy) [5]. Likewise, obesity increases the risk of a myriad of chronic diseases, including type 2 diabetes, coronary heart disease, hypertension, hyperlipidaemia, stroke and certain cancers. Obesity is linked to the development of additional comorbidities that can further complicate disease management, including mobility problems associated with osteoarthritis, obstructive sleep apnea and clinical depression [4].

Similar lifestyle interventions underpin the management of diabetes and obesity, and it is important to recall that intentional weight loss improves metabolic control in overweight and obese patients with type 2 diabetes. However, weight loss is more difficult to achieve in diabetic than nondiabetic individuals, at least partly due to the increased anabolic effects of insulin and reduced glycosuria with improved glycemic control [6].

Currently, there is a wide range of drug treatments to attack obesity, diabetes or a combination thereof. Conventional antidiabetic treatments include insulin, insulin sensitizers (e.g. TZDs), insulin secretagogues (e.g. sulphonylureas) and modulators of hepatic glucose production (metformin). Other treatments include peroxisome proliferator activated receptor-γ modulation by TZDs (rosiglitazone, pioglitazone), GLP-1 receptor agonists (exenatide and liraglutide) and DPP-4 inhibitors (saxagliptin) [4]. Drugs used in obesity treatment are classified in several groups, which include: anorexigenic drugs (decrease appetite or increase satiety), drugs that increase energetic expenditure and those that interfere with nutrients absorption and metabolism. Table 1 shows some antiobesity agents used currently.

In addition, since the benefits of weight loss on metabolic and cardiovascular parameters are well known, people with T2DM may elect to pursue more invasive interventions such as bariatric surgery to lose weight. The two most common procedures are laparoscopic adjustable banding of the stomach and the Roux-en-Y gastric bypass. There have been a number of recent studies showing that bariatric surgery produces a 25% reduction in mortality [4].

However, lifestyle habits and nutrition play a very important role to achieve success or failure during treatment, in addition to the use of pharmacological therapeutic agents to reduce weight and improve the normal course of the T2DM disease. There is a tendency for disease prevention through balanced diet and increased consumption of bioactive compounds in food, but more importantly, to manage them as auxiliaries in the treatment of diabesity.

The major objective of diet therapy in this group of individuals is weight reduction followed by maintenance of desirable body weight. Additional considerations include changes in the form, composition and distribution of meals as well as avoidance of certain items. For the most part, the earlier diet therapy is initiated, the greater are the potential benefits. Moreover, diet manipulation alone may reduce or eliminate the need for pharmacologic therapy of diabetes. Nutrition education should focus on the individual being able to make the appropriate changes to achieve weight loss and improve metabolic control [8].

Table 1. Antiobesity agents used currently

Drug classes	Therapeutics	Therapeutic function	Agents/drugs
Appetite suppressants	Adrenergic agonists	Increase release of NA-DA	Phendimetrazine Fenproporex Clobenzorex Benzphetamine Mazindol Phenylpropanolamine Phentermine
	Central serotonergic agonists	Increase release of 5HT	Fenfluramine Dexfenfluramine
		Selective inhibitors of serotonin reuptake	Fluoxetine Paroxetine Sestraline
	Central serotonergic and adrenergic agonists	Increase release of 5HT, NA, and reduce DA	Diethylpropion Fenproporex Clobenzorex Benfluorex
		Inhibit 5HT reuptake, NA and reduce DA	Sibutramine
	Cannabinoid receptor type 1 agonists		Rimonobant
Thermogenic drugs	Adrenergic	Increase NA	Ephedrine + caffeine
		β-3 Agonists	Research
Nutrient absorption inhibitors	Lipase inhibitors		Orlistat
	α -glucosidase inhibitor		Acarbose
	Fat substitutes		Olestra
	Fat binding		Fiber
	Gastric emptying inhibitor		Derivatives of amylins GLP-1
Hormones and peptides	Leptin analogue		Research
	CCK agonist		Research
	NPY agonist and galanine		Research
Hormones and peptides	Leptin analogue		Research
	CCK agonist		Research
	NPY agonist and galanine		Research
Type 2 Diabetes drugs	Sodium-glucose co-transporters (SGLT-2) inhibitors		Dapagliflozin/Forxiga Canagliflozin Tempagliflozin
	Glucagon-like peptide-1 (GLP-1) receptor agonists		Byetta/Exenatide Victoza/Liraglutide Degludec/Tresiba Semaglutide
	DPP-IV inhibitors		Sitagliptin Alogliptin Linagliptin
	Biosimilar insulin Humalog		Biosimilar insulin

NA: Noradrenaline; DA: Dopamine, 5HT: 5-Hidroxytriptamine; CCK: Colecistoquinine; NPY: Neuropeptide Y. Adapted from Salvador et al., (2002) [7].

In this context, the suggestion that whole grain foods might protect against the development of diabetes as well as being useful in the management of people who have already developed T2DM is relatively recent [9]. One of the key recommendations of the 2005 US Dietary Guidelines is "consume three or more ounce-equivalents of whole-grain products per day, with the rest of the recommended grains coming from enriched or whole-grain products. In general, at least half the grains should come from whole grains [10].

There are strong epidemiological evidence for the beneficial effect of whole grain and legume consumption on the risk for many chronic diseases, especially cardiovascular disease and diabetes. Along with whole grains, legumes constitute a food group that has been relatively understudied in the epidemiological context. The current evidence for recommendations about their inclusion in a healthy diet relate to their nutrient content (i.e., low in fat and a good source of soluble fiber and protein), rather than strong evidence for their role in chronic disease prevention. In relation to weight control, it may be that their generally low glycemic index (GI) value is the main benefit in weight control, since it may enhance satiety, although there have been suggestions that alpha amylase inhibitors in legumes may play a role as well [11]. Figure 1 lists the GI of three pulses and other types of foods.

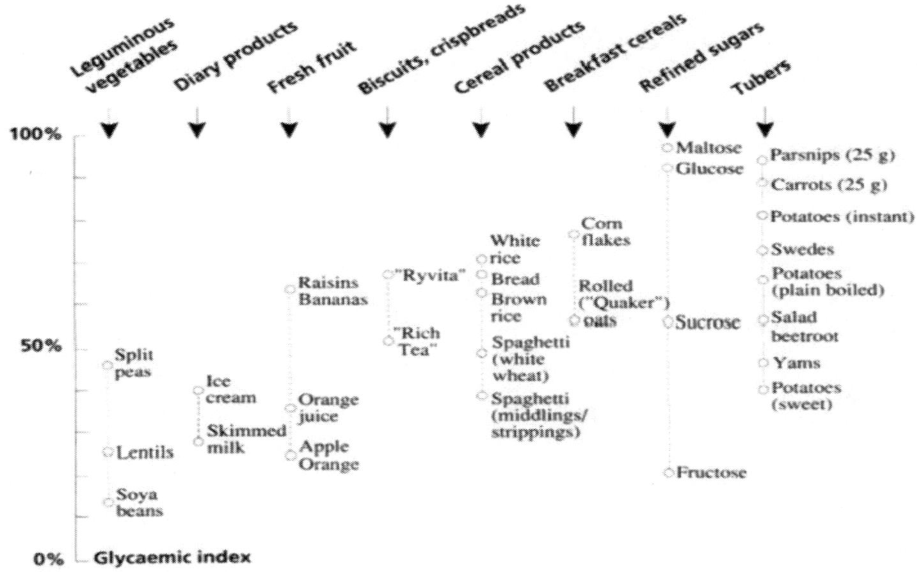

Figure 1. Glycaemic index of food and food categories Adapted from Bornet et al., [12].

The selection of low-GI foods, including legumes, allows efficient control of post-prandial insulin and glucose excursions. After intake of low-GI foods the moderate elevations in blood glucose are followed by a slow return to baseline values, without any sign of hypoglycaemia [13].

SCIENTIFIC EVIDENCE OF THE BENEFITS OF LEGUMES ON DIABESITY

Some studies have demonstrated that legumes provide different health benefits, including those related to obesity, diabetes and its complications. Forty eligible participants, men and women aged 35–55 years with a BMI of 27–39·9 kg/m^2, were studied by Mollard et al., [14]. Ten participants classified as having the MetSyn received a pulse diet, containing lentils (Nupak), chickpeas (Nupak), yellow split peas (Nupak) and navy beans (Ferma). Pulses provided about 70% of the energy in the diet. The results showed that regular consumption of pulses in an ad libitum diet was as effective at reducing risk factors of the MetSyn as energy restricted diet implemented through dietary counseling. Both the pulse and energy-restricted diets reduced energy intakes, waist circumference, systolic blood pressure, HbA1c, improved blood glucose control and insulin sensitivity (decreased insulin resistance), as well as improved blood lipids by the pulse diet indicated by HDL increase.

A low glycaemic index diet that included high consumption of legumes, including beans, chickpeas, and lentils, reduced glycosylated haemoglobin (HbA1c) levels and systolic blood pressure in patients with T2DM in a randomized controlled trial [15]. The researchers randomly assigned 121 patients with T2DM to receive a low glycaemic index legume diet or a whole wheat fibre diet for 3 months. The participants in the group consuming the low glycaemic index legume diet received dietary advice to increase their consumption of legumes to one cup per day of lentils, chickpeas or cooked beans, whereas those in the whole wheat fibre diet group were advised to increase consumption of whole wheat or whole grain carbohydrates. Levels of HbA1c fell by a mean 0.5% and 0.3% in patients who had consumed the low glycaemic index legume diet and the whole wheat fibre diet, respectively. Blood pressure, heart rate and estimated coronary heart disease risk also fell in the group consuming the low glycaemic index legume diet. Jenkins et al. [15] conclude that their findings support consumption of legumes as a specific component of a low glycaemic index diet in patients with T2DM, despite the modest improvements in glycaemic control observed in their study.

Nöthlings et al., [16] examined the associations of vegetables, legumes and fruit intake with all-cause and cause-specific mortality in a population (10,449 participants) with prevalent diabetes in Europe. The percentage of diabetic participants treated with insulin was inversely associated with intake of vegetables, legumes, and fruit, ranging from 32% in the lowest to 16% in the highest quartile. The percentage of diabetics with a self-reported heart attack at baseline was also inversely associated with intake of total vegetables, legumes, and fruit. The results indicated that consumption of vegetables and legumes is more beneficial for diabetes patients than consumption of fruit.

Sievenpiper et al. [17] performed a systematic meta-analysis to assess the evidence that dietary pulses may benefit glycaemic control. The meta-analysis consisted of three pooled analyses of 41 randomized controlled experimental trials, in a total of 1,674 participants, and they were supportive of diabetes association nutrition guidelines recommending the consumption of dietary pulses as a means of optimizing diabetes control. Pulses given alone or as part of low-GI or high-fibre diets improved the main markers of long-term glycaemic control, namely, GP (HbA1c or fructosamine) by pulses in low-GI and high-fibre diets and fasting blood glucose (FBG) by pulses given alone or as part of high-fibre diets. The strongest evidence for benefit was for pulse interventions containing chickpeas (*Cicer arietunum*), which decreased FBG in the pulses alone analysis with very low heterogeneity. Potential benefit was also seen for *Phaseolus vulgaris* (black, white, pinto, red and white kidney beans), but interpretation of these results was complicated by high heterogeneity, and for *Vicia faba* (fava beans), but the data were from only one study. Furthermore, although human clinical trial data are still forthcoming in examining the effects of pulse consumption on regional fat deposition, animal studies have suggested a beneficial effect. Chickpeas significantly reduced epididymal fat deposition as well as rats receiving chickpeas demonstrated similar levels of epididymal fat mass vs normal fat-fed rats [18]. On the other hand, rats fed diets containing mung beans demonstrated a 48% reduction in retroperitoneal adipose tissue mass, a visceral fat depot, compared with control diets [19]. A human clinical trial using whole pea flour as a dietary intervention, alongside an energy-controlled paradigm that prevented fluctuations in weight, demonstrated a significant reduction in android: gynoid fat ratios in women, indicating that fat deposition was directed away from the waistline area [20]. Marinangeli and Jones [21] have proposed the next hypotheses trying to explain how pulses and their fractions target satiety and related facets of metabolism that lead to excess weight gain (Figure 2):

Where: 1) Fermentation of pulse-derived fibres and resistant starch facilitates the production of short chain fatty acids such as butyrate. Butyrate has been shown to alter energy expenditure and fatty acid oxidation by promoting glycogen storage, decreasing glycolysis and increasing the expression of Pgc-1a. (2) Cellulose and hemicellulose, two insoluble fibres found in pulses mitigate excess adipose deposition and weight gain, causing an increase in hepatic Foxa2 and Pgc-1b expression. Cellulose and hemicelluloses also reduce the expression of adipose Gpr40. De novo lipogenesis could also be reduced via decreased expression of FAS secondary to low levels of refined sugars in pulse grains and blunt postprandial glucose responses. (3) Modulation of genes could also be secondary to the ability for pulses to induce changes in genes that facilitate energy expenditure and fat oxidation directly. (4) Pulse grains could contribute to the consumption of phyto-oestrogens that interact with ERa which increase SC adipose deposition and reduce visceral adipose deposition [21].

Moreover, several studies have demonstrated that pulse grains modulate satiety through increased and prolonged cholecystokinin (CCK) secretion and decreased rate of gastric emptying, respectively (Figure 3). Several studies demonstrate that legumes, such as chickpeas, navy beans and others reduce appetite compared with controls [21]. Each type of legumes may have differing potential beneficial effects on diabetes control and obesity, and therefore evidence for each legume type is presented below.

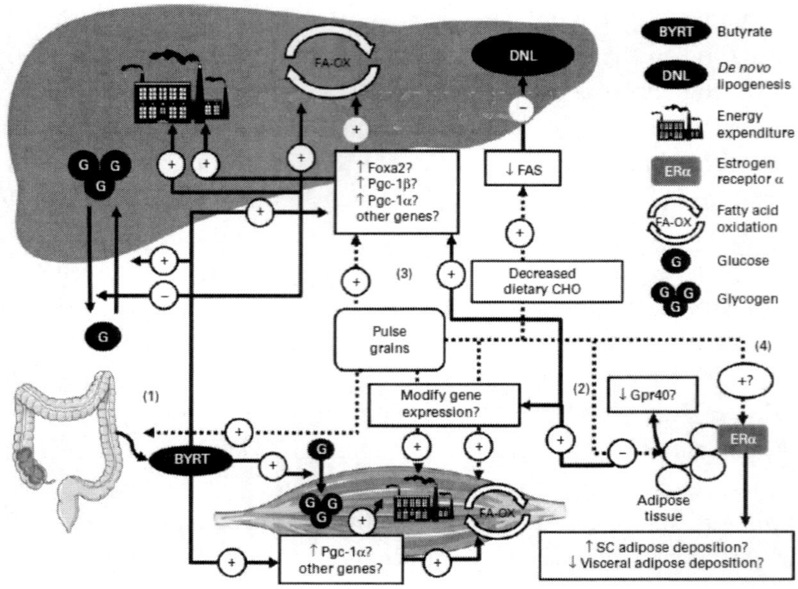

Figure 2. Pulse grains modulate thermogenesis, substrate trafficking ,oxidation and, adipose tissue deposition. Pgc-1α: PPARγ coactivator α; Foxa2: forkhead transcription factor; Pgc-1 β: PPARγ coactivator β; Gpr40: G-protein-coupled receptor- 40; FAS: fatty acid synthase; Erα: oestrogen receptor α; SC: facilitate subcutaneous; CHO: carbohydrates. Adapted from Marinangeli and Jones [21].

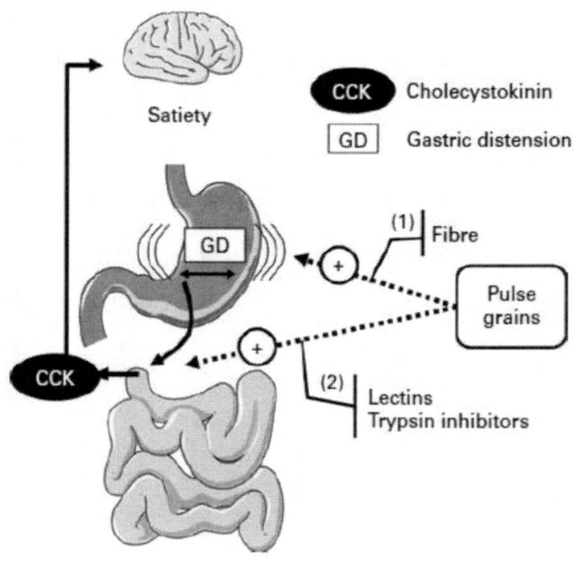

Figure 3. Mechanism proposed for induced secretion of cholecystokinin. CCK: cholecystokinin; GD: gastric distension by pulse grains. Adapted from Marinangeli and Jones [21]. Where: (1) Pulse grain fibres increase and prolong CCK secretion by inducing GD and decreasing the rate of gastric emptying, respectively. Decreasing the rate of gastric emptying can also prolong CCK secretion. (2) Pulse grain-derived lectins and trypsin inhibitors increase CCK secretion directly.

Beans

The common bean (*Phaseolus vulgaris* L.) is a legume predominantly found around Mexico and Central America. Common beans are rich in several phytochemicals such as polyphenols, carbohydrates, fiber and polysaccharides with potential health benefits [22]. It has been known for decades that certain bean extracts have the potential to inhibit the alpha amylase enzyme. They act as "carb blockers", which has the potential to slow CHO absorption through enzymatic inhibition [23] and is believed to cause weight loss by promoting the mobilization of the body's fat reserves as a result of energy restriction [24]. Pusztai et al. [25] observed a reduction in body fat in rats due to the consumption of raw kidney beans. The effect could also have arisen due to the presence of amylase inhibitors in the common beans since the lean body content of the obese rats was not affected. Hangen and Bennink [26] showed that black and navy bean diets reduced body weight and the fat percentage directly associated with anorexia and starch escape of digestion in the ileum of rats. It has been reported that the amylase inhibitor fed chronically to rats reduces feed intake.

Bean consumption has been associated with nutrient intakes and physiological parameters using the National Health and Examination Survey (NHANES) 1999-2002. Adults from NHANES were included, and three groups of bean consumers were identified for the present analyses 1) baked bean (BB); 2) variety bean (VB) (i.e., pinto beans, kidney beans, etc.).; and 3) variety bean and/or baked bean (VBBB). Body weight, BMI and waist circumference were lower in VB and VBBB consumers. Young adult VBBB consumers had lower body weights, decreased BMI values and smaller waist sizes than non-consumers (Figure 4 and 5). These differences typically led to a lower odds ratio for an increased waist. Young adults consuming VBBB were about 40% less likely to have an increased waist size or be obese than non-consumers.

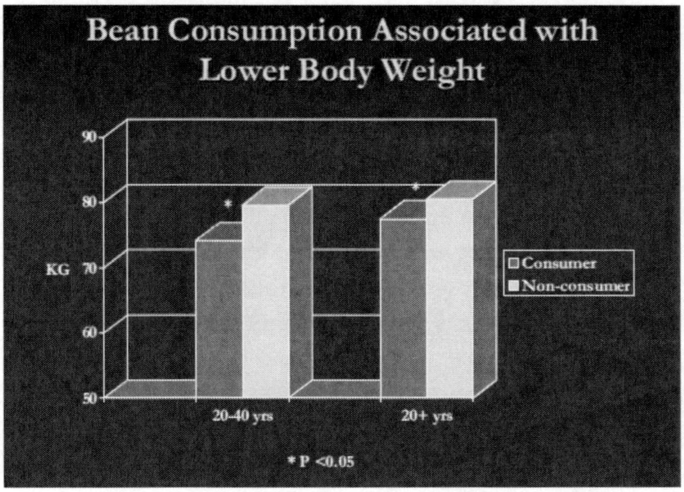

Figure 4. Bean consumption associated with lower body weight. Adapted from Papanikolaou and Fulgoni [27].

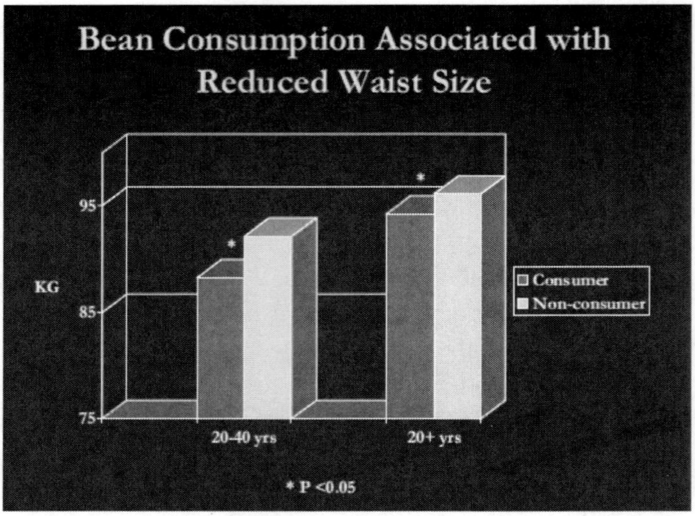

Figure 5. Bean consumption associated with reduced waist size. Adapted from Papanikolaou and Fulgoni [27].

The study concluded that adult bean consumers had lower body weights, smaller waist circumferences and a reduced risk for increased waist size and being obese relative to non-consumers [27].

Several types of beans have been recognized as potential anti-obesogenic food when added to the habitual diet. Small red, navy, and white kidney bean were added to diet of female Sprague-Dawley rats to determine if beans can modulate lipid metabolism, particularly modulation of cellular pathways responsible for the synthesis of many lipid metabolites, interconversion and/or degradation [28]. Dry red bean powder (small red bean) protected against obesogenic effect induced by high-fat diet and its comorbidity hypercholesterolemia in C57BL/6J male mice [29]. The impact on body weight was assessed for 7 weeks. The body composition of the bean-fed mice was shifted to a less obese state as indicated by the reduction in the Lee Index (a measure of adiposity) and lower plasma leptin, an indicator of adipocyte mass.

As mentioned before, diabetes can be triggered by obesity as one of its main consequences. In this regard, legumes, mostly beans, have been studied for its ability to improve diabetes. The hypoglycemic effect of the common bean has long been established and, in fact, bean pods are known as traditional remedies for diabetes mellitus [30].

Legumes, as beans, were the first class of foods recognized as having low GI values and have been recommended in many national DM guidelines [31, 32]. Several studies indicate that the long-term consumption of a diet with a high glycemic load is a significant independent predictor of the risk of developing type 2 diabetes. More recently, evidence has been accumulating that a low glycemic index (GI) diet might also protect against the development of obesity and therefore of diabetes [33]. Jenkins et al. [15] demonstrated that the increased legume consumption as part of a low-GI diet lowered HbA1c values, BP, heart rate, and estimated absolute CHD risk. Therefore, their study provides support for the use of legumes as a specific food option to lower the dietary GI in type 2 DM and for the recommendations to increase low-GI food consumption by many national diabetes associations.

The antihyperglycemic effect of *Phaseolus vulgaris* has also been studied on healthy rabbits. After the intragastric administration of kidney bean (decoction), a 50% dextrose solution was infused subcutaneously. The beans reduced (21%) the area under glucose tolerance curve, while the drug (tolbutamide) produced a 16% decrease [34]. The antidiabetic effects of beans could be explained by its fiber contents. Common beans are an important source of dietary fiber. Vergara-Castañeda et al., [35] reported 38.5% and 41.2% total fiber for raw and cooked bean, respectively, for cultivar Bayo Madero, while Campos-Vega et al., [22] obtained 37.6%, 25.6%, 28.3% and 30.1% polysaccharides in raw beans, and 48.1%, 55%, 36.5% and 42.2% in cooked Negro 8025, Bayo Madero, Pinto Durango and Azufrado Higuera, respectively.

In this regard, fiber is known to slow carbohydrate absorption. In another study, the oral administration of lyophilized flour of cooked Pinto Zapata bean showed a significantly hypoglycemic and antihyperglycemic effect (30 % and 33 %, respectively) at 300 mg/kg in diabetic rats. The results suggest that the treatment with bean might improve the conditions of diabetic patients [36].

In 2003, a study found that the glucose lowering effect of a hot aqueous extract of *P. vulgaris* (200 mg/kg) exceeded that of glibenclamide in streptozotocin-induced diabetic rats. The authors confirmed their results in a similar, additional study, finding a decrease in blood glucose and an increase in insulin levels comparable to that after glibenclamide [37].

In healthy rats, feeding of mung bean starch for 5 weeks reduced non-fasting plasma glucose and free fatty acid levels compared to rats fed wheat starch. In both healthy and diabetic rats the mung bean starch reduced plasma triacylglycerol concentrations and adipocyte volume. This indicates that the use of starch from mung beans may be beneficial for modifying both glucose and lipid metabolism [38].

Interestingly, a growing number of studies have indicated that the protein fraction of legumes has beneficial effects on health, improving the postprandial glycaemic response and reducing food intake and weight gain. Spadafranca et al. [39] evaluated the effects of a *P. vulgaris* extract (PVE) as a supplement in a mixed balanced meal (60% carbohydrates, 25% lipids and 15% protein), on glycometabolic and appetite control. PVE suplementation reduced the glucose response to the meal, in association with lower insulin and C-peptide excursions. Thus, PVE could be useful and beneficial in diabetic and glucose-intolerant subjects, both with insulin resistance and insulin secretory defect, in order to modulate their postprandial carbohydrate metabolism and prevent health complications. Furthermore, PVE suppressed ghrelin (a hunger signal peptide) secretion and affected satiety sensations, reducing the desire to eat. The hypothesis of this study was that PVE could have slowed gastric emptying via inhibition of carbohydrate hydrolysis, affecting ghrelin secretion and feelings of satiety.

Moreover, the study conducted by Rodgers et al., [40] presented evidence that a tepary bean diet combined with exercise, despite 37% of calories from fat, results in significantly less weight gain and mean body mass compared with fa/fa rats fed either a tepary bean diet alone, or control diet with or without exercise (Figure 6). Further, exercise and the consumption of the tepary bean diet results in significantly lower insulin concentration compared to control fed animals at weeks 4 and 8 of the intervention program (Figure 7). Tepary bean diet, with or without exercise, results in significantly lower serum triglycerides and cholesterol relative to animals fed the control diet.

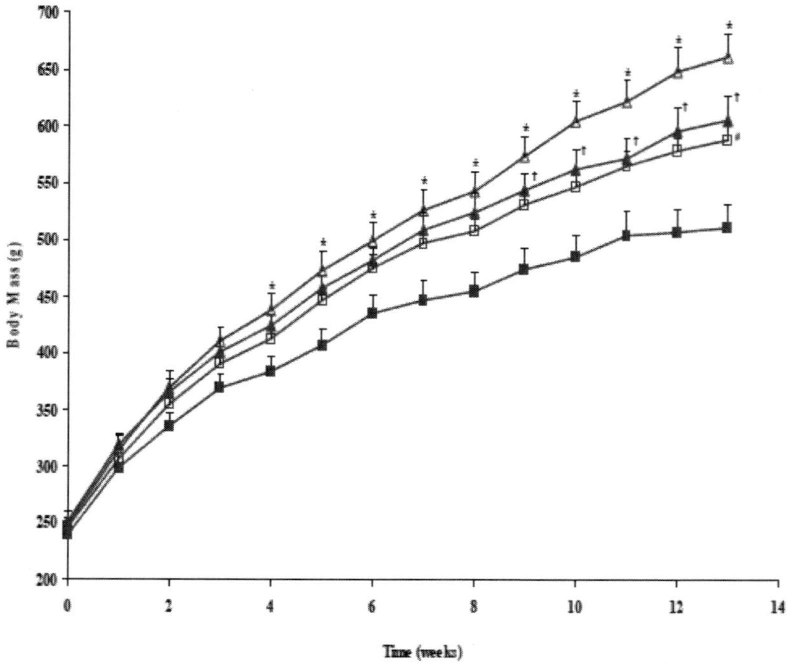

Figure 6. The effect of diet and exercise on mean body mass (g) of female fa/fa rats. Comparison of control diet (C) rats (; n=10) with control diet + exercise (CE) (▲; n=10), tepary bean diet (T) (▭; n=10) and tepary bean diet + exercise (TE) (■ ; n=10) rats. Values are expressed as mean ± sem. *p<0.05 C vs. TE; †p<0.05 CE vs. TE; # p<0.05 T vs TE. Adapted from Rodgers et al., [40].

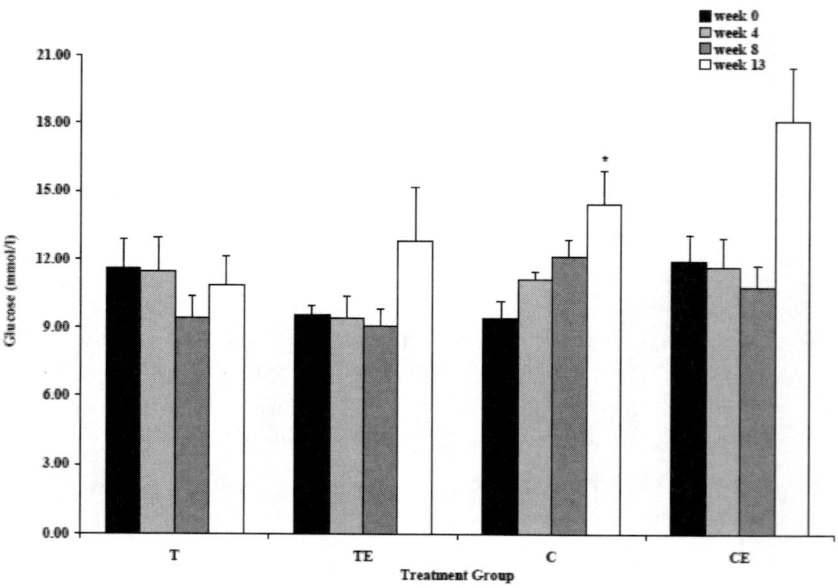

Figure 7. Effect of diet and exercise on mean fasting serum glucose concentration (mmol/L) over time. Values are expressed as mean + SEM, *n<0.05 study onset vs week 13. Adapted from Rodgers et al., [40].

Finally, one of the common consequences accompanying the course of diabetes is hypertension. Integral treatment of all pathologies inherent to diabetes should be taken into account to improve patient health. In this context, Oomah et al., [41] reported that an aqueous acetone extract of black bean hull had higher antioxidant and stronger inhibitory effect on both cyclooxygenases, COX-1 and COX-2, even outperforming aspirin. Moreover, bean hull water extracts were stronger inhibitors of lipoxygenase, 15-LOX. These results indicate the effectiveness of consuming beans as adjuvant in the treatment of diabetes.

Peas

Peas and yellow peas have also been proposed as chemoprotector food [9]. Peas provide an array of nutrients and phytochemicals that have beneficial health effects. Much of the protective function is due to the content of fiber and non-digestible grain fraction similar to those in beans. For example, yellow pea hulls consist primarily of cellulose (69%), hemicellulose (7.5%) and lignin (1.4%) [42] and cooked peas contain 16.26 g of total fiber, 5.04 g of soluble fiber and 11.23 g of insoluble fiber [43]. A recent study investigating the metabolic effects of insoluble fibres containing primarily cellulose and hemicellulose showed an increase in the expression of forkhead transcription factor (Foxa2) and PPARg coactivator b (Pgc-1b), two genes that facilitate hepatic fat oxidation [44].

Research is needed to understand how peas specifically may influence weight management, with body weight, BMI or waist circumference as primary endpoints, because data currently are inconclusive. For example, Lang et al., [45] showed no effect of pea protein on satiety, 24 h energy or macronutrient intakes, or on postprandial plasma glucose and insulin concentrations when compared with egg albumin, casein, gelatin, soy protein and wheat gluten.

The role of peas, together with chickpeas, and faba beans, in improving insulin resistance and post-meal blood glucose levels has been evaluated in numerous studies. Recently, Marinangeli and Jones [46] reported the health benefits of consuming peas in a human clinical trial with a controlled diet. The subjects consumed muffins made with whole (WPF) or fractionated pea flour (FPF) equivalent to half a cup per day of the pulse. The results showed lower (20%–25%), insulin concentrations among the WPF and FPF compared to those who received wheat based muffins. Other additional benefits were a significant reduction in Homeostatic Model Assessment-insulin resistance and android: gynoid fat ratio in women on a pea based diet. The authors concluded that under a controlled diet, the USDA's recommended intake of 50 g/d of pulse or half a cup of WPF and FPF decreased the fasting insulin concentrations and insulin resistance in hyper cholesterolemic and overweight individual fibers. Karlstrom et al., [47] demonstrated that lower mean postprandial and urine glucose excretions result from a diet composed of peas and beans in adults with type 2 diabetes mellitus.

Marinangeli et al., [48] investigated the use of whole yellow pea flour to create foods with a lower glycaemic index compared to foods made from whole wheat flour. The results demonstrated that foods containing whole yellow pea flour reduced postprandial glucose responses in individuals and, thus, may have a role in the management of type 2 diabetes. Lunde et al., [49] found that bread containing 17% pea hull fibre significantly reduced glycaemic response; however, the fibre breads also contained higher protein. These authors also found that pea fibre-enriched bread increased duration of satiety, compared with intake of regular bread.

The suppression of appetite and glycaemia by yellow peas (10 or 20 g of isolated yellow pea protein) was studied by Smith et al., [50] on blood glucose level and appetite of young healthy males (20–30 years). The results indicated that although the protein preloads did not affect subjective appetite, they suppressed food intake at 30 min, and reduced the glycaemic response to the treatment as well as the response to the second meal at 30 min. The suppression of food intake by yellow pea protein at 30 min corresponds with the peak in plasma amino acids following its consumption as well as other proteins with similar digestion kinetics.

Recently, a more complete study [51], compared the effects of feeding uncooked pea fractions (embryo vs. seed coat) for 4 weeks on glucose homeostasis in glucose intolerant rats and examined potential mechanisms influencing glucose homeostasis. Seed coats were used with low (non-colored seed coat; NSC) and high (colored seed coat; CSC) proanthocyanidin concentration. Both NSC and CSC groups had significantly reduced fasting blood glucose after 3 weeks of pea diets and peak glucose excursions in the oral glucose tolerance test. However, the mechanisms by which NSC and CSC feeding improved glucose homeostasis were different, because NSC reduced both fasting insulin and the insulin response to oral glucose, whereas CSC had no effect on insulin. The authors found that insulin signaling was improved in NSC rats, consistent with the reduction in insulin secretion and area of β-cells in pancreatic tissue. The NSC diet reduced the need for an adaptive response to developing insulin resistance (Figure 8). Moreover, in this study, the effects of pea seed coat feeding on AKT protein expression were restricted to CSC suggesting that the effects were not due to fibre or protein. In addition, improved glucose tolerance and lower insulin secretion in the NSC group could also be explained by slower or reduced glucose absorption from the gastrointestinal tract and colon thickness of the mucosa. Finally, the authors showed the reduction of stress on pancreatic islet cells as another benefit of feeding pea NSC. In conclusion, diets containing uncooked pea seed coats but not the embryos had beneficial effects on glucose tolerance in rats fed a high fat diet. However, the mechanisms by which pea seed coats exert their effects on blood glucose were not definitely identified but appeared to involve insulin signaling in skeletal muscle.

Finally, protein intake from peas may also have other beneficial effects associated with weight loss by the modulation of appetite sensations [52]. In this context, the study of Geraedts et al., [53] showed that the combination of pea protein with sucrose and the artificial sweetener sucralose induced stronger effects on GLP-1 release both *in vitro* (Figure 9) and *ex vivo* (Figure 10) compared to the compounds separately. Addition of sucrose or sucralose to STC-1 cells also stimulated the release of CCK and GLP-1, whereas addition of the sweeteners to human duodenal biopsies did not result in hormone release. The authors concluded that the combination of food ingredients plays an important role in regulating the release of satiety hormones. Abou-Samra et al., [54] recently reported that pea protein is effective in lowering short term food intake. But, when the preload was consumed 30 min before the *ad libitum* meal, total food intake was similar after pea protein compared to after water control, with a caloric compensation of around 100%. Mechanisms were not explored but gastric emptying might play a role. Other potential mechanisms can be related to the action of satiety hormones.

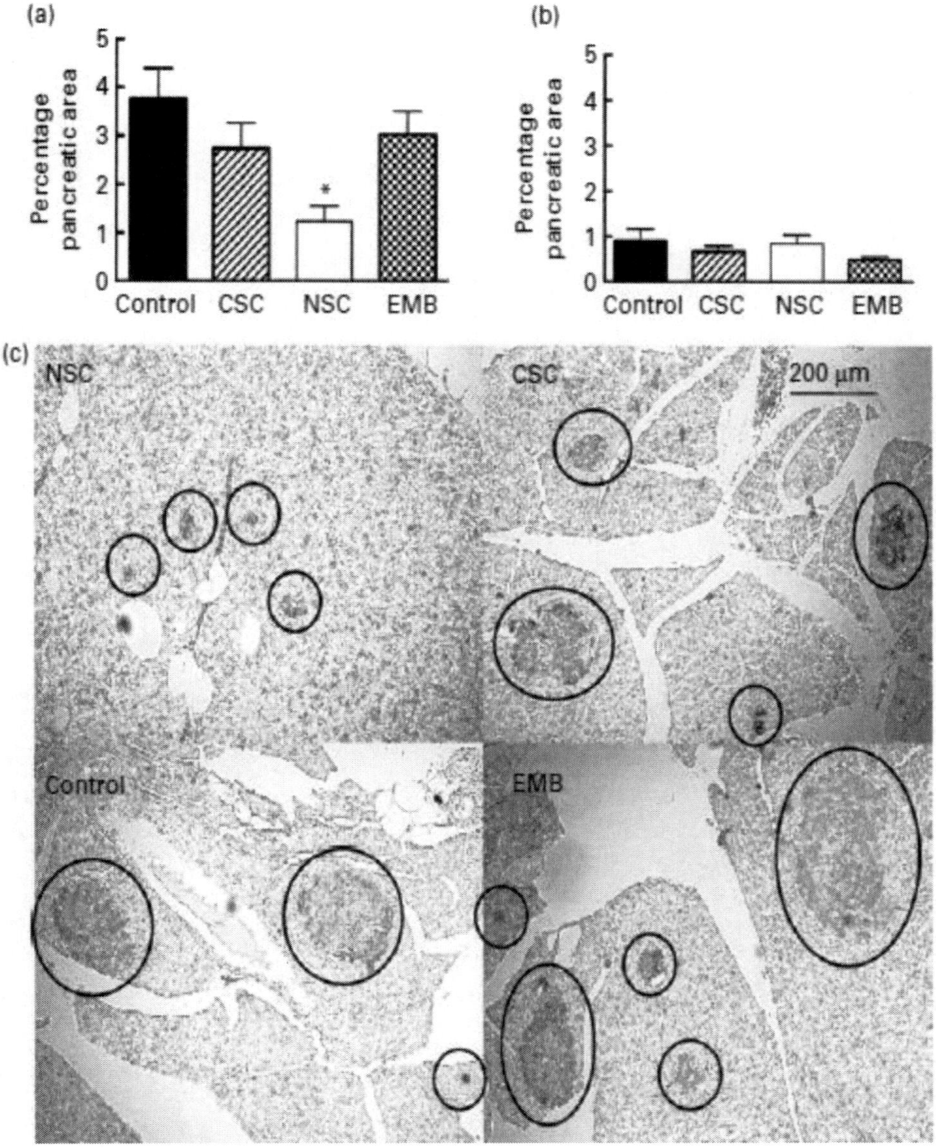

Figure 8. Morphology of pancreatic islets. The percentage of total pancreatic area occupied by cells immunostaining for (a) insulin and (b) glucagon was quantified. Values are means (n= 8 for all groups), with standard errors represented by vertical bars. *Mean value was significantly different from the control group fed the high-fat diet ($P < 0{\cdot}05$). (c) Representative photomicrographs. Insulin immunostaining is highlighted by the circles around each islet. CSC, colored seed coat diet; NSC, non-colored seed coat diet; EMB, embryo diet. Adapted from Whitlock et al., [51].

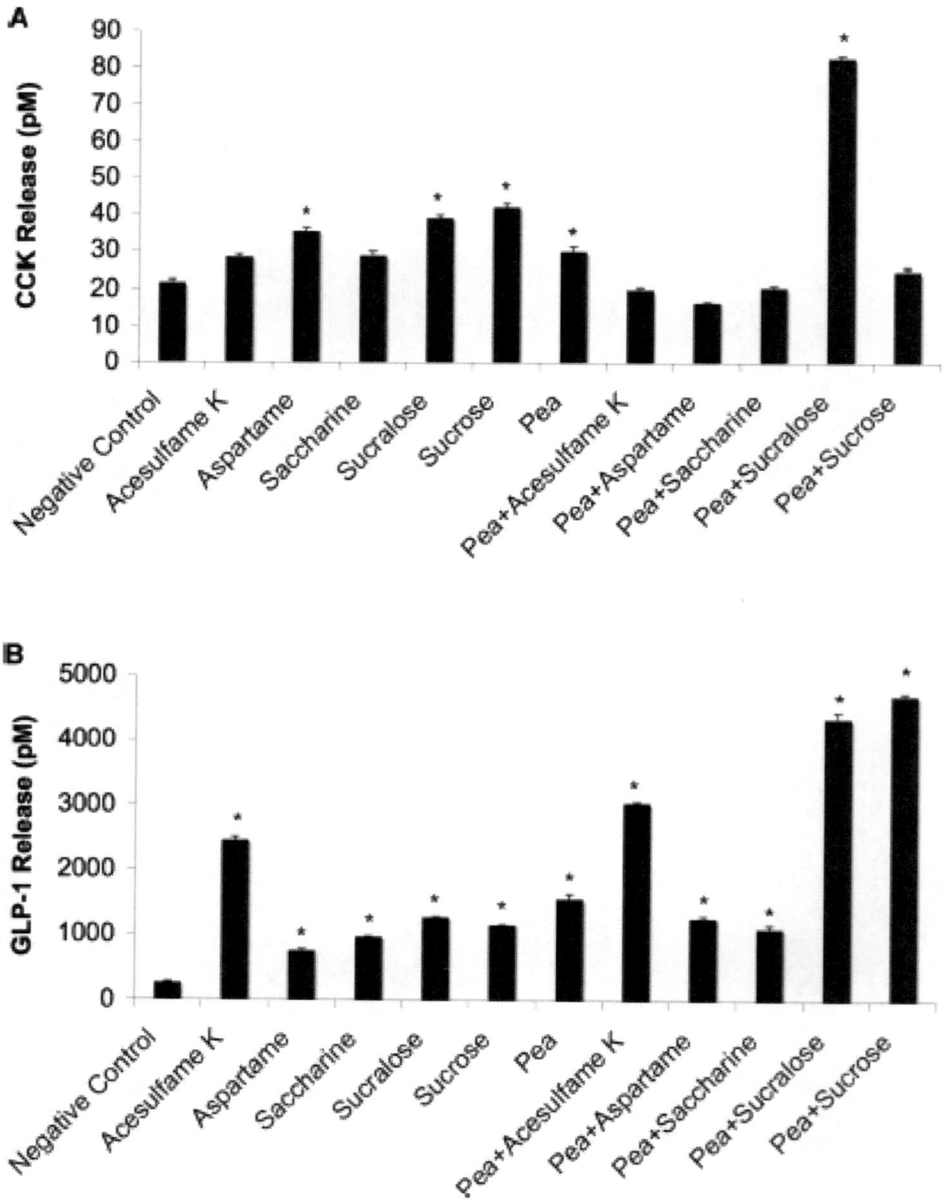

Figure 9. Hormone release from STC-1 cells after 2-h exposure to several sweeteners, pea protein, and combinations of pea protein with the sweeteners. After the incubation period, cholecystokinin (CCK) (A) and glucagon-like peptide 1 (GLP-1) (B) levels were measured in the supernatant. Results are expressed as mean ± SEM. *Significantly different from negative control, $p < 0.05$). Adapted from Geraedts et al., [53].

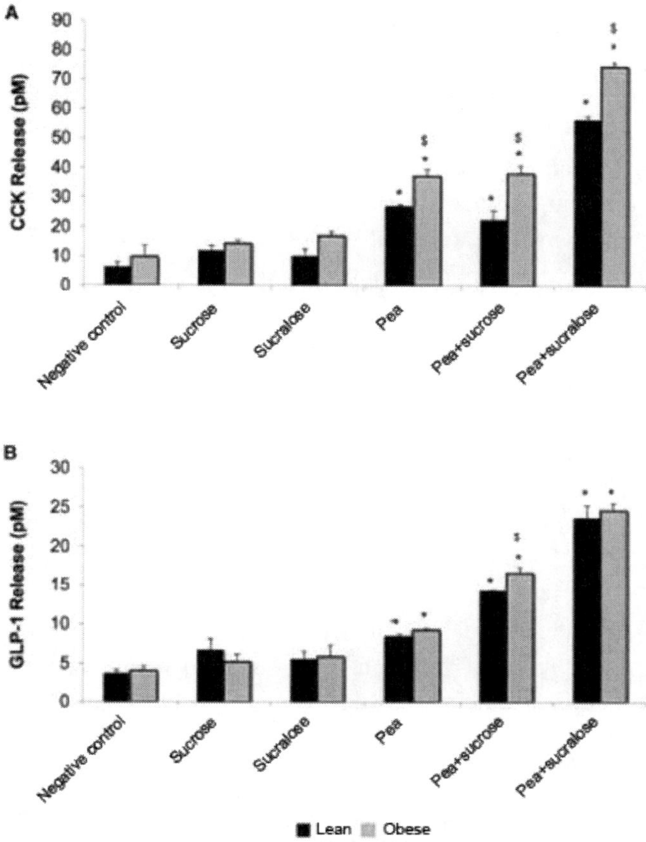

Figure 10. Hormone release from human duodenal biopsies in Ussing chambers after 2-h exposure to sucrose, sucralose, pea protein, and their combinations. Duodenal biopsies of lean and obese subjects were exposed to either, sucrose, sucralose, pea protein, or their combination. The levels of CCK (A) and GLP-1 (B) were measured in the supernatant of basolateral side of the biopsies in the Ussing chambers after being exposed to proteins for 2 h to the apical side. Results are expressed as mean ± SEM. *Significantly different from negative control, $p < 0.05$. $Significantly different from lean subjects. by Geraedts et al., [53].

Chickpeas

Chickpea (*Cicer arietinum* L.), also called garbanzo bean or Bengal gram, is the third most important pulse crop in production, next to dry beans and field peas. The major chickpea-producing countries include Pakistan, Turkey, Australia, Myanmar, Ethiopia, Iran, Mexico, Canada and the USA. In the semi-arid tropics, chickpea is an important component of the diets for people who cannot afford animal proteins or those who are vegetarian by choice. Chickpeas have high resistant starch and amylase contents [55] and the increased consumption of resistant starch is related to improved glucose tolerance and insulin sensitivity [56, 57]. Linoleic acid and β-sitosterol are the major PUFA and phytosterol, respectively, in chickpea seeds. Linoleic acid is biologically important due to its involvement in the production of PG. PG and β-sitosterol are involved in reducing blood pressure and smooth muscle constriction; therefore, chickpea seeds could be incorporated as part of a regular diet that may help reduce blood pressure, a major consequence of non-treated diabetes.

One study found a significant reduction in the blood glucose and insulin levels of healthy adults after consumption of whole chickpeas and black beans with white rice compared to a control meal containing solely white rice [58]. Incorporating chickpeas habitual *ad libitum* intake of 45 healthy participants for 12 weeks significantly increased dietary fiber content and energy supplied by protein and decreased energy supplied by saturated fat, and these led to improvements in fasting insulin and total cholesterol content [59]. Chickpeas lowered the post-pizza meal blood glucose area under the curve response compared to navy beans and yellow peas, used as treatment in a study performed with *ad libitum* pulse meal on healthy males [60].

In 1984, a group of research examined five leguminous foods, including chickpeas, and eight non-leguminous products considering their polyphenol concentrations and the blood glucose response in normal and diabetic volunteers. Their findings showed that for both normal and diabetic individuals, the leguminous foods reduced blood glucose response, measured as the glycemic index compared to non-leguminous foods. The presence of large polymeric polyphenols in legumes showed a higher correlation with glycemic index response [43].

Nestel et al., [61] conducted an acute and a long-term studies in healthy middle-aged men and women, measuring plasma glucose, insulin, and calculated homeostasis model assessment (HOMA; an index of insulin sensitivity) on 3 separated days over 3 h after the subjects consumed 50-g available carbohydrate loads from either chickpeas, wheat-based foods, or white bread, while the long-term comparison (*n*-20) was a randomized, crossover study in which chickpea-based and wheat-based foods were eaten for 6 wk each. The results of the study confirmed that a low–glycemic index food such as chickpeas reduce plasma glucose, insulin, and HOMA responses after a single meal than after meals of similar available carbohydrate content derived from wheat based or higher glycemic index foods. However, the results cannot be translated into longer-term because chickpeas did not improve the insulin sensitivity or glucose disposal.

The nutritional benefits of chickpea have led to its use in various culinary applications. In addition, chickpea is used in stews, soups and salads, and can be processed into flour. Chickpeas ability to improve blood glucose and insulin has been successfully used to develop products based on this legume. Johnson et al., [62] measured the glycemic and insulinemic responses and energy intake subsequent to consumption of breads made with normally processed chickpea flour, extruded chickpea flour, or white wheat flour. The extrusion process involves higher temperatures and shear compared with those used during normal processing. Gelatinized starch during extrusion processing may fragment and align differently than when normally processed and, by retrogradation, potentially increase the resistant starch content of the chickpea flour. The results showed a significantly higher insulin response and a non-significant tendency toward a lower glycemic response after ingestion of the chickpea bread (normal process). However, neither satiety nor energy intake differed among bread types at a buffet meal 2 h later, perhaps due to the variation in food form (chickpea flour) compared with other studies that have used whole pulses. Even when the hyperinsulinaemic effect of chickpea requires further investigation, all this evidence support the fact that chickpea is a good source of phytochemicals with biological effects.

Finally, phenolic compounds are considered the major component contributing to the total antioxidant activities of grains. These compounds have been associated with reduced risk of diabetes and other degenerative chronic diseases. Table 2 shows the total hydrophilic antioxidant activity (AoxA) or ORAC values of the studied chickpeas. Significant differences in hydrophilic AoxA were observed among desi chickpea seeds with ORAC values ranging from 43.9 (Red ICC14782 and Cream ICC3421) to 53.9 μmol (Brown ICC3512) Trolox equivalents/g sample, (dry weight basis). In general, these results show that the free phytochemicals were the primary contributors (59% - 65%) to the total hydrophilic AoxA [63].

There is growing evidence that excess generation of highly reactive free radicals, largely due to hyperglycemia, causes oxidative stress, which further exacerbates the development and progression of diabetes and its complications.

Overproduction and/or insufficient removal of these free radicals result in vascular dysfunction, damage to cellular proteins, membrane lipids and nucleic acids [64]. For these reasons, is important to include phenolic compounds into the biologically active compounds of chickpeas and all the legumes, specifically regarding diabetes.

Insoluble dietary fiber present in chickpea has been associated with reducing the incidence of colon cancer, whereas soluble fiber has been demonstrated to have a beneficial effect on weight loss and weight management [65]. Chickpea is also rich in protein considered as antinutritional compounds, with biological importance. In this regard, some researchers have extracted the proteinaceous alpha amylase inhibitor from chickpeas and measured its activity. The antinutritional protein inhibited 73.6% of the activity of human saliva, comparable to the α-amylase from kidney bean, which has been used as a dietary supplement in body weight management. Thus, the ability of the chickpea extract to hinder the digestion of complex carbohydrates could potentially be useful in controlling obesity in humans [66].

The study performed by Yang et al., [18] showed that rats consuming a high-fat diet with added chickpeas demonstrated similar energy intake compared with high fat-fed (HFD) rats and increased energy consumption compared with rats on normal-fat diets (NFD). Nonetheless, chickpeas significantly reduced epididymal fat deposition, a measure of visceral adipose tissue, compared with high fat-fed controls. Moreover, rats receiving chickpeas demonstrated similar levels of epididymal fat mass vs. normal fat-fed rats. A reduction in epididymal fat deposition in rats receiving chickpeas was also reflected in decreased leptin mRNA expression. Finally, the activities of lipoprotein lipase (LPL) in epididymal adipose tissue and hepatic TAG lipase in liver increased (40% and 23%, respectively) in HFD rats compared with those in NFD rats; dietary chickpeas completely normalized the levels. Both results, leptin and LPL mRNA expression, are illustrated in Figure 11. Therefore it is possible that some components of chickpea isoflavone may be responsible for the effects observed in the present study. In addition, the lipid fraction from chickpeas is high in unsaturated fatty acids, primarily linoleic and oleic acids, which may have synergistic effect of modifying body composition and plasma lipids [18].

In another recent study using a non-randomized design in which participants served as their own controls [67], individuals were followed for 4 wk on their usual diet, followed by 12 wk of chickpea supplementation (mean 104 g/d, which is just over 0.5 cup/d), then another 4 wk on their usual diet.

Table 2. Total phenolics and total hydrophilic antioxidant activity of desi chickpea cultivars

Cultivar	Phenolics content[a] (μg GAE/g)			Antioxidant activity[a] (μmol TE/g)		
	Free	Bound	Total	Free	Bound	Total
Red ICC14782	195 ± 4C	551 ± 8D	746 ± 7D	28.7 ± 0.8B	15.2 ± 0.5D	43.9 ± 1.2C
Green ICC5613	182 ± 5D	872 ± 10B	1.054 ± 12B	27.7 ± 0.5C	18.7 ± 0.3B	46.4 ± 1.1B
Brown ICC3512	211 ± 5B	1075 ± 12A	1.286 ± 10A	32.6 ± 0.5A	21.3 ± 0.4A	53.9 ± 1.2A
Cream ICC3421	250 ± 5A	690 ± 8C	940 ± 7C	26.2 ± 0.7D	17.7 ± 0.3C	43.9 ± 1.2C

[a]Data are expressed as means ± standard deviations; μg GAE: micrograms gallic acid equivalents; μmol TE: micromols trolox equivalents; [A-D]Means with different superscripts in the same column are significantly different (Duncan, $p \leq 0.05$). Adapted from Garzón-Tiznado et al., [63].

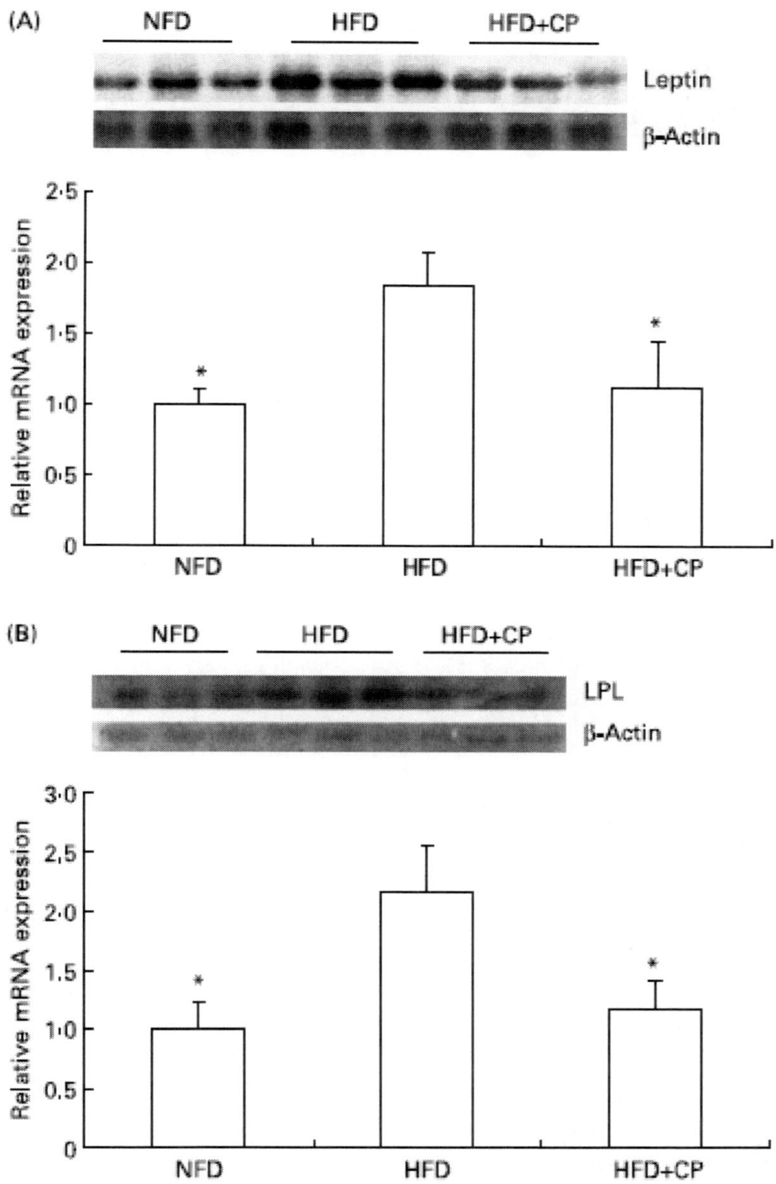

Figure 11. Representative Northern blots for mRNA of (A) leptin and (B) lipoprotein lipase (LPL) in epididymal adipose tissue of rats fed a normal-fat diet (NFD), a high-fat diet (HFD) or the HFD plus chickpeas (HFD + CP) for 8 months. Total RNA (20 mg per lane) was analysed. The corresponding blot for the β-actin RNA is also shown. Values are means, with standard deviations represented by vertical bars. * Mean value was significantly different from that of the HFD group (P<0.05). Adapted from Yang et al., [18].

Participants reported a significant increase in feelings of satiety in the chickpea phase relative to the first habitual phase and a significant decrease in satiation in the second habitual phase relative to the chickpea phase. However, it is unclear when these ratings were collected during the study.

Based on the few studies that have been conducted on chickpea, there is some indication that this legume may help to increase satiety, at least in the short term and weight loss during intentional energy restriction over a few weeks. Moreover, the hypoglycemic effect of chickpea has been recognized, so we could think of a therapeutic use.

Lentils

The lentil (*Lens culinaris Medikus* subsp. culinaris) is a lens-shaped grain legume, rich source of protein, minerals (potassium, phosphorus, iron, zinc) and vitamins for human nutrition. Furthermore, its consumption with wheat or rice provides a balance in essential amino acids for human nutrition because of its high lysine and tryptophane content. Currently lentil is primarily grown in the developing world with a particular concentration in Asia where India and Turkey are the largest producing countries [68].

Several non-nutritive carbohydrates including the raffinose-family oligosaccharides, α-galactosides, β-glucans and other polysaccharides, together with resistant starch constitute the fibre fraction of lentil. Dietary fibre is important for general functioning of the gastrointestinal tract, as it assists faecal motility, helps maintain proper faecal water balance [69], and the non-digestible fraction from legumes has a direct role in obesity and diabetes management, due to its slow digestion, fat binding, and other physiological functions. Lentils are high in fiber, containing approximately 8 g of fiber per 100 g of cooked lentils [70].

A diet containing lentils can serve as a prebiotic source that provides needed dietary fibre for the maintenance of a balanced gut ecosystem [71, 72]. According to a new hypothesis, gut microbes have a role in the host's metabolic homeostasis. As type 2 diabetes mellitus and obesity are associated with low-grade inflammation and an altered composition of the gut microbiota, a bacterial compound might act as a triggering factor in the development of obesity, diabetes mellitus and inflammation induced by a high-fat diet. Some of the molecular pathways involved in the development of these diseases are presented in Figure 12. Prebiotics, following fermentation by bacteria, can drive qualitative and selective changes in the composition of the gut microbiota. Certain fermentable carbohydrates with prebiotic properties can counteract the overexpression of several host targets that are involved in the development of adiposity, metabolic disorders and inflammation [73, 74]. The molecular actions offered by lentils on preventing obesity and diabetes, could be linked to it prebiotic effect, supporting this theory.

The results of Johnson et al., [75] on the characterization of lentil indicate that this legume may be a good source of prebiotic carbohydrates. Total prebiotic carbohydrate concentrations suggest that a 100 g serving of lentils may provide over 13 g of prebiotics. The authors compared prebiotic compounds among different lentil cultivars (Table 3).

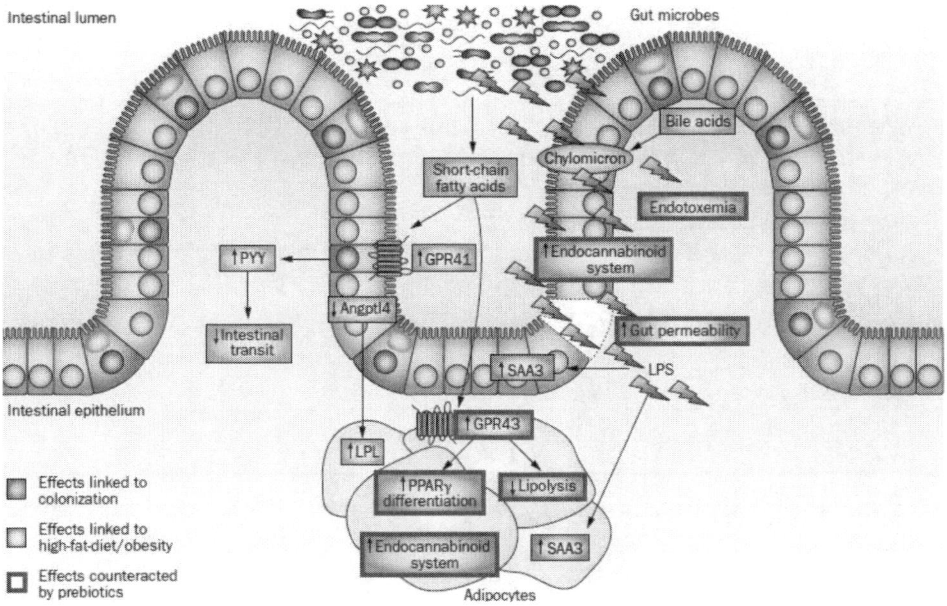

Figure 12. The host's metabolic responses to changes affecting its gut microbiota, including colonization of the gut in germ-free mice, obesity and high-fat diet, and administration of prebiotics. The short-chain fatty acids produced by the fermentation of carbohydrates bind to GPR41 in the intestine and promote the expression of PYY, which slows down the intestinal transit. Some short-chain fatty acids also activate GPR43—the expression of which is increased by a high-fat diet —in the adipose tissue. This activation decreases lipolysis, increases PPARγ-related differentiation and thereby increases adiposity. The gut microbiota promotes LPL-controlled fatty acid storage in the adipose tissue by blunting the intestinal expression of ANGPTL4. The increased LPS level in the blood is linked to the intestinal activation of the endocannabinoid system that increases the gut's permeability. LPS activates the production of SAA3 peptide in the gut and the adipose tissue. In obese animals, prebiotics increase the production of GLP-1 and GLP-2, decrease the intestinal wall's permeability, inhibit the endocannabinoid system in the gut and the adipose tissue, blunt the overexpression of GPR43 and increase lipolysis in the adipose tissue, thereby decreasing adiposity. Abbreviations: ANGPTL4, angiopoietin-related protein 4; GLP, glucagon-like peptide; GPR, G-protein coupled receptor; LPL, lipoprotein lipase; LPS, lipopolysaccharide; PPARγ, peroxisome proliferator-activated receptor γ; PPY, peptide YY; SAA3, serum amyloid A3 protein. Adapted from Delzenne et al., [74].

The results indicated that total soluble prebiotic carbohydrates were 5753 mg/100g in green and 5260 mg/100g in red lentils; besides, resistant starch was slightly higher in green (7.8 g/100g) than in red (7.4 g/100g) lentils. Relative concentrations of prebiotic carbohydrates may be more closely linked to green- or red-cotyledon traits than seed size. The authors analyzed commercial lentil market classes that were relatively high and uniform in total prebiotic carbohydrate concentrations. The concentrations of total prebiotic carbohydrates of these two types are consistent with their respective market classes, small red and large green, respectively.

The important role of lentils on disease risk reduction may be associated with the effect of some probiotic, generating a synergism between both [76]. Such is case when the number of probiotic bacteria in yogurt at the initial stages of storage was enhanced and overall microbial counts maintained from polysaccharides of green, red, and dehulled red lentils, which also exhibited strong antioxidant potential suggesting that they play an important role in free radical scavenging. These results propose the use of legumes as functional food additives for the population.

Table 3. Mean concentration of prebiotic carbohydrates of 10 lentil varieties grown in North Dakota, USA, in 2010 and 2011

Variety	mg/ 100 [a]				
	Sorb	Mann	Raff+Stach[b,c]	Verb	Nys[b]
CDC Greenland	1109[c]	221[c]	2426	1770[b]	57
CDC Lemay	1039[c]	163[d]	2497	1495[d]	57
CDC Red Rider	1036[c]	160[d]	2419	1586[cd]	52
CDC Redberry	1226[b]	176[d]	2349	1481[d]	61
CDC Richlea	1295[ab]	294[a]	2319	1731[bc]	62
CDC Rosetown	1325[ab]	158[d]	2586	922[e]	62
CDC Rouleau	1304[ab]	199[c]	2793	1082[e]	63
CDC Viceroy	1285[ab]	215[c]	2530	1800[b]	79
Pennell	1231[b]	204[c]	2684	1968[a]	57
Riveland	1349[a]	249[b]	2492	1784[b]	68
Mean	1220	203	2509	1562	62
SE	11.6	2.2	17	18	0.6

SE, standard error of combined data (n=120). Sorb, sorbitol; Mann, mannitol; Raff, Raffinose; Stach, stachyose; Verb, verbascose; Nys, nystose.

[a] Means within a column followed by different letters are significantly different at $p<0.05$.

[b] Mean concentration of varieties are not significantly different.

[c] Raffinose and stachyose are reported as total raffinose and stachyose concentration due to similar elution times for the separation method. Adapted from Johnson et al., [75].

Other mechanisms have been proposed on glycaemic response from lentils. For example, Liu et al., [77] isolated and characterized the starch fractions of different pea, lentil and chickpea cultivars grown in Canada under identical environmental conditions. The *in vitro* digestibility and physicochemical properties were investigated and the correlations between physicochemical properties and starch digestibility were determined. Lentil starch exhibited lower rapidly digestible starch (RDS) content and hydrolysis rate. Consequently, the expected GI of the starch fraction from lentils was the lowest among the tested starch fractions. Inclusion of lentils as part of a healthy diet can confer a significant reduction of the dietary glycaemic load (GL), which reflects dietary carbohydrate quality and quantity, and is directly related to the reduced risk for type 2 diabetes. Previous studies have shown that a blood glucose profile levels or flattens after eating lentils even in insulin dependent diabetics [78].

Lentils and other pulses, besides their high fiber content (total and soluble) and resistant starch, present anti-nutrients and polyphenols that have been recommended as key components of weight loss and diabetic reduction strategies [79]. Studies have indicated that most phenolics in lentils are essentially located in their seed coat or hull [80]. The seed coat was found to be very rich in catechins, procyanidins dimers and trimers, but contained small amounts of glycosides of quercetin, myricetin, luteolin, and apigenin, whereas the cotyledon contained mainly hydroxybenzoic and hydroxycinnamic acids in low concentration.

In black lentils, the major anthocyanin is delphinidin 3-O-(2-β-*D*-glucopyranosyl-α-*L*-arabinopyranoside). Amarowicz et al., [81] identified 24 phenolic compounds including phenolic acids and flavonoids in the extract of lentils. Moreover, phenolics may provide a protective effect against hyperglycemia-induced chronic diseases through dual protection of antioxidant and inhibition of starch digestion; phenolic compounds are relevant for type 2 diabetes management and cyclooxygenase inhibition for inflammation as catechins has demonstrated anti-diabetic effects in human studies [82, 83]. A recent study [84] indicates that lentil hull extract exert anti-inflammatory activity, partly via lipoxygenase (15-LOX) inhibition. This inhibition may influence the inflammation process since 15-LOX participates in oxidative modifications of low-density lipoproteins (LDL) and in the development of atherosclerotic lesions.

Wolever et al., [85] showed that lentils fed as a single food decreased glucose area under the curve at a subsequent meal relative to white wheat bread, whereas lentils as part of a complete meal did not. This data is further convoluted by the fact that the subsequent meals were different: for the lentils as a single food study, the subsequent meal was a drinkable glucose solution, whereas the second meal for the lentils as part of a complete meal study was also a complete, mixed meal. Therefore, the form of the subsequent meal could also have influenced the difference in response to lentils in this series of experiments. Jenkins et al., [86] confirmed the observation that the blood glucose response was markedly flatter after eating lentils compared with bread. This difference was reduced when the lentils were dried in an oven for 12 h, whereas boiling for 1 h was without effect.

The effects of an *ad libitum* pulse meal on food intake (FI), appetite and blood glucose (BG) in healthy males, using chickpeas, lentils, navy beans and yellow beans as treatments, was determined by Mollar et al., [60]. The results indicated that the effects on satiation at the treatment meal and BG after the second meal were dependent on pulse type. Specifically, at the treatment meal, lentils exhibited the strongest satiating properties, resulting in lower FI compared to chickpeas and pasta and sauce, whereas navy beans also resulted in lower FI compared to chickpeas. At 60 min, lentils and navy beans resulted in lower BG concentrations compared to yellow peas. At 80 min, chickpeas and lentils reduced BG further compared to yellow peas and pasta and sauce. These studies illustrate the unmet need in research on the use of lentils and in general legumes to prevent or decrease the diabesity rate.

Conclusion

It has long been recognized that legumes are functional foods that promote good health and have concurrent therapeutic properties. Legumes are a good source of nutraceutical compounds that act on several molecular pathways related to degenerative chronic diseases. Available evidence suggests that legume consumption can impact metabolic profiles associated with obesity and may be an effective dietary measure to aid in weight loss and/or maintenance, such as diabetes and resistance insulin control. Moreover, legumes intake has beneficial effects on satiety, modeling hormones involved in appetite sensations, offering a new range of possibilities for the generation of adjuvants in the comprehensive treatment of obesity. The latest reports actually recommend bean supplementation to control diabetes [87] and as a preventive tool in curbing obesity and metabolic syndrome [88]. This information is very important to increase consumption of this food around the world and consider the possibility of including fractions thereof in the development of functional foods.

REFERENCES

[1] WHO. World Health Organization. Fact Sheet Number 311, 2008. Available at URL: http://www.who.int/mediacentre/factsheets/ fs311/en/index.html. Last accessed on 23 September 2009.

[2] Astrup, A; Finer, N. Redefining Type 2 diabetes: Diabesity' or 'Obesity Dependent Diabetes Mellitus'?. *Obesity Reviews*, 2000 1, 57–59.

[3] Han, TS; Richmond, P; Avenell, A; Lean, MEJ. Waist circumference reduction and cardiovascular benefits during weight loss in women. *International Journal of Obesity*, 1997, 21, 127–134.

[4] Colagiuri, S. Diabesity: therapeutic options. *Diabetes, Obesity and Metabolism,* 2010, 12, 463–473.

[5] ADA. American Diabetes Association. Complications of diabetes in the United States, 2008. Available from URL: http://www.diabetes.org/diabetesstatistics /complications. jsp.

[6] Bailey, CJ. New Therapies for Diabesity. *Current Diabetes Reports*, 2009, 9, 360–367.

[7] Salvador, J.; Silva, C.; Santos, E. Tratamiento farmacológico de la obesidad. Anales del Sistema Sanitario de Navarra. 2002, 25, 143-161.

[8] Griver, K; Henry, RR. Nutritional management of obesity and diabetes. *Nutrition Research*, 1994, 14, 465-483.

[9] Venn, BJ; Mann, JI. Cereal grains, legumes and diabetes. *European Journal of Clinical Nutrition*, 2004, 58, 1443–1461.

[10] USDA. US Department of Agriculture Department of Health and Human Services (HHS). Dietary Guidelines for Americans 2005. Available at: www.healthierus.gov/dietaryguidelines. Accessed 20 February 2006.

[11] Williams, PG; Grafenauer, SJ; O'Shea, JE. Cereal grains, legumes, and weight management: a comprehensive review of the scientific evidence. *Nutrition Reviews*, 2008, 66(4), 171–182.

[12] Bornet, FR; Billaux, MS; Messing, B. Glycaemic index concept and metabolic diseases. *International Journal of Biology Macromolecules*, 1997, 21, 207–219.

[13] Rizkalla, SW; Bellisle, F; Slama, G. Health benefits of low glycaemic index foods, such as pulses, in diabetic patients and healthy individuals. *British Journal of Nutrition*, 2002, 88, S255–S262.

[14] Mollard, RC; Luhovyy, BL; Panahi, S; Nunez, M; Hanley, A; Anderson, GH. Regular consumption of pulses for 8 weeks reduces metabolic syndrome risk factors in overweight and obese adults. *British Journal of Nutrition*, 2012, 108, S111–S122.

[15] Jenkins, DJA; Kendall, CWC; Augustin, LSA; Mitchell, S; Sahye-Pudaruth, S; Blanco Mejia, S; Chiavaroli, L; Mirrahimi, A; Ireland, C; Bashyam, B; Vidgen, E; de Souza, RJ; Sievenpiper, JL; Coveney, J; Leiter, LA; Josse RG. Effect of Legumes as Part of a Low Glycemic Index Diet on Glycemic Control and Cardiovascular Risk Factors in Type 2 Diabetes MellitusA Randomized Controlled Trial. *Archives of Internal Medicine*, 2012, 172(21), 1653-1660.

[16] Nöthlings, U; Schulze, MB; Weikert, C; Boeing, H; van der Schouw, YT; Bamia, C; Benetou, V; Lagiou, P; Krogh, V; Beulens, JWJ; Peeters, PHM; Halkjær, J; Tjønneland, A; Tumino, R; Panico, S; Masala, G; Clavel-Chapelon, F; de Lauzon, B; Boutron-

Ruault, MC; Vercambre, MN; Kaaks, R; Linseisen, J; Overvad, K; Arriola, L; Ardanaz, E; Gonzalez, CA; Tormo, ML; Bingham, S; Khaw, KT; Key, TJA; Vineis, P; Riboli, E; Ferrari, P; Boffetta, P; Bueno-de-Mesquita, HB; van der A, DL; Berglund, G; Wirfält, E; Hallmans, G; Johansson, I; Lund, E; Trichopoulo, A. Intake of Vegetables, Legumes, and Fruit, and Risk for All-Cause, Cardiovascular, and Cancer Mortality in a European Diabetic Population. *Journal of Nutrition*, 2008, 138: 775–781.

[17] Sievenpiper, JL; Kendall, CW; Esfahani, A; Wong, JM; Carleton, AJ; Jiang, HY; Bazinet, RP; Vidgen, E; Jenkins, DJ. Effect of non-oil-seed pulses on glycaemic control: a systematic review and meta-analysis of randomised controlled experimental trials in people with and without diabetes. *Diabetologia,* 2009, 52(8), 1479-1495.

[18] Yang, Y; Zhou, L; Gu, Y; Zhang, Y; Tang, J; Li, F; Shang, W; Jiang, B; Yue, X; Chen, M. Dietary chickpeas reverse visceral adiposity, dyslipidaemia and insulin resistance in rats induced by a chronic high-fat diet. *British Journal of Nutrition*, 2007, 98, 720–726.

[19] Morris, KL; Zemel, MB.) Effect of dietary carbohydrate source on the development of obesity in agouti transgenic mice. *Obesity Research,* 2005, 13, 21–35.

[20] Marinangeli, CP; Jones, PJ. Whole and fractionated yellow pea flours reduce fasting insulin and insulin resistance in hypercholesterolaemic and overweight human subjects. *British Journal of Nutrition*, 2011, 105, 110–117.

[21] Marinangeli, CPF; Jones, PJH. Pulse grain consumption and obesity: effects on energy expenditure, substrate oxidation, body composition, fat deposition and satiety. *British Journal of Nutrition*, 2012, 108, S46–S51.

[22] Campos-Vega, R; Reynoso-Camacho, R; Pedraza-Aboytes, G; Acosta-Gallegos, JA; Guzman-Maldonado, SH; Paredes- Lopez, O; Oomah, BD; Loarca-Piña, G. Chemical composition and in vitro polysaccharide fermentation of different beans (Phaseolus vulgaris L.). *Journal of Food Sciences*, 2009, 74, 59–65.

[23] Preuss, HG. Bean Amylase Inhibitor and Other Carbohydrate Absorption Blockers: Effects on Diabesity and General Health. *Journal of the American College of Nutrition*, 2009, 28(3), 266–276.

[24] Obiro, WC; Zhang, T; Jiang, B. The nutraceutical role of Phaseolus vulgaris a-amylase inhibitor. *British Journal of Nutrition*, 2008, 100, 1–12.

[25] Pusztai, A; Grant, G; Buchan, WC; Bardocz, S; De Carvalho, AF; Ewen; SW. Lipid accumulation in obese Zucker rats is reduced by inclusion of raw kidney bean (Phaseolus vulgaris) in the diet. *British Journal of Nutrition*, 1998, 79, 213–221.

[26] Hangen, L; Bennink, MR. Consumption of black beans and navy beans (Phaseolus vulgaris) reduced azoxymethaneinduced colon cancer in rats. *Nutrition and Cancer*, 2002, 44, 60–65.

[27] Papanikolaou, Y; Fulgoni, VL. Bean Consumption Is Associated with Greater Nutrient Intake, Reduced Systolic Blood Pressure, Lower Body Weight, and a Smaller Waist Circumference in Adults: Results from the National Health and Nutrition Examination Survey. 1999-2002. *Journal of the American College of Nutrition*, 2008, 27(5), 569–576.

[28] Mensack, MM; McGinley, JN; Thompson, HJ. Metabolomic analysis of the effects of edible dry beans (Phaseolus vulgaris L.) on tissue lipid metabolism and carcinogenesis in rats. *British Journal of Nutrition,* 2012, 108, S155–S165.

[29] Zhu, Z; Jiang, W; Thompson, HJ. Edible dry bean consumption (Phaseolus vulgaris L.) modulates cardiovascular risk factors and diet-induced obesity in rats and mice. *British Journal of Nutrition*, 2012, 108, S66–S73.

[30] Helmstadter, A. Beans and diabetes: Phaseolus vulgaris preparations as antihyperglycemic agents. *Journal of Medicinal Food*, 2010, 13, 251 – 254.

[31] Bantle, JP; Wylie-Rosett, J; Albright, AL; Apovian, CM; Clark, NG; Franz, MJ; Hoogwerf, BJ; Lichtenstein, AH; Mayer-Davis, E; Mooradian, AD; Wheeler, ML. Nutrition recommendations and interventions for diabetes: a position statement of the American Diabetes Association. *Diabetes Care*, 2008, 31(1), S61-S78.

[32] Misra, A; Sharma, R; Gulati, S; Joshi, SR; Sharma, V; Ghafoorunissa; Ibrahim, A; Joshi, S; Laxmaiah, A; Kurpad, A; Raj, RK; Mohan, V; Chandalia, H; Krishnaswamy, K; Boindala, S; Gopalan, S; Bhattiprolu, SK; Modi, S; Vikram, NK; Makkar, BM; Mathur, M; Dey, S; Vasudevan, S; Gupta, SP; Piru, S; Joshi, P; Khanna, K; Mathur, P; Krishnaswamy, S; Madan, J; Karmarkar, M; Seth, V; Passi, SJ; Chadha, D; Bhardwaj, S; National Dietary Guidelines Consensus Group. Consensus dietary guidelines for healthy living and prevention of obesity, the metabolic syndrome, diabetes, and related disorders in Asian Indians. *Diabetes Technology and Therapeutics*, 2011, 13(6), 683-694.

[33] Ludwig, D. Dietary glycemic index and obesity. *Journal of Nutrition*, 2000, 130, 280S–3S.

[34] Roman-Ramos, R; Flores-Saenz, FJ; Alarcón, A. Anti-hyperglycemic effect of some edible plants. *Journal of Ethnopharmacoloy*, 1995, 48, 25-32.

[35] Vergara-Castañeda, HA; Guevara-González, RG; Ramos-Gómez, M; Reynoso-Camacho, R; Guzmán-Maldonado, H; Feregrino-Pérez, AA; Oomah, BD; Loarca-Piña, G. Non-digestible fraction of cooked bean (Phaseolus vulgaris L.) cultivar Bayo Madero suppresses colonic aberrant crypt foci in azoxymethane-induced rats. *Food and Function,* 2010, 1(3), 294-300.

[36] Ambriz, T; Ramos, M; Salgado, L.M; Reynoso, R. Hypoglycemic effect of cooked bean (Phaseolus vulgaris) and its relationship with renal failure in diabetic rats. *Food Science and Food Biotechnology in Developing Countries*. 2006.

[37] Helmstädter A. Beans and Diabetes: Phaseolus vulgaris Preparations as Antihyperglycemic Agents. *Journal of Medicinal Food*, 2010, 13(2), 251–254.

[38] Madar, Z; Stark, AH. New legume sources as therapeutic agents. *British Journal of Nutrition*, 2002, 88, S287–S292.

[39] Spadafranca, A; Rinelli, S; Riva, A; Morazzoni, P; Magni, P; Bertoli, S; Battezzati, A. Phaseolus vulgaris extract affects glycometabolic and appetite control in healthy human subjects. *British Journal of Nutrition,* 2012, 9, 1-7.

[40] Rodgers, CD; Burch, PE; Van Heest, JL; Suzuki, K; Bussiere CT; Schemmel, RA. A tepary bean diet and exercise delays indices of type 2 diabetes in female fa/fa rats. *Inernational Journal of Diabetes and Metabolism,* 2007, 15, 38-45.

[41] Oomah, BD; Corbé, A; Balasubramanian, P. Antioxidant and Anti-inflammatory Activities of Bean (Phaseolus vulgaris L.) Hulls. *Journal of Agricultural and Food Chemistry*, 2010, 58, 8225–8230.

[42] Reichert, RD. Quantitative isolation and estimation of cell-wall material from dehulled pea (Pisum sativum) flours and concentrates. *Cereal Chemistry*, 1981, 58, 266–270.

[43] Udahogora, M. Health benefits and bioactive compounds in field peas, faba beans, and chickpeas. In: Yu L, Tsao R, Shahidi F. *Cereals and Pulses: Nutraceutical Properties and Health Benefits.* Oxford, UK. John Wiley and Sons, Inc.; 2012; 199-215.

[44] Isken, F; Klaus, S; Osterhoff, M; Pfeiffer, AF; Weickert, MO. Effects of long-term soluble vs. insoluble dietary fiber intake on high-fat diet induced obesity in C57BL/6J mice. *The Journal of Nutritional Biochemistry*, 2010, 278–284.

[45] Lang, V; Bellisle, F; Oppert, JM; Craplet, C; Bornet, FR; Slama, G; Guy-Grand, B. Satiating effect of proteins in healthy subjects: a comparison of egg albumin, casein, gelatin, soy protein, pea protein, and wheat gluten. *American Journal of Clinical Nutrition*, 1998, 67, 1197–1204.

[46] Marinangeli, CPF; Jones, PJH. Chronic Intake of Fractionated Yellow Pea Flour Reduces Postprandial Energy Expenditure and Carbohydrate Oxidation. *Journal of Medicinal Food*, 2011, 14(12), 1654–1662.

[47] Karlstrom , B; Vessby, B; Asp, NG; Boberg, M; Lithell, H; Berne, C. Effects of leguminous seeds in a mixed diet in noninsulin-dependent diabetic patients. *Diabetes Research*, 1987, 5, 199–205.

[48] Marinangeli, CP; Kassis, AN; Jones, PJ. Glycemic responses and sensory characteristics of whole yellow pea flour added to novel functional foods. *Journal of Food Sciences*, 2009, 74, S385–S389.

[49] Lunde, MS; Hjellset, VT; Holmboe-Ottesen, G; Hostmark, AT. Variations in postprandial blood glucose responses and satiety after intake of three types of bread. *Journal of Nutrition and Metabolism*, 2011.

[50] Smith, CE; Mollard, RC; Luhovyy, BL; Anderson, GH. The effect of yellow pea protein and fibre on short-term food intake, subjective appetite and glycaemic response in healthy young men. *British Journal of Nutrition*, 2012, 108, S74–S80.

[51] Whitlock, CA; Kozicky, L; Jin, A; Yee, H; Ha, C; Morris, J; Field, CJ; Bell, RC; Ozga, JA; Chan, CB. Assessment of the mechanisms exerting glucose-lowering effects of dried peas in glucose-intolerant rats. *British Journal of Nutrition*, 2012, 108, S91–S102.

[52] Eisenstein, J; Roberts, SB; Dallal, G; Saltzman, E. High-protein weight-loss diets: are they safe and do they work? A review of the experimental and epidemiologic data. *Nutr Rev*, 2002, 60, 189-200.

[53] Geraedts, MC; Troost, FJ; Saris, WH. Addition of sucralose enhances the release of satiety hormones in combination with pea protein. *Molecular Nutrition and Food Research*, 2012, 56(3), 417-424.

[54] Abou-Samra, R; Keersmaekers, L; Brienza, D; Mukherjee, R; Macé, K. Effect of different protein sources on satiation and short-term satiety when consumed as a starter. *Nutrition Journal*, 2011, 10, 139.

[55] Jukanti, AK; Gaur, PM; Gowda, CLL; Chibbar, RN. Nutritional quality and health benefits of chickpea (Cicer arietinum L.): a review. *British Journal of Nutrition*, 2012, 108, S11–S26.

[56] Tharanathan, RN; Mahadevamma, S. Grain legumes – a boon to human nutrition. Trends of Food Sciences Technology, 2003, 14, 507–518.

[57] James, SL; Muir, JG; Curtis, SL; Gibson, PR. Dietary fibre: a roughage guide. *Internal Medicine Journal*, 2003, 33, 291–296.

[58] Thompson, SV; Winham DM; Hutchins, AM. Black bean and chickpea consumption reduce glycemic response as part of a rice meal. *The FASEB Journal*, 2009, 23, 540.2.

[59] Pittaway, JK; Robertson, IK; Ball, MJ. Chickpeas may influence fatty acid and fiber intake in an ad libitum diet, leading to small improvements in serum lipid profile and glycemic control. *Journal of the American Dietetic Association*, 2008, 108, 1009–1013.
[60] Mollard, RC; Zykus, A; Luhovyy, BL; Nunez, MF; Wong, CL; Anderson, GH. The acute effects of a pulse-containing meal on glycaemic responses and measures of satiety and satiation within and at a later meal. *British Journal of Nutrition*, 2012, 108, 509–517.
[61] Nestel, P; Cehun, M; Chronopoulos, A. Effects of long-term consumption and single meals of chickpeas on plasma glucose, insulin, and triacylglycerol concentrations. *American Journal of Clinical Nutrition*, 2004, 79, 390 –5.
[62] Johnson, SK; Thomas, SJ; Hall, RS. Palatability and glucose, insulin and satiety responses of chickpea flour and extruded chickpea flour bread eaten as part of a breakfast. *European Journal of Clinical Nutrition*, 2005, 59, 169–76.
[63] Garzón-Tiznado, JA; Ochoa-Lugo, MI; Heiras-Palazuelos, MJ; Domínguez-Arispuro, DM; Cuevas-Rodríguez, EO; Gutiérrez-Dorado, R; Milán-Carrillo, J; Reyes-Moreno, C. Acceptability Properties and Antioxidant Potential of Desi Chickpea (Cicer arietinum L.) Cultivars. *Food and Nutrition Sciences*, 2012, 3, 1281-1289.
[64] Schultz-Johannsen, J; Harris, AK; Rychly DJ; Ergul, A. Oxidative stress and the use of antioxidants in diabetes: Linking basic science to clinical practice. *Cardiovasc Diabetol*. 2005; 4, 5.
[65] Agriculture and Agri-Food Canada. Chickpeas: Situation and outlook. *Bi-weekly Bulletin*, 2006, 19, 13.
[66] Hao, X; Jiangui, L; Qinghua, S; Jusong, Z; Xiaoling, HE; Hao, MA. Characterization of a novel legumin apha–amylase inhibitor from chickpeas (Cicer arietinum l.) seeds biosci. *Biotechnology and Biochemistry*, 2009, 73, 1200–1202 .
[67] Murty, CM; Pittaway, JK; Ball, MJ. Chickpea supplementation in an Australian diet affects food choice, satiety and bowel health. *Appetite,* 2010, 54, 282–8.
[68] Ramdath, DD; Tsao, R. Bioactives and health benefits of lentils (Lens culinaris L.). In: Yu L, Tsao R, Shahidi F. Cereals and Pulses: *Nutraceutical Properties and Health Benefits*. Oxford, UK. John Wiley and Sons, Inc.; 2012; 217-228.
[69] Gallaher, DD; Schneeman, BO. *Present Knowledge in Nutrition*. 7th ed., Washington, DC, ILSI Press, 1966.
[70] USDA (United States Department of Agriculture). *Agriculture Research Services*. Products and Services SR 20. 2009. Available at:http://www.ars.usda.gov /Services/docs.htm?docid ¼1747. Accessed on 06 May 2011.
[71] Guillon, F; Champ, MMJ. Carbohydrate fractions of legumes: uses in human nutrition and potential for health. *British Journal of Nutrition*, 2002, 88, S293–S306.
[72] Blaut, M; Clavel, T. Metabolic diversity of the intestinal microbiota: implications for health and disease. *Journal of Nutrition*, 2007, 137, 751S–775S.
[73] Cani, PD; Delzenne, NM. The role of the gut microbiota in energy metabolism and metabolic disease. *Current Pharmaceutical Design*, 2009, 15, 1546–1558.
[74] Delzenne, NM; Neyrinck, AM; Bäckhed, F; Cani, PD. Targeting gut microbiota in obesity: effects of prebiotics and probiotics. *Nature Reviews Endocrinology*, 2011, 7, 639-646.
[75] Johnson, CR; Thavarajah, D; Combs Jr., GF; Thavarajah, P. Lentil (Lens culinaris L.): A prebiotic-rich whole food legume. *Food Research International,* 2013, 51, 107–113.

[76] Agil, R; Gaget, A; Gliwa, J; Avis, TJ; Willmore, WG; Hosseinian, F. Lentils enhance probiotic growth in yogurt and provide added benefit of antioxidant protection. LWT - *Food Science and Technology*, 2013, 50, 45-49.

[77] Liu, S; Willett, WC; Stampfer, MJ; Hu, FB; Franz, M; Sampson, L; Hennekens, CH; Manson, JE. A prospective study of dietary glycemic load, carbohydrate intake, and risk of coronary heart disease in US women. *American Journal of Clinical Nutrition*, 2000, 71, 1455–1461.

[78] Jenkins, DJA; Wolever, TMS; Taylor, RH; Barker, HM; Fielden, H. Exceptionally low blood glucose response to dried beans: comparison with other carbohydrate foods. *British Medical Journal*, 1980, 281, 578-80.

[79] Wylie-Rosett, J; Segal-Isaacson, CJ; Segal-Isaacson A. Carbohydrates and increases in obesity: does the type of carbohydrate make a difference? *Obesity Research*, 2004, 12, 124S–129S.

[80] Dueñas, M; Hernández, T; Estrella, I. Phenolic composition of the cotyledon and the seed coat of lentils (Lens culinaris L.). *European Food Research Technology*, 2002, 215, 478–483.

[81] Amarowicz, R; Estrella, I; Hernández, T; Dueñas, M; Troszynska, A; Agnieszka, K, and Pegg, R.B. Antioxidant activity of a red lentil extract and its fractions. *International Journal of Molecular Sciences*, 2009, 10, 5513–27.

[82] Rani, MP; Padmakumari, KP; Sankarikutty, B; Cherian, OL; Nisha, VM; Raghu, KG. Inhibitory potential of ginger extracts against enzymes linked to type 2 diabetes, inflammation and induced oxidative stress. *International Journal of Food Sciences and Nutrition*, 2011, 62, 106-110.

[83] Asgar, A. Anti-diabetic potential of phenolic compounds: A review. *International Journal of Food Properties*, 2013, 16, 91–103.

[84] Boudjou, S; Oomah, BD; Zaidi, F; Hosseinian F. Phenolics content and antioxidant and anti-inflammatory activities of legume fractions. *Food Chemistry*, 2013, 138, 1543–1550.

[85] Wolever, TMS; Jenkins, DJA; Ocana, AM; Rao, VA; Collier, GR. Second-meal effect: low-glycemic-index foods eaten at dinner improve subsequent breakfast glycemic response. *American Journal of Clinical Nutrition*, 1988, 48(4), 1041–1047.

[86] Jenkins, DJA; Thorne, MJ; Camelon, K; Jenkins, A; Rao, V; Taylor, RH; Thompson, LU; Kalmusky, J; Reichert, R; Francis, T. Effect of processing on digestibility and the blood glucose response: a study of lentils. *American Journal of Clinical Nutrition*, 1982, 36, 1093-1101.

[87] Hernández-Saavedra, D; Mendoza-Sánchez, M; Hernández-Montiel, HL; Guzmán-Maldonado, HS; Loarca-Piña, GF; Salgado, LM; Reynoso-Camacho, R. Cooked common beans (*Phaseolus vulgaris*) protect against β-cell damage in streptozotocin-induced diabetic rats. *Plant Foods for Human Nutrition*, 2013, DOI 10.1007/s11130-013-0353-1.

[88] Nilsson, A; Johansson, E; Ekström, L; Björck, I. Effects of a brown beans evening meal on metabolic risk markers and appetite regulating hormones at a subsequent standardized breakfast: A randomized cross-over study. *PLoS ONE*, 2013, 8(4): e59985. doi:10.1371/journal.pone.0059585

In: Legumes
Editors: Hiroto Satou and Ren Nakamura
ISBN: 978-1-62808-280-7

Chapter 4

SOYBEAN: OVERVIEW OF THE NUTRITIONAL PROFILE AND THE IMPLICATIONS FOR NUTRITION AND HEALTH EFFECTS

I. Mateos-Aparicio*[*], *A. Redondo-Cuenca and M. D. Tenorio-Sanz
Departamento de Nutrición y Bromatología II. Bromatología. Facultad de Farmacia.
Universidad Complutense de Madrid. Madrid, Spain

ABSTRACT

Soybean represents an excellent source of high quality protein, it has a low content in saturated fat, contains a great amount of dietary fiber and the isoflavone content makes soybean singular among other legumes. Most of the studies have been focused on the positive effect of soybean protein over hyperlipidemia and the prevention of cardiovascular disease. Besides, isoflavones focused many attention due to they are associated with the prevention and the treatment of different chronic diseases. Some studies have shown also the role of the soy dietary fiber and the galacto-oligosaccharides in different diseases such as hypercholesterolemia, diabetes and obesity.

Soybean is a source of functional ingredients interesting from the nutritional point of view and because their potential health effects. Indeed, soybeans and soy foods are widely promoted and eaten based on assumed relationships between its consumption and beneficial health effects. This chapter is focused on the latest results of research on health properties and uses of soybean as well as an overview of the nutritional profile of this legume.

1. INTRODUCTION

Functional foods are one of the most promising fields concerning nutritional sciences. This type of food is interesting from the consumer point of view with the prospect of maintaining health and preventing diseases by using natural foods as part of the habitual diet,

[*] E-mail: inmateos@farm.ucm.es.

and also from the industry point of view, for the added value of the products. There are several raw materials that can be used for healthy purposes and soybeans are among those with the greatest potential [1-3].

The potential of soybeans as a functional food is being currently explored by the food industry. Soybeans and soy foods, like soymilk, tofu and miso, are widely promoted and eaten based on assumed relationships between its consumption and beneficial health effects in humans including chemoprevention of breast and prostate cancer, osteoporosis, cardiovascular disease as well as relieving menopausal symptoms. The evidence provided by epidemiological and interventional studies that show a lower incidence of these health conditions in Asian countries, which have high soy consumption, is the basis of this relationship [4-9].

It is widely believed that the origins of soybean are in China, probably in the north and central regions, 4000-5000 years ago. Soybean was introduced in Europe about 1712 by a German botanist, Engelbert Kaempfer. Later, Carl von Linné gave soybeans a genetic name, *Glycine max*. Until the early 80s, the European Union (EU) countries showed limited interest for the cultivation of soybean. During the 80s, especially in France and Italy, an extensive research was carried to promote this crop. Only these two countries hold a significant production area since in the whole EU is a minor crop due to the poor climate and soil conditions [1, 10]. In Spain there have been numerous attempts to promote the cultivation of soybean. After different stages in relation with the interest for this crop, the production now is about 1700 tones, being the main producers regions Andalusia and Extremadura [11]. The water requirements are the main cause of the reduced expansion of soybean in Spain [12]. The introduction in United States of America (USA) dates from the mid-eighteenth century, and since the last decade of the nineteenth century, soybean has been intensively investigated. First introduced as forage, the success as oilseed in Europe between 1900 and1910 promoted the interest in U.S., but during the 20s and 30s it was grown as forage. The production of soybeans increased in the twentieth century becoming one of the most economically important crops, the second after the maize [1, 12]. In 1983, the US Department of Agriculture approved the use of soy as a partial substitute for animal protein in the diets of some schools. However, only about 2.1% of the American society daily protein intake is based on soy, while the Japanese consume about 13% of this legume protein [13].

Botanically, soybean belongs to the order *Rosaceae*, family *Leguminosae* or *Papillonaceae* or *Fabaceae*, subfamily *Papilionoidae*, the genus *Glycine* and the cultivable specie *Glycine max*. Soybean is a singular food because of its rich nutrient content. Soybean contains vegetable protein, dietary fibre, oligosaccharides, phytochemicals (especially isoflavones), and minerals (Table 1) [14].

Table 1. Proximate composition of yellow soybean from intensive crop [14]

g/100g	SOYBEAN (fresh matter)	SOYBEAN (dry matter)
MOISTURE	10.02	-
PROTEIN	43.13	39.86
FAT	18.80	19.86
DIETARY FIBER	24.36	22.00
MINERALS	4.80	5.20

Soybean plant presents a variable height ranging 50-130 cm. The root system has a main root and a great and variable number of secondary roots. The leaves are alternate and the inflorescences are axillary raceme in varying numbers. They consist of 2-35 white or purple butterfly-like flowers. Each inflorescence can provide from one to more than twenty pods. The fruits are dehiscent pods and each pod in normal conditions contains 2-3 seeds. The color shape and dimensions of soybeans are greatly variable but generally they are spherical and straw-colored [12, 15, 16].

The soybean is harvested firstly for seed production to obtain protein meal for animal feed production. Although the market for soy products for human consumption is much lower than the market of flour and oil feed, there is some optimism regarding these products because different studies show how certain foods containing soy may help prevent and treat chronic diseases such as cardiovascular disease or cancer. Global soybean production is projected at 269.4 million tons and the soybean production in U.S., the first producer, is estimated at more than 83 millions of tons [17]. Most of which is processed for animal feed and also, although in minor extension, for human consumption food [1, 15]. These foods are typically grouped in two categories: non-fermented foods (soy meal, tofu, soy milk and others) and fermented (tempeh, miso, soy sauce and others) (Table 2).

Table 2. Main foodstuff from soybeans

Main Industrial Products Soybean oil Soybean meal (whole, total or partially defatted, toasted, texturized)
Traditional Food Without fermentation: green soybean (with or without pods), germinated seed and buds, toasted seed, soybean milk and tofu Fermented: tempeh, miso, soy sauce and others
Protein products Flakes, concentrate, isolate, texturized
New Generation Food Ice-cream, yogurt, hamburger, sausage
Meal-enriched products Bread, breakfast cereals, pasta, snacks and others
Dietary supplements Lecithin, isoflavones, etc.

The main processed soybean lines for human consumption are oil extraction and soy milk and tofu obtaining. The oil is extracted from the soybean seeds with solvents and subsequently refined to erase the crude oil bad odor. Soy milk obtaining essentially consists of an aqueous extract of soybean seeds. The seed is soaked for about 12 hours, although the time required for adequate hydration depends on the size and variety of soybean and the water temperature. The Japanese method submits the slurry to cook for about 10 minutes at 100-110 °C while Chinese method just applies a wet milling without heating. After the extraction, the pulp or okara is removed and soy milk obtained. Soy milk is a creamy consistency liquid with raw-legume flavor (Figure 1) [18].

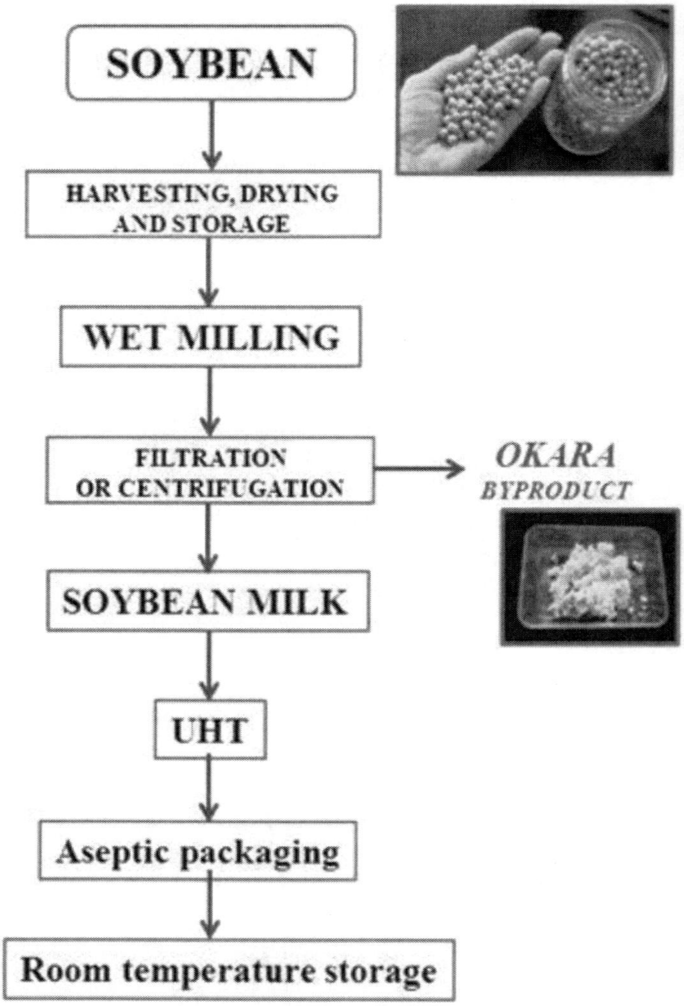

Figure 1. Scheme of the soybean beverage obtaining [18].

In May 1998, the U.S. company Protein Technologies International demanded an authorization from the Food and Drug Administration (FDA) to label their soybean products with a health claim based on the role of soybean protein in reducing the risk of coronary heart disease. In October 1999, FDA approved this claim and established that every food with a minimum of 6.25 g of soybean protein and included in a low-in-saturated-fat diet and cholesterol may reduce the risk of coronary heart disease. Moreover, this claim just can figure in food which contains less than 3g of fat, less than 1g of saturated fat and less than 20 mg of cholesterol per serving. This authorization had very important consequences in the developing and intake of new products, trade, research and soybean health effects [19]:

- Many food companies have tried to break into the soybean products market
- The consumer interest have increased about soybean product
- The food manufacturers have incorporated soy protein to many traditional food and have developed new products with that protein
- The researches about soybean health properties have increased

On the other hand, the results from recent researches suggest that soybean dietary fiber and isoflavones plays a role in the prevention and treatment of chronic diseases such as heart disease, cancer, diabetes, and hypertension as well as other clinical pathologies [8,10,20-28], although there is not any approved health claim.

2. Composition and Nutritional Profile

2.1. Protein

Soybean is a protein source with a content major than the most legumes. Beans averages 20-25%, whereas the soybean protein content is around 40% independently of the crop type (conventional: 40.4 g/100 g; ecological: 41.8 g/ 100 g; transgenic: 38.9 g/100 g; non-transgenic: 39.5 g/100 g), but significantly minor if the harvesting is before maturation (conventional: 37.1 g/100 g; ecological: 36.8 g/ 100 g) [29].

Soybean protein provides "full protein" that means soybean present every essential aminoacids, although it is low in sulfur amino acids, being methionine the most significant limiting amino acid, followed by cysteine and threonine. This difference is actually not really high and soybean protein foods are equivalent in quality to animal protein. Soybean protein contains sufficient lysine, which is deficient in most cereal proteins. Therefore, soybean amino acid profile is complementary to cereal amino acid profile. On the other hand, there are other natural components present in soybeans such as trypsin inhibitors, lectins and phytic acid, which reduce protein quality [10, 23] (see Antinutrients).

Table 3. Relation of the essential and non-essential aminoacids for humans

Essential aminoacids	Non-essential aminoacids
Isoleucine	Alanine
Leucine	Arginine
Lysine	Asparagine
Methionine	Aspartic acid
Phenylalanine	Cysteine
Threonine	Glutamic acid
Tryptophan	Glycine
Valine	Proline
Histidine (children)	Serine
	Tyrosine
	Histidine (adults)

Storage proteins are predominant, such as the 7S globulin (β-conglycinin) and 11S globulin (glycinin), which represent about 80% of total protein content, as well as less abundant storage proteins such as 2S, 9S, and 15S globulins [30]. Interestingly, β-conglycinin but not glycinin is capable of improving serum lipid profiles, in the absence of phytoestrogens [31, 32].

The substitution of animal protein by dietary soy protein has been shown to reduce plasma total and LDL cholesterol and triglyceride levels [22, 23, 26, 28, 31, 32]. Although soybean is cited as one of the eight most frequent human allergenic ingredients, a substantial number of clinical and animal studies indicate that soy protein is less immunologically reactive than many other food proteins, such as cow's milk; however, it should be avoided in individuals with a high risk of food allergy [33].

2.2. Carbohydrates

Soybeans contain about 35% carbohydrates being the second largest component after protein; approximately half of these carbohydrates are structural polysaccharides [18, 29]. The seed coat, also known as soybean hull, constitutes the 8% of the whole soybean, containing about 86% complex carbohydrates [1]. The starch, which is present in cotyledons, is predominant when maturation begins representing between 13% and 50% of the total carbohydrate fraction [34]; however, in mature soybean, starch content falls to approximately 1% [29].

Cell wall structural components cover a large variety of polysaccharides excluding α-glucans (starch) and compose the dietary fiber. Dietary fiber plays an important role in many physiological processes and in the prevention of diseases of different origin. Increased dietary fiber intake can be achieved by changing dietary habits, increasing the consumption of high fiber foods and consuming fiber-fortified foods and fiber supplements. Dietary fiber has all the characteristics required to be considered as an important ingredient in the formulation of functional foods [18, 35, 36]. Soybean seeds fiber is supposed to have good nutritional and functional properties. The neutral taste and the absence of color of such fibers make them suitable for incorporation into food products without any change in their quality. Dietary fiber of soybean (24.4%) is classified as soluble (3.5%) and insoluble (20.9%) fiber. The insoluble dietary fiber is characterized by its high content in glucose, followed by galactose, arabinose and xylose. These monosaccharides are from cellulose, xyloglucans, xylogalacturonans and non-solubilized pectic material in strong association with cellulose. The most representative monomers of soluble fiber are galactose, uronic acids and arabinose from rhamnogalacturonan regions very substituted with arabinans, galactans and/or arabinogalactans [35,37,38]. Soybean seeds soluble polysaccharides are considered to have a globular structure with long neutral side chains of α-(1,3) and α-(1, 5) arabinan and β-(1,4) galactan [39].

The non-starch available carbohydrates include sucrose (2.5-8.2%), raffinose (0.1-0.9%), stachyose (1.4-4.1%), galactose, fructose, glucose and arabinose [18]. The α-galactosides, raffinose and stachyose, are well known as antinutritional factors because their presence has been linked to flatulence and abdominal discomfort [1]. Flatulence occurs because mammals (including humans) have no α-galactosidase present in their intestinal mucosa which is required to hydrolyze these compounds, thus they are non-digestible oligosaccharides, constituents of dietary fiber and can act as a prebiotic [40-42]. The prebiotic concept was coined by Gibson and Roberfroid in 1995 [43] and more recently revisited by Gibson et al. (2004) [44] and Roberfroid in 2007 [45]. A prebiotic can be defined as selectively fermented ingredient that allows specific changes, both in the composition and/or activity in the gastrointestinal microflora that confers benefits upon host well-being and health. Hence, ingestion results in α-galactosides passing into the large intestine where bacteria anaerobically ferment them with the production of hydrogen, CO_2, traces of methane gas and short chain fatty acids (SCFA) [46]. They can increase the colonic population of bifidobacteria, known to contribute positively to human health by suppressing intestinal putrefaction, reducing both constipation and diarrhoea, stimulating the immune system and increasing resistance to infection [47]. The α-galactosides are presented in all parts of legumes plant but accumulate in the seeds and roots during development. Soybean oligosaccharides have been widely studied in Japan where they are considered "FOSHU" (Foods Specified for Health Use). In USA, their safety and bifidogenic capacity have been recognized by FFDCA (Federal Food, Drugs and Cosmetic Act) and DSHEA (Dietary Supplements Health and Education Act) and they are used as supplements and functional ingredients [48]. In Europe, by now although there is an increasing interest in prebiotics, the health claims about them have been denied by EFSA (European Food Safety Agency) following the Regulation (EC) No 1924/2006 [49], because the Agency did not consider the increase of some specific microbiota populations enough to establish a beneficial effect on gut health.

2.3. Lipids

Most beans are very low in fat but the exception is soybean. Soybean contains 20% fat, which are mostly polyunsaturated fatty acid triglycerides, about 2-5% of phospholipids, mainly as lecithin, and small amounts of other lipids in the form of sterols such as stigmasterol and sitosterol [50, 51]. Casale (2004) [52] showed soybean fat contents between 22-23% depending on the year of harvest, very similar to those of Redondo et al. (2007) [29], 18-22 % for yellow soybean and less than 1%, for immature seeds. This low content is due to the lipid synthesis from carbohydrates has not yet occurred in immature seeds.

The fatty acid profile of soybean oil include oleic acid (18:1), the main representative of monounsaturated fatty acids, and two polyunsaturated fatty acids, linoleic acid (18:2, ω-6) and α-linolenic acid (18:3, ω-3) [50, 53]. Linoleic acid is the substrate for the synthesis of different fatty acids of the ω-6 series, like arachidonic acid, which form functional compounds of enormous importance for our organism, such as prostaglandins. The α-linolenic acid is the substrate for the synthesis of fatty acids of the ω-3 series as eicosapentaenoic acid (EPA) and docosahexaenoic acid (DHA) [54]. Polyunsaturated fatty acids, especially eicosapentaenoic acid (EPA) and docosahexaenoic acid (DHA), are being studied for their health benefits [55-58]. The consumption of full-fat soy foods contributes significantly to α-linolenic acid intake [10]. The biochemical composition and physical appearance of soybean seeds determine the quality of various soy foods. Indeed, according to the location Kumar et al. (2006) [59] noted that fat content varies between 18.5 and 20.9%, oleic acid between 24.5 and 31.3%, linoleic acid between 46.0 and 50.9% and the linolenic acid between 6.0 and 7.9%.

Soybeans have a high content of phospholipids such as lecithin (phosphatidylcholine) (1.5-2.5%) and these compounds may contribute to its hypocholesterolemic effect [51, 60]. Lecithin is a phospholipid that also contains glycerol, phosphate and choline. It is present in every cell of our body, particularly in nervous tissue, heart, kidneys and endocrine glands [50]. There have been several benefits of lecithin [61, 62], among which we can cite the protective and strengthening the nervous system as the hill becomes lecithin in acetylcholine, a neurotransmitter. In liver, choline from lecithin promotes fat metabolism, preventing infiltration. In fact, a choline deficiency is associated with development of fatty liver, and even cirrhosis. Lecithin facilitates the mobilization and endogenous fat oxidation, helping to prevent accumulation. Lecithin helps to correct the deficiencies of choline can cause hemorrhagic degeneration of the kidneys. By its action emulsifier, lecithin promotes intestinal fat absorption. Other compounds of interest in the lipid fraction of soybean are the tocopherols, which act as natural antioxidants and have vitamin E functions [51].

2.4. Minerals

Soybean seed quality is often determined by seed protein, oil, fatty acid, and mineral content [63]. The global mineral content of soybeans, determined as ash, is about 5-6 %. The major mineral constituents are potassium (1.7-2.3 g% dry matter), phosphorus (480-940 mg/100g DM), calcium (200-310 mg/100g DM) and magnesium (220-310 mg/100g DM). Na content in soy seeds is low (2-10 mg/100g) and for that it is interesting for hipertension. The minor constituents comprise trace elements of nutritional importance, such as iron (8-17 mg/100g DM), zinc (4.2- 6.8 mg/100g DM) or copper (1.3-1.8 mg/100g DM) [64-69]. Soybean contain a higher amount of minerals as K, Ca, Mg, P and Fe than others leguminous seeds or most of cereals [50].

Mineral content in soybean crops changes widely in relation to factors such as varieties, climate, cultural methods (rates of fertilizer application) and mineral content of the soil [50]. However, Redondo et al. (2007) [29] showed that the mineral level is non-dependent on the crop type for the same variety (conventional: 4.8 g/ 100 g, ecological: 4.7 g/100 g, transgenic: 5.3 g/100 g, non-transgenic 4.8: g/100 g) and it is higher than in immature seeds (conventional: 3.4 g/100 g, ecological: 3.1 g/100 g). In many areas soybeans are grown in rotation with corn (*Zea mais*). Crops rotation has been shown to improve nutrient use efficiency [70]. Besides, the application of phosphorus as fertilizer can improve the phosphorus content in the grain legumes. Phytic acid is the primary phosphate reserve in most seeds accounting for 60% to 90% of total phosphorus [71]. Their content can vary with genotype, climate, soil and year [72]. In seeds, it is stored in the protein bodies and globoids in the endosperm as mixed salts of cations. These salts called phytins contain mainly K and Mg, together with smaller amounts of Ca, Fe, Zn and Cu and these forms play an important role in mineral storage [73, 74].

Soybeans and soybean products serve as a dietary source of minerals but their bioavailability may be reduced as a result of the presence of phytates that could chelate divalent cations (Ca, Mg, Fe, Zn or Cu) and limit their intestinal absorption [70]. Similarly, phosphorus phytate, which is the predominant phosphorus in legumes, is biologically unavailable to humans [67, 70]. Mineral bioavailability is not the same for each cation. In this sense, Messina (1999) [10] found poor bioavailability for iron and in contrast relatively good for zinc. Calcium bioavailability is reasonably good and could be attributed to the relatively low sulfur aminoacid (SAAs) content of beans and also to the presence of non-digestible oligosaccharides in soy seeds that provide an advantage in terms of calcium retention [75]. The reported hypercalciuric effect of protein is at least partially due to the metabolism of SAAs. Human studies showed that the consumption of soy protein is associated with a markedly lower urinary calcium excretion compared with the consumption of similar amounts of whey protein or a mixture of animal proteins [10]. Soybeans contains galacto-oligosaccharides (GOS), which promote the growth of beneficial lactic acid bacteria in the colon and their acidic fermentation in the caecum could be related with an increase in mineral absorption [42].

2.5. Isoflavones

Polyphenols constitute one of the largest and ubiquitous group of phytochemicals [76]. Several thousand polyphenolic molecules have been identified in higher plants and several hundred are found in edible plants [77]. Polyphenols are also an essential part of the human diet; in fact it is suggested that more than 8000 different dietary polyphenols exist with flavonoids and phenolic acids being the most common ones in food [78]. Among the interesting biological properties exhibited by plant polyphenols there has been an increase interest regarding cancer prevention. Indeed, a number of *in vitro* and *in vivo* studies have shown that plant polyphenols could be used as chemopreventive agents against different cancer types [79, 80]. The major structural characteristic of the polyphenols is one or more hydroxyl groups binding to one or more aromatic rings and they are divided into two major groups, flavonoids and non-flavonoids [77]. Flavonoids share a common flavan core formed with 15 carbon atoms and this class can be divided into flavanols (e.g. catechins), flavonols (e.g. quercetin, myricetin, kaempferol), anthocyanidins (e.g. cyaniding, delphinidin), flavones (e.g. apigenin, diosmin), flavanones (e.g. naringenin, hesperitin) and chalcones (e.g. phloretin). The non-flavonoids contain an aromatic ring with one or more hydroxyl group. Non-flavonoids include stilbene (e.g. trans-resveratrol), phenolic acids (e.g.caffeic acid, gallic acid), saponins (e.g. ginsenoside), and other polyphenols such as curcumin and tannins [77].

Several classes of polyphenols have been identified in soybeans, including saponins, phenolic acids and isoflavones [82, 83]. Isoflavones are non-nutritive substances that possess health protective benefits. They are a subclass of flavonoids and the basic structure of flavonoid compounds is the flavone nucleus, which has two benzene rings linked with heterocyclic pyrane (Figure 2). They are described as phytoestrogen compounds because they exhibit estrogenic activity (similar effects to estradiol hormones). Soybeans are recognised as a major dietary source of isoflavones, and soy-based products have been shown to contain significant amounts of these compounds; fermented soy products, including miso and tempeh, also contain high concentrations of isoflavones [84].

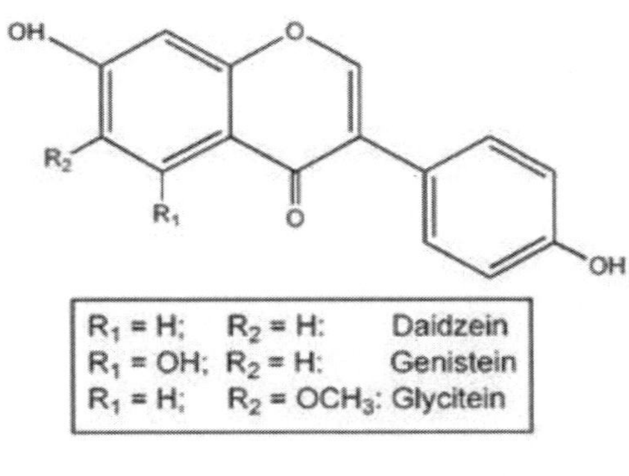

Figure 2. Chemical structures of soybean isoflavones. Aglycone forms.

In soybean, the isoflavones are tightly associated with protein. The association of high-quality protein and phytochemicals, especially isoflavones, is unique among plant-based proteins because isoflavones are not widely distributed in plants other than legumes. Soybeans contain 0.1 to 5 mg/g total isoflavones, primarily genistein, daidzein, and glycitein [85]. These nonsteroidal compounds, commonly known as soy phytoestrogens, are naturally present as the ß-glucosides genistin, daidzin, and glycitin, representing 50% to 55%, 40% to 45%, and 5% to 10% of the total isoflavone content, respectively [86]. In their naturally occurring glycosidic form, isoflavones are conjugated at the 7 position and are often esterified with acetyl or malonyl groups at the 60 position of the sugar residue. In plants, the most widely studied isoflavones, genistein and daidzein, are usually encountered as conjugates (acetyls, glycosides or malonyls) [87].

The main isoflavones found in soybeans are genistein (4,5,7-trihidroxyisoflavone), daidzein (4,7-dihidroxyisoflavone), glycitein (4,7-dihidroxy-6-metoxi-isoflavone) and their respective acetyl, malonyl and aglycone forms thus implying 12 isoflavonic compounds in total. Genistein and daidzein are the primary isoflavones in soybean [8, 88]. Isoflavones bioavailability depends on the conjugate form, indicating the aglycone as the biologically active compound [89]. Inactive precursors of isoflavones are activated by enzymes in the small intestinal, which favors absorption and subsequent hepatic biotransformation and enhances their biological activity. Genistein is processed to be converted into dihydrogenistein and 6'-hydroxy-O-desmethylangolensin, which are more easily absorbable, and have greater bioavailability. Daidzein is converted into equol and O-desmethylangolensin [50].

Isflavones have primarily attracted attention because of their possible function in ameliorating postmenopausal symptoms, such as hot flashes and osteoporosis. Isoflavones have also been reported to potentially affect cardiovascular diseases, cognitive function, and breast and prostate cancer [90-93]. Several animal and human clinical studies as well as epidemiological studies (many of which compared the occurrence of a certain disease in Asian countries with that in Western countries), have suggested that soy consumption has preventative effects on several hormone-dependent diseases [94]. There are several possible mechanisms of action by which isoflavones may act on disease prevention, including estrogenic/ anti-estrogenic activity, cell anti-proliferation, induction of cell cycle arrest and apoptosis, prevention of oxidation, anti-inflammatory, regulation of the host immune system, and changes in cellular signaling [95-99]. However, there is still controversy and an unanimous position about if isoflavones, other soy phytochemicals or components are responsible for the health benefits of soy consumption is still far from being reached [8].

2.6. Antinutrients

Soybean contains a number of bioactive substances that have been traditionally classified as antinutritional factors (ANFs). Among those factors, there are heat-labile components that exert a negative impact on the nutritional quality of the protein: protease inhibitors and lectins, and heat stable compounds, such as phytates, flatulence-producing oligosaccharides or saponins with relatively less significant antinutritional effects. ANFs decrease the nutritional value of food by lowering the digestibility or bioavailability of nutrients, but also could be classified as bioactive compounds with significant potentials benefits to human health. All of them play both beneficial and adverse roles in animal and human nutrition [100].

Protease inhibitors are known to defend the plant against predators (microorganisms or insects). In humans, they inhibit pancreatic proteases impairing dietary protein digestion and decreasing their nutritional value [72, 101]. The main adverse effects in humans are growth inhibition due to a reduced proteolysis of the exogenous protein, and hypertrophy and hyperplasia of the pancreas that may enhance endogenous sulphur-rich proteins loss [72]. There are two families of these antinutrients isolated from soybeans referred as Kunitz and Bowman-Birk families. The first present a molecular weight of 20.00-25.000 Da and few disulphide bonds that combine specifically with trypsin. The second one have low molecular weight, seven disulphide bonds and capacity to inhibit trypsin and chymotrypsin at two independent binding sites, and the formed complexes have low dissociation constant. In contrast with these adverse effects, Bowman-Birk inhibitors has been studied as an anticancer agent [102]. Other potentially beneficial attributes include anti-inflammatory activity and probable effects against obesity, several degenerative and autoimmune diseases [103]. The activity of these inhibitors, mainly Kunitz inhibitor, is reduced or destroyed when the soybean is heated during processing (soaking, boiling, autoclaving or extrusion cooking), but the thermal treatment could also destroy other essential nutrients and it is necessary to apply moderate heat treatment in order to obtain beneficial effects on the nutritional value [101].

Lectins (or phytohemagglutinins) are sugar-binding proteins or glycoproteins that have the ability to agglutinate red blood cells. These compounds are found in most plant foods; however grain legumes are the main source in the human diet [100]. Lectin in raw soybeans also known as soybean agglutinin (SBA) can resist the gastrointestinal tract digestion. This allows that SBA bind to glycoprotein receptors on the epithelial cells of the intestinal mucosa, which causes disruption of the brush borders and could inhibits growth by interfering non-specifically with the absorption of nutrients [104]. Douglas et al. (1999) [105] detected that SBA produced approximately 15% of growth depression in chicks, whereas Liener (1996) [106] estimated that lectins accounted for approximately 50% of the growth-inhibiting effect of raw soybean meal in rats. Lectins are rapidly degraded upon heating, but dry heat treatment is not as effective as moist heat in deactivating SBA [107]. It has seen that soybean lectins have both anti-nutritional and anti-carcinogenic properties [108].

Phytic acid (myo-inositol 1,2,3,4,5,6-hexakisphosphate, I6) is naturally present in soybeans and can make up to 1-1,5g [65,66]. It is considered an antinutrient, mainly due to this ability to bind essential minerals and also to interact non-specifically with protein and starch and consequently reduce their uptake in the human intestine [109]. The phytate (IP6), and some of their degradation products, are well-known inhibitors of absorption of multivalent metal ions, especially Zn, Ca and Fe [110]. During food processing and digestion, phytates can be dephosphorylated by phytases to produce degradation products such as myo-inositol pentakis (IP5), tetrakis (IP4), tris- (IP3), bis- (IP2) and monophosphates (IP1) [100]. The high ability to chelate minerals is directly correlated with the number of reactive phosphate groups [101]. The main inosytol phosphate in legumes is IP6, and the highest content correspond to soybeans (1.8% dry matter) [72,110]. The controlled degradation of phytic acid has showed to improve the uptake of essential dietary minerals [110]. On the other hand, phytic acid may play a role in reducing cancer risk, possibly because of its antioxidant effects [111].

Soybean oligosaccharides, mainly stachyose and raffinose, are galatooligosaccharides (GOS) and they are non-digestible oligosaccharides because there is no α-galactosidase in the human intestinal mucosa. Thus, they pass into the large intestine where anaerobic bacteria ferment them with the production of carbon dioxide, hydrogen, and traces of methane causing discomfort, flatulence and sometimes diarrhea [10]. However, GOS promote the growth of beneficial lactic acid bacteria in the colon and are recognized as prebiotics [40, 41, 42].

In the case of saponin, they are glucosides of a steroid or a triterpenoid group that may interact with the intestinal mucosal cells and reduce the absorption of nutrients [103]. On the other hand, hypocholesterolemic and anticarcinogenic effects have been attributed to these non-flavonoids polyphenols [10, 96].

3. Implications For Nutrition and Health Effects

The health benefits of legumes such as chickpeas, beans, lentils and soybean, among others have been much studied. Soybean foods represent an excellent source of high quality protein, are low in saturated fat, and are cholesterol free. But, moreover, soybean contains a great amount of dietary fiber; it is the second largest component in soybeans [18, 29, 38, 88]. Based in this composition, it has shown to reduce risk of heart disease, cancer and osteoporosis and has a role in other diseases such as diabetes or obesity.

3.1. Soybean Intake, Cholesterol and Heart Disease

Cholesterol is a lipid synthesized in the liver, intestine and other tissues. Numerous dietary components contain cholesterol, and although an excess of these compounds can increase cholesterol blood levels, they depend mainly on liver synthesis. Cholesterol is transported associated with other proteins and lipids as lipoproteins in blood. High density lipoproteins (HDL) transport cholesterol from peripheral tissues, deposed there for low density lipoproteins (LDL), to the liver where it is excreted bile duct. HDL reduces tissue cholesterol and avoids the blood vessel deposit while LDL contributes to atheroma plaque formation. Nevertheless, cholesterol is part of cellular membranes and essential for the synthesis of bile acids which are involved in the intestinal digestion of fat, vitamin D and other steroids hormones as estrogens and testosterone [50].

The causes of cardiovascular disease (CDV) are diverse being atherosclerosis and/or hypertension are the most common, so high cholesterol levels particularly LDL constitute just one of the risk factors in the CDV. This disease is one of the major health problems in most developed countries, and although CDV usually affects elderly people, the antecedents of this disease, notably atherosclerosis, begin in early life. Because of that there is an increased emphasis on preventing atherosclerosis by modifying risk factors, such as diet, exercise and no smoking [50, 112, 113].

Several studies corroborate that the regular intake of low-fat, high-fiber diet including whole grains, fresh fruit, vegetables and legumes significantly decreases CVD. The tobacco interruption, limited alcohol consumption, low blood pressure, decrease body fat and moderate physical activity also contribute to diminish and prevent the CVD [112].

Epidemiologic studies have shown that CVD incidence is lower in Asiatic countries than in western countries [114], so in addition to genetic factors, it has been attributed to dietary factors, especially to the elevated soybean products consumption. Soybean specific studies remark that the substitution in the diet of animal protein for soybean protein reduces the concentration of total cholesterol, LDL and triglycerides, so the consumption of soybean protein can help to reduce cardiovascular disease risk [22, 23, 26, 28, 31]. However, the mechanisms responsible for the hypocholesterolemic effect have not been identified yet. The lack of understanding about this mechanism remains an obstacle for a better acceptance of soybean protein by clinical community. There are different hypothesis to explain these mechanisms. One of these hypothesis is that amino acid composition or distribution in soybean change the cholesterol metabolism, possibly, due to changes in endocrine status, because there are alterations in insulin:glucagon ratio and thyroid hormone concentrations [115], as well as an increase in plasma thyroxin concentrations which is related with a reduction in plasma cholesterol [116]. Also, it has been studied that soybean diet could block the inflammatory processes associated with atherogenesis because of the control of the adhesive functions of vascular endothelium. This is due to the regulation of key endothelial cell adhesion molecules, monocyte integrin function, cytokines and chemokines that control migration of monocytes [117]. Another hypothesis proposes that non-protein components such as fiber or isoflavones associated with soybean protein affect cholesterol metabolism [21, 118]. Different clinical research indicates that to achieve a cholesterol-lowering effect, it is important to consume soybean protein with its natural isoflavones fraction [119, 120]. It is thought that soybean isoflavones may inhibit atherosclerotic development because they have antioxidant properties against LDL oxidation, which generates a cascade of events producing atherosclerotic plaques [24]. In addition, isoflavones possess a hypocholesterolemic effect, although this effect is still under investigation. It might be due to the interaction of isoflavones with estrogenic receptors, because of the structural similitude between these compounds and their metabolites and estrogens [21, 118].

Dietary fiber is the second largest compound in soybeans [29, 38, 88] and it has health benefits including laxation, attenuation of blood cholesterol and glucose, and increased fecal excretion of bile acids. Early epidemiologic research on dietary fiber from cereals and legumes indicated an inverse relationship between dietary fiber intake and coronary disease risk [26, 121]. Legumes constitute a source of dietary fiber, relatively rich in soluble fiber [26]. Results from intervention trials have shown that soluble dietary fiber may diminish blood cholesterol concentrations [122, 123], reduce blood pressure [124], promote body-weight loss [125], and improve insulin sensitivity [126], thereby reducing the risk of CVD. Dietary fiber is a potential substrate for colonic fermentation that can beneficially affect the host by selective stimulation of growth and/or activity of a limited number of colonic bacteria capable of producing short-chain fatty acids (SCFA), mainly acetate, propionate, and butyrate. Small amounts of branched SCFA may also be formed from indigestible protein [127]. Those SCFA contribute in the lowering-LDL effects, indeed propionate has been shown to inhibit hepatic cholesterol synthesis [128].

3.2. Soybean Intake and Obesity

Obesity is usually defined as fat body excess. It is a medical concern because body fat excess may have an adverse effect on health, leading to reduced life expectancy and/or increasing health problems. To assess the obesity classes the body mass index (BMI) or Quetelet index is used. This is an approach for estimating the human body fat as the individual's body mass divided by the square of the height (Kg/m^2) [129].

Table 4. OMS Criteria to define overweight and obesity classes based in BMI [129]

	BMI ranges
Normal weight	18.5-24.9
Overweight	25.0-29.9
Class I Obesity	30.0-34.9
Class II Obesity	35.0-39.9
Class III Obesity	>40

The body fat percentage is the total weight of fat divided by total weight and this is the only body measurement which directly calculates a person's relative body composition without regard to height. López-Jiménez & Cortés-Bergoderi (2011) [130] have shown that when the BMI is normal the prevalence of different diseases depend on the fat body percentage.

Obesity is most commonly caused by a combination of excessive food energy intake, lack of physical activity, and genetic susceptibility, although some cases are caused primarily by genes, endocrine disorders, medications or psychiatric illness [131]. The major concern about obesity is that can increase the likelihood of various diseases, particularly heart disease, type 2 diabetes, obstructive sleep apnea, certain types of cancer, and osteoarthritis. Thus obesity is a leading preventable cause of death worldwide, with increasing prevalence in adults and children, being one of the most serious public health problems in 21st century [132]. In fact, the World Health Organization (WHO) has predicted that overweight and obesity may soon replace the traditional public health concerns such as undernutrition and infectious diseases as the most significant cause of poor health.

The low frequency of obesity and related metabolic disorders in Asian populations has drawn attention towards soybean, which is a characteristic component in Asiatic diets. Current evidence from animal and human studies suggests that diets rich in soybean and phytoestrogens have beneficial effects on many aspects of diabetes and obesity [133]. Therefore, the inclusion of soybean products in the diet of obese people is very useful and covers two important goals. First, soybean is a food with a limited energy value, therefore facilitates the maintaining caloric restriction with a high quality protein intake [1]. The other input is the ability to improve metabolic abnormalities responsible for major comorbidities of obesity: hyperlipidemia, hyperinsulinemia, insulin resistance, type II diabetes and / or metabolic syndrome [134].

Most studies about the relation between obesity and the soybean intake are about protein. The soybean protein intake in dietetic therapy for obesity may be related with the effect over the lipid metabolism. As it was showed in the previous section, soybean protein seems to diminish the cholesterol levels [22, 23, 26, 28, 31]. For years, the question has been whether the active component is protein itself or a combination with other components. Kim et al. (2012) [135] showed that black soybean anthocyanins contribute to antiobesity and antidiabetic soybean properties. However, isoflavones focused much research about their potential antiobesity properties by themselves or in combination with soybean protein. In fact, there is some evidence that soybean dietary phytoestrogens, mainly isoflavones, favorably alter glycemic control, improve weight and fat loss, lower triglycerides, LDL and total cholesterol. Their structural similarities to endogenous estrogens produce weak estrogenic effects by binding to the intranuclear estrogen receptors (ERs). Additionally, the isoflavones have been proved to exert non-ER–mediated effects through numerous other pathways. Epidemiologic and laboratory data suggest that genistein and daidzein, the main soy isoflavones, could have health benefits in human obesity [27]. However, these results must be interpreted with care, and additional evidence is needed before a firm conclusion can be drawn. In particular, since not all activities related to soybean can be assigned to the estrogenic-like activity, further studies are needed to identify firstly which soy constituent(s) improve metabolic parameters when ingested and secondly, which are the mechanisms whereby dietary soy improves metabolic-related conditions.

3.3. Soybean Intake and Diabetes

Nutritional intervention studies performed in animals and humans suggest that the ingestion of soy protein associated with isoflavones improves glucose control and insulin resistance. In studies of human subjects with or without diabetes, soy protein also appears to moderate hyperglycemia and reduce body weight, hyperlipidemia, and hyperinsulinemia, supporting its beneficial effects on diabetes [136].

Soy protein is reported to have numerous beneficial effects in humans, including improvements in body composition and insulin secretion [137]. Soy proteins are rich in arginine and glycine, which are involved in insulin and glucagon secretion from the pancreas. Decreased plasma insulin by soy protein may be due to decreased release from the pancreas or increased hepatic removal. Soy proteins may improve diabetes by reducing insulin resistance and reduce adiposity by inhibiting insulin secretion from the pancreatic cells or by inhibiting lipogenesis and enhancing lipolysis in liver and adipocytes [136]. On the other hand, Vedavanam et al. (1999) [138] suggest that soy isoflavones may be beneficial for diabetic subjects because of their estrogenic activity and their ability to prevent glucose-induced lipid peroxidation. Moreover, they inhibit intestinal glucose uptake by decreasing sodium-dependent glucose transporter, which results in a reduction in postprandial hyperglycemia.

Aging plays an important role in insulin resistance and ß-cell dysfunction in both men and women, however the prevalence and progression are lower in premenopausal women in comparison to men, but the incidence markedly increases with high fat accumulation after menopause [139, 140]. This phenomenon suggests that estrogen prevents insulin resistance and ß-cell dysfunction. Estrogen is a key mediator for glucose-stimulated insulin secretion and the expansion and maintenance of ß-cell mass in rats and mice [141, 142]. Isoflavones may improve antidiabetic actions through estrogen receptors. The estrogen receptor α is emerging as a key molecule involved in glucose and lipid metabolism. An inhibitory effect of soy isoflavones on intestinal glucose transport may help reduce postprandial hyperglycemia in diabetes [136]. Ademiluyi and Oboh (2013) [143] indicate that soybean phenolic rich extracts possess potential antidiabetic activities existent as the ability to inhibit α-amylase and α-glucosidase activities *in vitro*. The soybean phenolic compounds may regulate the glucose uptake from the intestinal lumen by inhibiting carbohydrate digestion and absorption, leading to normal glucose homeostasis. *In vitro* studies have shown that a soybean phytochemical extract containing genistein and daidzein inhibits glucose uptake into rabbit intestinal brush border membrane vesicles in a dose-dependent manner and also protects against glucose-induced oxidation of human LDL [138].

Glyceollins are phytoalexins converted from the isoflavon daidzein in unboiled soybeans with fungi infection, and they often act as antifungal, antibacterial, or anticancer compounds [144, 145]. It is likely that glyceollins have positive health effects on metabolic diseases such as obesity, type 2 diabetes, and cardiovascular diseases. Results suggest that glyceollins may be beneficial in improving glucose homeostasis by potentiating β-cell function and mass and improving glucose utilization in diabetic animal models [146].

Soybeans are well-known for improving energy and glucose homeostasis due to their isoflavonoid and soy protein content [147], whereas their fermented products, such as chungkookjang, natto, and meju, have demonstrated in previous studies the capacity to improve homeostasis better than soybeans due to the increase in isoflavone aglycones and small peptides [148]. Fermentation is an excellent processing method for improving nutritional and functional properties of soybeans due to the increased content of small bioactive compounds. The large protein, lipid, and carbohydrate molecules in raw soybean are broken down by enzymatic hydrolysis during fermentation to small molecules such as peptides, amino acids, fatty acids, and sugars [148]. Isoflavones, which are mostly present as 6-o-malonylglucoside and ß-glucoside conjugates and associated with proteins in soybean, are also broken down by heat treatment and fermentation [149]. During preparation of fermented soy foods, 6-o-malonylglucosides, the most prevalent soybean isoflavones, are converted to 6-o-acetylglucosides or ß-glucosides by heating, and ß-glucosides are unconjugated by the action of ß-glucosidases secreted by fermentation microorganisms [150]. Although no studies have evaluated the effects of specific smaller peptides made from soybean fermentation on glucose regulation, these peptides may have significant effects on glucose metabolism because the changes in isoflavonoids cannot explain the improvement of glucose homeostasis. Thus, further studies are needed to determine the antidiabetic effects of specific peptides [148].

Therefore, phytoestrogens and proteins in soybeans seem to have beneficial actions both on glucose metabolism, and additional micronutrients such as saponins, phytosterols, trypsin inhibitors, as well as the amino acid and protein composition may have additive or synergistic effects. Fermentation of soybeans leads to structural changes in proteins and phytoestrogen, which may contribute to more beneficial effects on glucose metabolism. It remains unclear if soy phytoestrogens mediate the antidiabetic effect through estrogenic pathways or if the effect is mediated by non-estrogenic activity of isoflavones, isoflavone metabolites, soy peptides, or other bioactive soy compounds [148]. In diabetes, one or more components may have a beneficial effect singly, synergistically, or additively with other active components. Researches performed in diabetic patients with soybean diets show several potential advantages, but at the moment very much work is required to define the exact role of soybean in the control of diabetes mellitus [136].

3.4. Soybean Intake and Cancer

Many studies have attempted to establish the relationship between soy intake and the incidence of various types of cancer, since it has been observed that the incidence is much lower in Asian countries where soy consumption is high. It was also determined that in this incidence may play an important role lifestyle, diet and intake of fats and fibers [151].

The lower risk of prostate and breast cancer in areas of high soybean and isoflavone intake, especially in Asia [152], as well as the increased risk observed for Asian people who migrated to Western countries or adopted a westernized lifestyle, are well known [153]. High soy food intake can be related to a reduction in prostate cancer risk [154] and for breast cancer; a smaller risk reduction is also found [155]. Asian women whose soy intake was high during puberty experience lower risk for breast cancer than women who did not consume soy products or did only as adults [156]. Thus the consumption of soybean is associated with the relatively low levels of different cancers in countries that include soybean in their diets. Soybeans contain high levels of compounds with anticancer activity, such as phytates, protease inhibitors, saponins, phenolic acids, and isoflavones. Most of the data suggest that isoflavones are the responsible ones for the hypothesized anticancer effects of soybean.

Isoflavones and their metabolites have several hormonal and non-hormonal activities and thus they have been reported to modulate steroid biosynthesis, transport and metabolism, as well as carcinogen activation and detoxification, to inhibit cell proliferation induced by growth factors, to induce cell cycle arrest or apoptosis, to favor cell differentiation, to reduce oxidative stress or to inhibit angiogenesis, cell invasiveness and metastasis [151, 157]. They may act through modulation of cell signaling, regulation of gene expression and/or specific inhibition of some key enzyme activities. The major focus in isoflavone research so far lies on the modulation of steroid hormone receptor signaling. Estrogens regulate many physiological processes in hormone-dependent tissues, including cell growth and differentiation, apoptosis and tissue-specific gene regulation, but also influence the pathological processes of hormone-dependent diseases, such as breast and prostate cancers. The biological actions of estrogens are mediated by the binding to one of the two specific nuclear receptors, ERα and ERβ [158], which can induce gene transcription of estrogen-responsive target genes. A general hypothesis is that estrogens acting via ERα exert strong proliferation stimulatory effects while those interacting with ERβ tend to reduce this stimulation. Furthermore, ERβ has been shown to repress ERα-controlled transcription [159]. Cell response would therefore depend on the balance between ERα and ERβ expression levels within a given cell type. This balance is altered in favor of ERα during tumor progression [160], emphasizing the protective role of ERβ signaling and the usefulness of ERβ-selective agonists in cancer prevention. Eight *in vitro* and *in vivo* studies consistently reported a down-regulation of ERα and an up-regulation of ERβ mRNA and protein levels by genistein at 1–10 mM in breast cancer cells [161, 162]. In addition to the inhibition of cell proliferation, isoflavones can induce apoptosis in human breast and prostate cells [151].

Furthermore, the beneficial effects of isoflavones may in part be attributable to their epigenetic properties, including 1) changes in the DNA methylation pattern, 2) regulation of histone modifications and 3) changes in the expression of some microRNAs.

1) The hypermethylation induced transcriptional silencing of tumor suppressor genes constitutes a frequent epigenetic defect in many human cancers. Reversal of gene

hypermethylation, which may in part be achieved by inhibiting DNA methyltransferase activity in cancer cells, is a plausible and promising avenue for developing epigenetic drugs. Genistein, one of the many phytoestrogens contained in soybeans, has been lately studied as a demethylating agent. Although *in vitro* studies have shown that genistein apparently induces DNA demethylation through DNA methyltransferase inhibition, animal studies have observed rather increased DNA methylation following the treatment. The effect of genistein on methylation in humans has been recently tested, and in agreement with the animal data, the results showed that hypermethylation was increased after treatment and correlated with the genistein level [163].

2) Evidence indicates that dietary polyphenols can also regulate gene expression through changes in histone modifications. So, genistein decreases the histone deacetylase activity (HDAC) in squamous esophageal and renal cancer cell lines, but not in prostate cancer cells [164].
3) MicroRNAs (miRNAs) have been recently discovered as key regulators of gene expression. Although there is still limited evidence, recent reports have identified that dietary polyphenols can also modulate gene expression by targeting various oncogenic or tumor suppressive miRNAs. Soybean isoflavones have been found to regulate the miRNA expression in pancreatic cancer cells. Natural compounds like a combination of genistein, daidzein, and glycitein could function as miRNA regulators which are likely to be important for designing novel therapies for pancreatic cancer [165].

Many studies have been conducted to understand the effects of soy isoflavones on breast and prostate cancer; however the mechanisms of action are still not entirely elucidated [166] and it should be shed more light in this field.

3.5. Soybean Intake and Menopause

There is scientific evidence that isoflavones may alleviate post-menopausal symptoms derived from the lack of estrogen, mainly hot flushes, but also other symptoms such as vaginal dryness and sleep disturbances, anxiety, depression, loss of libido among other symptoms [167, 168]. Hormone replacement therapy with estrogens improves the condition but the compliance to such medications is generally very low. Soy-derived products have been proposed to provide comparable benefits to estrogens [169, 170].

It is well known that Asian women have considerably fewer menopausal symptoms than Western women, while the rate of hot flushes ranged from 70 to 80% in European women, this decreased by 18 and 14% in China and Singapore, respectively [171]. These differences diminished when oriental population migrated to the U.S.A. and adopted their alimentary habits. It has been hypothesized that this difference is due to their high intake of phytoestrogen, particularly isoflavones [172].

Phytoestrogens are thought to have estrogenic and non-estrogenic effects like selective estrogen receptor modulators, depending on the environment of circulating hormone and receptor availability [173]. In menopausal women phytoestrogen effects appear to be related with their coupling to β-receptor [169, 171]. Genistein and daidzein are the most abundant phytoestrogens in soy products but equol, a daidzein metabolite, is thought to have the most important estrogenic effect [174]. Daidzein is converted to equol in the gut by bacterial flora, whose converting capacity may be influenced by the individual diet habits [175], and there also great interethnic variations in this conversion capacity [176].

The hot flush is the most common and frequent symptom of menopause. The etiology and mechanisms of this vasomotor disturbance are not yet completely understood. Apparently, this climacteric symptom may be linked to central noradrenergic responses [177] with accompanying dysfunction of the thermoregulatory center of the hypothalamus, core body temperature elevation and changes in plasma catecholamine levels [178]. It results in tachycardia and increase temperature in skin, with a compensatory activation of the mechanisms of heat loss through sweating and skin vasodilation [179]. Positive phytoestrogens effect on the hot flushes could be explained through the hypothesis of an agonist estrogenic action on the thermoregulatory hypothalamic center [169].

The main non-vasomotor climacteric symptom caused by physiological responses to low concentrations of estrogens is the vaginal atrophy. These responses include reduced vaginal blood flow and secretions, tissue changes, and a change in the pH of vaginal fluid, from acidic to neutral. The vagina is covered by a mucosa with a Malpighian stratified epithelium with marked sensitivity to sex steroids, particularly to estrogens. Isoflavones have the ability to bind to estrogen receptors and exert the biological responses, also on vaginal parameters [180].

Numerous investigations have been done to assess whether phytoestrogens are effective in menopause symptoms obtaining contradictory results. Epidemiological studies have demonstrated the reduction of the severity of hot flushes and the improvement of other symptoms such as vaginal cell maturation and dryness [172, 181, 182]. Some clinical trials with phytoestrogens in postmenopausal women showed moderate estrogenic effects on hot flushes and other menopausal related disorders [167, 183], but one feature that stands out in nearly all studies is a large placebo effect [184, 185]. Despite this, a few studies have found significant decreases in number or severity of hot flushes within the soy treatment group beyond placebo [167, 169, 170, 182], although the clinical significance of some of the changes observed is debatable. Howes et al. (2006) [91] suggest that the benefit appeared to be positively associated with the frequency of the flushes and possibly the used isoflavones dose. Williamson-Hughes et al. (2006) [186] support the notion that higher concentrations of genistein may be more effective. Jou et al. (2008) [188] concluded that isoflavones supplementation improves menopausal symptoms only in women with the ability to produce equol. Nevertheless, the most recently results described in several meta-analysis and systematic reviews of clinical trials on the impact of isoflavones on menopausal vasomotor symptoms give inconsistent results [173, 180, 188].

Different studies asses the efficacy of soy isoflavones on the morphologic changes at vaginal epithelium. Some investigations have found an increase of the vaginal cell maturation with phytoestrogens [189], an estrogenic effect on vaginal cytology in postmenopausal women [190] and vaginal dryness [169], but the vaginal cell maturation has not been confirmed in other studies [167, 183]. Chiechi et al. (2003) [168] conclude that a soy-rich diet is effective increasing the maturation indices of vaginal cells and the results obtained indicate a positive estrogenic effect on vaginal epithelia attributable to prolonged exposure to dietary phytoestrogens.

3.6. Soybean Intake and Osteoporosis

Osteoporosis is a gradual loss of the number of cross bridges in the organic bone matrix, combined with a reduction in bone mineral density and loss of bone mass. This fact increase risk of bone fractures occurring during aging. There are two forms of osteoporosis: post-menopausal osteoporosis that affects at woman and senile osteoporosis that occurs in both aging women and men [191]. The risk of osteoporosis is most relevant to post-menopausal women, since the decline in estrogen production that occurs with menopause is a major contributing factor [192].

Estrogen activity is crucial to prevent bone loss in women. The relative estrogen deficiency after the menopause is the main factor in the genesis of osteoporosis and subsequent fractures in the elderly women. Physiological doses of estrogens prevent or retard bone loss associated with menopause, but also there are side effects such as fluid retention, nausea, migraine, weight gain and slightly increased risk of hormone-depending tumors, including breast and uterus cancer. Moreover, to obtain bone health benefit hormone replacement therapy should be given continuously for many years, but approximately 70% of the women drop out after the first year of treatment [168, 191].

Studies have shown that Asian populations habitually consuming soybean products have a lower risk of osteoporosis when compared with western populations [88, 118, 171]. Given the estrogenic properties of isoflavones and their structural similarity to estradiol or to ipriflavone, a synthetic isoflavone used in the treatment of the osteoporosis, soy isoflavones were suggested to prevent bone loss associated with menopause and this has been the starting point for formulating the hypothesis that soy or soy phytoestrogens could constitute an alternative to the hormone replacement therapy [10, 193].

The estrogen receptor ER-α predominates in the uterus whereas ER-β predominates in bone (Gustafsson, 2000). It would be reasonable to think that in postmenopausal women, soy isoflavones (genistein, daidzein and the bacterial conversion product of daidzein: equol) bind to free estrogenic receptors ER-β in the bone [194]. Isoflavones show preference in binding to ER-β with slightly less affinity than estradiol [195].

The mechanism of how isoflavones acts on bone is unknown. It has been postulated that genistein stimulate osteoprotegerin activity and has a direct inhibitory effect on bone resorption [10, 196]. Osteoprotegerin is secreted by osteoblasts that are responsible for building bone. If the activity of the osteoclasts responsible for resorption is greater than the osteoblasts, there will be a gradual reduction of bone structure, leading to osteoporosis. Osteoprotegerin ligand system neutralizes regulating nuclear κß osteoclast differentiation and activity [193]. Moreover, daidzein shows an increase of bone mass in postmenopausal women [10]. It is also possible that other factors can act together with phytoestrogens such as calcium levels, galacto-oligosaccharides and/or soy proteins presented in soy products. Calcium increases bone density, GOS allows better calcium absorption and soy proteins can decrease urinary calcium excretion, which may support bone preservation [42, 191, 195, 197].

Preliminary reports of research using animal models indicate that isoflavonoids are able to prevent bone loss in ovariectomized rats [198, 199], but in contrast there are negative results in research with monkey models [200]. The first human study that found a positive effect of soy on bone mineral density was published by Potter et al. (1998) [201] and from this date, many human observational and intervention studies have been made. Numerous observational studies relating soy intake of pre- and post-menopausal women to bone mineral density have generally reported significant positive associations [202-204], with only some studies reporting no association [205, 206]. Subsequently, human intervention studies were made to assess effects of phytoestrogen consumption on bone mineral density at the lumbar spine or femoral neck and on biomarkers of bone turnover and the results are contradictory. Some authors [207-209] have showed that a treatment with isoflavone-rich soy is important in maintaining bone integrity in postmenopausal women, but other [193, 210-212] did not confirm these results, so the matter is still controversial. Possibly, these discrepancies are the result of the use of different extracts and doses of isoflavones and / or soy foods, differences in metabolism between individuals in the trial (or not equol producers), differences in duration of clinical trials, potential problems in the inclusion criteria and a possible involvement of different diets and lifestyles of the individuals participating in the trials [193]. Additional research is needed to determine if isoflavones are an effective alternative to hormone replacement therapy for the prevention and treatment of osteoporosis.

REFERENCES

[1] Liu, K. *Soybeans as Functional Foods and Ingredients*. Champaign, Illinois, USA: AOCS Press; 2004

[2] Roberfroid, MB (2000) Concepts and strategy of functional food science: the European perspective. *American Journal of Clinical Nutrition*, 71, 1660S-1664S

[3] Schwager, J; Mohajeri, MH; Fowler, A; Weber, P (2008). Challenges in discovering bioactives for the food industry. *Current Opinion in Biotechnology*, 19, 66-72

[4] Clair, RS and Anthony, M (2005). Soy, isoflavones and atherosclerosis. *Handbook of Experimental Pharmacology,* 170 301-323

[5] Sarkar, FH and Li, Y (2003). Soy isoflavones and cancer prevention. *Cancer Investigation,* 21, 744-757

[6] Ikeda, Y; Iki, M; Morita, A; Kajita, E; Kagamimori, S; Kagawa, Y; Yoneshima, H (2006). Intake of fermented soybeans, natto, is associated with reduced bone loss in postmenopausal women: Japanese population-based osteoporosis (JPOS) study. *Journal of Nutrition*, 136, 1323–1328

[7] Moriguchi, EH; Moriguchi, Y; Yamori, Y (2004). Impact of diet on the cardiovascular risk profile of Japanese immigrants living in Brazil: contributions of World Health Organization CARDIAC and MONALISA studies. *Clinical and Experimental Pharmacology and Physiology*, 31, S5-7

[8] Rostagno, MA; Villares, A; Guillamón, E; García-Lafuente, A; Martínez, JA (2009). Sample preparation for the analysis of isoflavones from soybeans and soy foods. *Journal of Chromatography A*, 1216, 2–29

[9] Wu, AH; Yu, MC; Tseng, CC; Pike, MC (2008). Epidemiology of soy exposures and breast cancer risk. *British Journal of Cancer, 98*, 9-14

[10] Messina, MJ (1999). Legumes and soybeans: overview of their nutritional profiles and health effects. *American Journal of Clinical Nutrition,* 70, 439S– 450S.

[11] FAOSTAT: http://faostat3.fao.org/home/index.html#VISUALIZE

[12] López-Bellido L. Soja. In: López-Bellido L (Ed). *Cultivos Industriales*. Madrid, Spain: Mundi-Prensa; 2002

[13] Erdman, JW and Fordyce, EJ (1989). Soy products and the human diet. *American Journal of Clinical Nutrition*, 49, 725-737

[14] Mateos-Aparicio, I. *Caracterización de la fibra alimentaria en distintos tipos de grano de soja*. Tesina. Madrid, Spain: Universidad Complutense de Madrid: 2004

[15] Guerrero, A. Soja. In: Guerrero A (Ed). *Cultivos herbáceos extensivos*. Madrid, Spain: Mundi-Prensa; 1998

[16] Silveira, JM and Durán, JM (1998). La soja: un cultivo poco conocido en España. *Vida rural*, 75, 28-33

[17] USDA: http://www.usda.gov/oce/commodity/wasde/latest.pdf

[18] Mateos-Aparicio, 2009. Aprovechamiento de subproductos de leguminosas para la obtención de ingredientes funcionales. Comparación de metodologías para la caracterización de fibra alimentaria. PhD Thesis. *Colección Digital de Tesis de la UCM (http://eprints.ucm.es/)*

[19] Food and Drug Administration (FDA) (1999). FDA approves new health claim for soybean protein and coronary heart disease. FDA Talk Paper: www.fda.gov

[20] Adlercreutz, H (2002). Phyto-oestrogens and cancer. *Lancet Oncology*, 3, 364-373

[21] Adlercreutz, H; Heinonen, S-M; Penalvo-Garcia J (2004). Phytoestrogens, cancer and coronary heart disease. *Biofactors*, 22, 229-236

[22] Anderson, JW; Johnstone, BM; Cook-Newell, ME (1995). Meta-analysis of the effects of soybean protein intake on serum lipids. *The New England Journal of Medicine*, 333, 276-282

[23] Anderson, JW; Smith, BM; Washnock, CS (1999). Cardiovascular and renal benefits of dry bean and soybean intake. *American Journal of Clinical Nutrition*, 70, 464-474.

[24] Engelman, HM; Alekel, DL; Hanson, LN; Kanthasamy, AG; Reddy, MB (2005). Blood lipid and oxidative stress reponses to soy protein with isoflavones and phytic acid in postmenopausal women. *American Journal of Clinical Nutrition*, 81, 590-596

[25] Jenkins, DJA; Kendall, CWC; Augustin, LSA; Franchesci, S; Hamidi, M; Marchie, A; Jenkins AL; Axelsen, M (2003). Glycemic index: overview of implications in health and disease. *American Journal of Clinical Nutrition*, 76, 266-273

[26] Kushi, LH; Meyer, KM; Jacobs, DR (1999). Cereals, legumes, and chronic disease risk reduction: evidence from epidemiologic studies. *American Journal of Clinical Nutrition*, 70, 451-458

[27] Ørgaard, A and Jensen, L (2008). The Effects of Soy Isoflavones on Obesity. *Experimental Biology and Medicine*, 233, 1066-1080

[28] Reynolds, K; Chin, A; Lees, KA; Nguyen, A; Bujnowski, D; He, J (2006). A Meta-Analysis of the Effect of Soybean Protein Supplementation on Serum Lipids. *American Journal of* Cardiology, 98, 633-640

[29] Redondo-Cuenca, A; Villanueva-Suárez, MJ; Rodriguez-Sevilla, MD; Mateos-Aparicio, I (2007). Chemical composition and dietary fibre of yellow and green commercial soybeans (*Glycine max*). *Food Chemistry*, 101, 1216-1222

[30] García, MC; Torre, M; Marina, ML; Laborda, F; Rodríguez AR (1997). Composition and characterization of soyabean and related products. *Critical Reviews in Food Science and Nutrition*, 37, 361-391

[31] Ferreira, ES; Silva, MA; Demonte, A; Neves, VA (2011). Soy β-Conglycinin (7S Globulin) Reduces Plasma and Liver Cholesterol in Rats Fed Hypercholesterolemic Diet. *Journal of Medicinal Food*, 14, 94-100

[32] Kohno, M; Hirotsuka, M; Kito, M; Matsuzawa, Y (2006). Decreases in serum triacylglycerol and visceral fat mediated by dietary soybean β-conglycinin. *Journal of Atherosclerosis and Thrombosis*, 13, 247–255

[33] Cordle, CT (2004). Soy Protein Allergy: Incidence and Relative Severity. *Journal of Nutrition*, 134, 1213S-1219S

[34] Macrae, R; Robinson, RK; Sadler, MJ. *Encyclopaedia of food science, food technology and nutrition*. London, UK: Academic Press; 1993

[35] Rodríguez, R; Jiménez, A; Fernández-Bolaños, J; Guillén, R; Heredia, A (2006). Dietary fibre from vegetable products as source of functional ingredients. *Trends in Food Science and Technology*, 17, 3–15

[36] Mateos-Aparicio, I (2012). Beans by-products, potential sources for functional ingredients. In: Popescu E and Golubev I (eds.) *Beans: Nutrition, Consumption and Health*. Nova Publishers Inc., NY, USA

[37] Mateos-Aparicio, I; Redondo-Cuenca, A; Villanueva-Suárez, MJ (2010). Isolation and characterization of cell wall polysaccharides from legume by-products: okara, pea pod and broad bean pod. *Food Chemistry*, 122, 339-345.

[38] Redondo-Cuenca, A; Villanueva-Suárez, MJ; Mateos-Aparicio, I (2008). Soybean seeds and its by-products okara as sources of dietary fibre. Measurement by AOAC and Englyst method. *Food Chemistry*, 108, 1099-1105

[39] Nakamuraa, A; Furutab, H; Katob, M; Maedaa, H; Nagamatsuc, Y. Effect of soybean soluble polysaccharides on the stability of milk protein under acidic conditions. *Foods Hydrocolloids*, 17, 333-343

[40] American Association of Cereal Chemist (AACC) (2001). The Definition of Dietary Fiber. *Cereal Foods World*, 46, 112-126

[41] Gray, J. Dietary Fibre. Definition, Analysis, Physiology and Health. *ILSI Europe Concise Monograph Series*. Brussels, Belgium: ILSI; 2006

[42] Tenorio, M.D.; Préstamo Martínez, G.; Espinosa Martos, I.; Rupérez Antón, P. (2010). Soybean whey enhances mineral balance and caecal fermentation in rats. *European Journal of Nutrition*, 49, 155-163.

[43] Gibson, GR; Roberfroid, MB (1995). Dietary modulation of human colonic microbiota: introducing the concept of prebiotics. *Journal of Nutrition*, 125, 1401-12

[44] Gibson, GR; Probert, HM; Van Loo, J; Rastall, RA; Roberfroid, MB. Dietary modulation of the human colonic microbiota: updating the concept of prebiotics. *Nutrition Research Reviews*, 17, 259-275

[45] Roberfroid, MB (2007). Prebiotics: the concept revisited. *Journal of Nutrition*, 137, 830-837

[46] Rochfort, S and Panozzo, J (2007): Phytochemicals for health, the role of pulses. *Journal of Agricultural and Food Chemistry*, 55, 7981–7994

[47] Eiwegger, T; Stahl, B; Haidl, P; Schmitt, J; Boehm, G; Dehlink, E; Urbanek, R; Szepfalusi, Z (2010). Prebiotic oligosaccharides: in vitro evidence for gastrointestinal epithelial transfer and immunomodulatory properties. *Pediatric Allergy and Immunology*, 21,1179-1188

[48] Tuohy, KM; Rouzaud, GCM; Brück, WM; Gibson, GR (2005). Modulation of the human gut microflora towards improved health using prebiotics-Assesment of efficacy. *Current Pharmaceutical Design*, 11,75-90

[49] Regulation (EC) no 1924/2006 of the European Parliament and of the Council of 20 December 2006 on nutrition and health claims made on foods. *Official Journal of European Union*, 404, 9-25
[50] Pastor, V and Perote, A. *La salud y la soja*. Madrid, Spain: Grupo Leche Pascual. EDIMSA Editores médicos S.A.; 2007
[51] Ridner, E. Soja. Propiedades nutricionales y su impacto en la salud. 1[a] ed. Buenos Aires. Argentina: Grupo QSA. Sociedad Argentina de Nutrición; 2006.
[52] Casale, MG. Composición de variedades argentinas. Base de datos de cultivos agroalimentarios de ILSI. In: Harris, S; Kovalskys, I; López Musi, JC; Rubinstein, C (Eds). *Soja y Nutrición. Serie de Informes Especiales. Volumen I.* ILSI Argentina. Buenos Aires, Argentina: Editorial Publitec SAECyM; 2004
[53] Luna, A (2007). Composición y procesamiento de la soja para consumo humano. *Investigación y Ciencia de la Universidad Autónoma de Aguascalientes*, 37, 35-44
[54] Herrera, E. Formación de las grasas: biosíntesis de ácidos grasos y triacilgliceridos. In: Herrera, E (Ed). Bioquímica. Aspectos estructurales y vías metabólicas. Vol 3. Nueva York, USA: McGraw-Hill; 1991
[55] Caygill, CPJ; Charlett, A; Hill, MJ (1996). Fat, fish, fish oil and cancer. *Brithis Journal of Cancer*, 74, 159–164
[56] Nair, SSD; Leitch, JW; Falconer, J; Garg, ML (1997). Prevention of cardiac arrhythmia by dietary (n23) polyunsaturated fatty acids and their mechanism of action. *Journal of Nutrition*, 127, 383–393
[57] Simopoulos, AP (1999). Essential fatty acids in health and chronic disease. *American Journal of Clinical Nutrition,* 70, 560S–569S
[58] Stone, NJ (1997). Fish consumption, fish oil, lipids, and coronary heart disease. *American Journal of Clinical Nutrition*, 65, 1083–1086
[59] Kumar, V; Rani, A; Solanki, A; Hussain, SM (2006). Influence of growing environment on the biochemical composition and physical characteristics of soybean seed. *Journal of Food Composition and Analysis,* 19, 188–195
[60] Anderson, JW. Soja: salud, nutrición y seguridad. In: Harris, S; Kovalskys, I; López Musi, JC; Rubinstein, C (Eds). *Soja y Nutrición. Serie de Informes Especiales. Volumen I.* ILSI Argentina. Buenos Aires. Argentina: Editorial Publitec SAECyM; 2004
[61] Sachan, DS; Hongu, N; Johnsen, M (2005). Decreasing Oxidative Stress with Choline and Carnitine in Women. *Journal of the American College of Nutrition*, 24, 172-176
[62] Wilson, TA; Meservey, CM; Nicolosi, RJ (1998). Soy lecithin reduces plasma lipoprotein colesterol and early atherogenesis in hypercholesterolemic monkeys and hamsters: beyond linoleate. *Atherosclerosis*, 140, 147-153
[63] Bellaloui, N.; Bruns, A. H.; Guillen, A.M.; Abbas, H.K.; Zablotowicz, R.M.; Mengistu, A.; Paris, R.L. (2010). Soybean seed protein, oil, fatty acids and mineral composition as influenced by soybean-corn rotation. *Agricultural Sciences*, 1, 3, 102-109
[64] FAO. Technology of production of edible flours and protein products from soybeans: http://www.fao.org/docrep/t0532E/t0532E00.ht
[65] ILSI (2010), Crop Composition Database Version 4.0, International Life Science Institute. Washington D.C., website: www.cropcomposition.org
[66] OECD. Revised consensus document on compositional considerations for new varieties of soybean (*Glycine max* (l.) Merr): key food and feed nutrients, antinutrients, toxicants

and allergens. *Series on the Safety of Novel Foods and Feeds No. 25.* Paris, France: ENV/JM/MONO; 2012

[67] Sandberg, A.S. (2002). Bioavailability of minerals in legumes. *British Journal of Nutrition*, 88, S281–S285

[68] Souci, SW; Fachmann, W; Kraut, H. Food Composition and Nutrition Tables, 7th revised and completed edition. MedPharm Scientific Publishers, Boca Raton, FL; 2008

[69] USDA-ARS: http://www.ars.usda.gov/Services/docs.htm?docid=8964

[70] Mathews, RH. *Legumes. Chemistry, Technology and Human Nutrition.* New York, USA: Marcel Dekker, Inc.; 1989

[71] Khokhar, S and Apenten, RKO. Anti-nutritional factors in food legumes and effects of processing. In: Squires, Victor R (ed) *The role of food, agriculture, forestry and fisheries in human nutrition.* Encyclopedia of Life Support Systems, EOLSS Publishers Ltd.; 2003

[72] Muzquiz, M; Varela, A; Burbano, C; Cuadrado, C; Guillamón, E; Pedrosa, M (2012). Bioactive compounds in legumes: pronutritive and antinutritive actions. Implications for nutrition and health. *Phytochemistry Review*, DOI 10.1007/s11101-012-9233-9

[73] Raboy, V; Young, KA; Dorsch, JA; Cook, A (2001). Genetics and breeding of seed phosphorus and phytic acid. *Journal of Plant Physiology*, 158, 489–497

[74] López, W; Leenhardt, F; Coudray, C; Remesy, C (2002). Minerals and phytic acid interactions: is it a real problem for human nutrition? *International Journal of Food Science and Technology*, 37, 727-739

[75] Mateos-Aparicio, I; Redondo-Cuenca, A.; Villanueva-Suárez, MJ.; Zapata-Revilla, MA; Tenorio-Sanz, MD (2010). Pea pod, broad bean pod and okara, potential sources of functional compounds. *LWT-Food Science and Technology*, 43, 1467-1470

[76] Link, A; Balaguer, F; Goel, A (2010). Cancer chemoprevention by dietary polyphenols: Promising role for epigenetics. *Biochemical Pharmacology*, 80, 1771–1792

[77] Manach, C; Scalbert, A; Morand, C; Rémésy, C; Jiménez, L (2004). Polyphenols: food sources and bioavailability. *American Journal of Clinical Nutrition*, 79, 727–747

[78] Bravo, L (1998) Polyphenols: chemistry, dietary sources, metabolism, and nutritional significance. *Nutrition Review*, 56, 11, 317–333

[79] Korkina, LG; De Luca, C; Kostyuk, VA; Pastore, S (2009). Plant polyphenols and tumors: from mechanisms to therapies, prevention, and protection against toxicity of anti-cancer treatments. *Current Medical Chemistry*, 16, 3943–3965

[80] Ramos, S (2008). Cancer chemoprevention and chemotherapy: dietary polyphenols and signaling pathways. *Molecular Nutrition and Food Research*, 52, 507–526

[81] Stagos, D; Amoutzias, GD; Matakos, A; Spyrou, A; Tsatsakis, AM; Kouretas, D (2012). Chemoprevention of liver cancer by plant polyphenols. *Food and Chemical Toxicology*, 50, 2155–2170

[82] Anderson, RL and Wolf, WJ (1995). Compositional changes in trypsin inhibitors, phytic acid, saponins and isoflavones related to soybean processing. *Journal of Nutr*ition, 125, 581S-588S

[83] Romani, A; Vignolini, P; Galardi, C; Aroldi, C; Vazzana, C; Heimler, D (2003). Polyphenolic content in different plant parts of soy cultivars grown under natural conditions. *Journal Agricultural and Food Chemistry,* 51, 5301-5306

[84] Boniglia, C; Carratu, B; Gargiulo, R; Giammarioli, S; Mosca, M; Sanzini, E (2009). Content of phytoestrogens in soy-based dietary supplements. *Food Chemistry*, 115, 1389–1392

[85] Velasquez, MT and Bhathena, SJ (2007). Role of dietary soy protein in obesity. *International Journal of Medical Science,* 4,72-82

[86] Murphy, P; Song, T; Buseman, G; Barua, K; Beecher, G; Trainer, D; Holden, J (1999) Isoflavones in retail and institutional soy foods. *Journal of Agricultural and Food Chemistry*, 47, 2697-2704

[87] Taylor, JI; Grace, PB; Bingham, SA (2005). Optimization of conditions for the enzymatic hydrolysis of phytoestrogen conjugates in urine and plasma. *Analytical Biochemistry*, 341, 220–229

[88] Mateos-Aparicio, I; Redondo Cuenca, A; Villanueva-Suárez, MJ; Zapata-Revilla, A (2008). Soybean, a promising health source. *Nutrición Hospitalaria*, 23, 305-312

[89] Sarkar, FH and Li, Y (2003). Soy isoflavones and cancer prevention. *Cancer Investigation,* 21, 744-757

[90] Espin, JC; Garcisa-Conesa, MT; Tomas-Barberan, FA (2007). Nutraceuticals: facts and fiction. *Phytochemistry*, 67, 2986–3008

[91] Howes, LG; Howes, JB; Knight, DC (2006). Isoflavone therapy for menopausal flushes: a systematic review and meta-analysis. *Maturitas*, 55, 203–211

[92] Ikeda, Y; Iki, M; Morita, A; Kajita, E; Kagamimori, S; Kagawa, Y; Yoneshima, H (2006). Intake of fermented soybeans, natto, is associated with reduced bone loss in postmenopausal women: Japanese population-based osteoporosis (JPOS) study. *Journal of Nutrition*, 136, 1323–1328

[93] Verheus, M; van Gils, CH; Keinan-Boker, L; Grace, PB; Bingham, SA; Peeters, PHM (2007). Plasma phytoestrogens and subsequent breast cancer risk. *Journal of Clinical Oncology*, 25, 648–655

[94] Otieno, DO; Ashton, JF; Shah, NP (2007). Isoflavone phytoestrogen degradation in Fermented soymilk with selected b-glucosidase producing L. Acidophilus strains during storage at different temperatures. *International Journal of Food Microbiology,* 115, 79–88

[95] Kao, TH; Wu, WM; Hung, CF; Wu, WB; Chen, BH (2007). Anti-inflammatory effects of isoflavone powder produced from soybean cake. *Journal of Agricultural and Food Chemistry,* 55, 11068-11079

[96] Messina, M; Kucuk, O; Lampe, JW (2006). An overview of the health effects of isoflavones with an emphasis on prostate cancer risk and prostate-specific antigen levels. *Journal of The AOAC International*, 89, 1121-1134

[97] Sarkar, FH and Li, Y (2003). Soy isoflavones and cancer prevention. *Cancer Investigation,* 21, 744-757

[98] Steiner, C; Arnould, S; Scalbert, A; Manach, C (2008). Isoflavones and the prevention of breast and prostate cancer: new perspectives opened by nutrigenomics. *British Journal of Nutrition,* 99, ES78-108

[99] Zhang, EJ; Ng, KM; Luo, KQ (2007). Extraction and purification of isoflavones from soybeans and characterization of their estrogenic activities. *Journal of Agriculture and Food Chemistry,* 55, 6940-6950

[100] Champ, M.M.J. (2002). Non-nutrient bioactive substances of pulses. British *Journal of Nutrition*, 88, S307-S319

[101] Morales, ML and Troncoso, AM. Sustancias antinutritivas presentes en alimentos. In Camean, AM and Repetto (Eds): *Toxicología alimentaria*. Madrid, Spain: Díaz de Santos; 2006

[102] Balisteiro, DM; Rombaldi, CV; Genovese, MI (2013). Protein isoflavones, trysin inhibitory and in Vitro antioxidante capacities: Comparison among conventionally and organically Brown soybeans. *Food Research International*, 51, 8-14

[103] Dilis, V and Trichopoulou, A (2009). Nutritional and health properties of pulses. *Mediterranean Journal of Nutrition and Metabolism*, 1,149–157

[104] Vasconcelos, IM; Tadeu, J; Oliveira, A (2004). Antinutritional properties of plant lectins. *Toxicon*, 44, 385-403

[105] Douglas, MW; Parsons, CM; Hymowitz, T (1999). Nutritional evaluation of lectin-free soybeans for poultry. *Poultry Science*, 78, 91–95

[106] Liener, I (1996). Effects of processing on antinutritional factors in legumes: the soybean case. *Archivos Latinoamericanos de Nutrición*, 44, 48S-54S

[107] Yating M and Wang J (2010). Deactivation of Soybean Agglutinin by Enzymatic and Other Physical Treatments. *Journal of Agricultural and Food Chemistry*, 58, 11413–11419

[108] Gonzalez de Mejia, E and Prisecaru, V (2005). Lectins as bioactive plant proteins: a potential in cancer treatment. *Critical Review in Food Science and Nutrition*, 45, 445–455

[109] Khattab, RY and Arntfield, SD (2009). Nutritional quality of legume seeds as affected by some physical treatments. 2. Antinutritional factors. *LWT-Food science and Technology*, 42, 1113-1118

[110] Sandberg, AS (2002). Bioavailability of minerals in legumes. *British Journal of Nutrition*, 88, S281–S285

[111] Kumar, V; Sinha, AK; Makkar, HPS; Becker, K (2010). Dietary roles of phytate and phytase in human nutrition: A review. *Food Chemistry*, 120, 945-959

[112] Ignarro, LJ; Balestrieri, ML; Napoli, C (2007). Nutrition, physical activity, and cardiovascular disease: an update. *Cardiovascular research*, 73, 326–40

[113] McGill, HC; McMahan, CA; Gidding, SS (2008). Preventing heart disease in the 21st century: implications of the Pathobiological Determinants of Atherosclerosis in Youth (PDAY) study. *Circulation,* 117, 1216–27

[114] Ueshima, H; Sekikawa, A; Miura, K; Chowdhury T; Takashima, N; Kita, Y; Watanabe, M; Kadota, A; Okuda, N; Kadowaki, T; Nakamura, Y; Okamura, T (2008). Cardiovascular Disease and Risk Factors in Asia. A Selected Review. *Circulation*, 118, 2702-2709

[115] Potter, SM (1995). Overview of proposed mechanism for the hypocholesterolemic effect of soybean. *Journal of Nutrition*, 125, 606-611.

[116] Forsythe, WA (1995). Soy protein, thyroid regulation and cholesterol metabolism. *Journal of Nutrition*, 125, 619-623

[117] Nagarajan, S (2010). Mechanisms of anti-atherosclerotic functions of soy-based diets. *The Journal of Nutritional Biochemistry*, 21, 255-260

[118] Brouns, F (2002). Soya isoflavones: a new and promising ingredient for the health foods sector. *Food Research International*, 35, 187-193

[119] Lichtenstein, AH (1998). Soybean protein, isoflavones and cardiovascular disease risk. *Journal of Nutrition*, 128. 1589-1592

[120] Wagner, JD; Schwenke, DC; Greaves, KA; Zhang, L; Anthony, MS; Blair, RM; Shaodan, MK; Williams, JK (2003). Soy protein with isoflavones, but not an isoflavone-rich supplement, improves arterial low-density lipoprotein metabolism and atherogenesis. *Arteriosclerosis, Thrombosis and Vascular Biology*, 2241-2246

[121] Flight, I and Clifton, P (2006). Cereal grains and legumes in the prevention of coronary heart disease and stroke: a review of the literature. *European Journal of Clinical Nutrition*, 60,1145–59

[122] Brown, L; Rosner B, Willett WW, Sacks FM (1999). Cholesterol-lowering effects of dietary fiber: a meta-analysis. *American Journal of Clinical Nutrition*, 69, 30–42

[123] Anderson, JW and Major, AW (2002). Pulses and lipaemia, short- and long-term effect: potential in the prevention of cardiovascular disease. *British Journal of Nutrition*, 88, 263-271

[124] Whelton, SP; Hyre, AD; Pedersen, B; Yi, Y; Whelton, PK; He, J (2005). Effect of dietary fiber intake on blood pressure: a meta-analysis of randomized, controlled clinical trials. *Journal of Hypertension*, 23, 475–81

[125] Howarth, NC; Saltzman, E; Roberts, SB (2001). Dietary fiber and weight regulation. *Nutrition Review*, 59, 129–39

[126] Anderson, JW; Randles, KM; Kendall, CW; Jenkins, DJ (2004). Carbohydrate and fiber recommendations for individuals with diabetes: a quantitative assessment and meta-analysis of the evidence. *Journal of the American College of Nutrition*, 23, 5–17

[127] Roberfroid, MB (2001). Prebiotics: preferential substrates for specific germs? *American Journal of Clinical Nutrition,* 73, 406S–9S

[128] Delzenne, NM and Kok, N (2001). Effects of fructans-type prebiotics on lipid metabolism. *American Journal of Clinical Nutrition*, 73, 456-458

[129] WHO: http://www.who.int/mediacentre/factsheets/fs311/es/

[130] López-Jiménez, F and Cortés-Bergoderi, M (2011). Obesidad y Corazón. *Revista Española de Cardiología*, 64, 140-9

[131] Swinburn, BA; Sacks, G; Hall, KD; McPherson, K; Finegood, DT; Moodie, ML; Gortmaker, SL (2011). The global obesity pandemic: shaped by global drivers and local environments. *Lancet*, 378, 804-814

[132] Barness, LA; Opitz, JM; Gilbert-Barness, E (2007). Obesity: genetic, molecular, and environmental aspects. *American Journal of Medicine and Genetic*, 143A (24), 3016–34

[133] Cederroth, CR and Nef, S (2009). Soy, phytoestrogens and metabolism: A review. *Molecular and Cellular Endocrinology*, 304, 30-42

[134] Merritt JC (2004). Metabolic syndrome: soybean foods and serum lipids. *Journal of National Medical Association*, 96, 1032–1041

[135] Kim, H-K; Kim, J-N; Hanb, S-N; Nam, J-H; Nac, H-N; Had, TJ (2012). Black soybean anthocyanins inhibit adipocyte differentiation in 3T3-L1 cells. *Nutrition Research*, 32, 770-777

[136] Bhathena, SJ and Velasquez, MT (2002). Beneficial role of dietary phytoestrogens in obesity and diabetes. *American Journal of Clinical Nutrition,* 76, 1191–1201

[137] Sites, CK; Cooper, BC; Toth, MJ; Gastaldelli, A; Arabshahi, A; Barnes, S (2007). Effect of daily supplement of soy protein on body composition and insulin secretion in postmenopausal women. *Fertility and Sterility*, 88, 1609-1617.

[138] Vedavanam, K; Srijayanta, S; O'Reilly, J; Raman, A; Wiserman, H (1999) Antioxidant action and potential antidiabetic properties of an isoflavonoid-containing soyabean phytochemical extract (SPE). *Phytotherapy Research*, 13, 601–608

[139] González, C; Alonso, A; Grueso, NA; Díaz, F; Esteban, MM; Fernández S, Patterson, AM (2001). Effect of treatment with different doses of 17ß-estradiol on insulin receptor substrate-1. *Journal of Pancreas*, 2, 140-149

[140] Zethelius, B; Hales, CN; Lithell, HO; Berne, C (2004). Insulin resistance, impaired early insulin response, and insulin propeptides as predictors of the development of type 2 diabetes: a population-based, 7-year follow-up study in 70-year-old men. *Diabetes Care*, 27, 1433-1438

[141] Ferrannini, E; Camastra, S; Gastaldelli, A; Sironi, AM; Natali, A; Muscelli E; Mingrone, G; Mari, A (2004). Beta-cell function in obesity: effects of weight loss. *Diabetes,* 53, S26-33

[142] Nadal, J; Rovira, M; Laribi, O; Leon-Quinto, T; Andreu, E; Ripoll, C; Soria, B (1998) Rapid insulinotropic effect of 17ß-estradiol via a plasma membrane receptor. *The FASEB Journal*, 12:1341-8

[143] Ademiluyi, AO and Oboh, G (2013). Soybean phenolic-rich extracts inhibit key-enzymes linked to type 2 diabetes (α-amylase and α-glucosidase) and hypertension (angiotensin I converting enzyme) *in vitro. Experimental and Toxicologic Pathology*, 65, 305-309

[144] Paxton, JD (2000). Biosynthesis and accumulation of legume phytoalexins. In: Sharma, RP; Salunkhe, DK (Eds). *Mycotoxins and phytoalexins*. Boca Raton: CRC Press

[145] Payton-Stewart, F; Schoene, NW; Kim, YS; Burow, ME; Cleveland, TE; Boue, SM; Wang, TTY (2009). Molecular effects of soy phytoalexin glyceollins in human prostate cancer cells LNCaP. *Molecular Carcinogenesis*, 48, 862–71

[146] Park, S; Kim, DS; Kim, JH; Kim, JS; Kim, HJ (2012). Glyceollin-containing fermented soybeans improve glucose homeostasis in diabetic mice. *Nutrition*, 28, 204-211

[147] Gudbrandsen, OA; Wergedahl, H; Berge, RK (2009). A casein diet added isoflavone enriched soy protein favorably affects biomarkers of steatohepatitis in obese Zucker rats. *Nutrition*, 25, 574–80

[148] Kwon, DY; Daily III, JW; Kim, HJ; Park, S (2010). Antidiabetic effects of fermented soybean products on type 2 diabetes. *Nutrition Research,* 30, 1–13

[149] Choi, HK; Yoon, JH; Kim, YS; Kwon, DY (2007). Metabolomic profiling of cheonggukjang during fermentation by 1H NMR spectrometry and principal components analysis. *Process Biochemistry,* 42, 263-266

[150] Murphy, P; Song, T; Buseman, G; Barua, K; Beecher, G; Trainer, D; Holden, J (1999). Isoflavones in retail and institutional soy foods. *Journal of Agricultural and Food Chemistry*, 47, 2697-2704

[151] Magee, PJ and Rowland, IR (2004). Phyto-oestrogens, their mechanism of action: current evidence for a role in breast and prostate cancer. *British Journal of Nutrition*, 91, 513–531

[152] Hsing, AW; Tsao, L; Devesa, SS (2000). International trends and patterns of prostate cancer incidence and mortality. *International Journal of Cancer,* 85, 60–67

[153] Sim, HG and Cheng, CWS (2005). Changing demography of prostate cancer in Asia. *European Journal of Cancer*, 41, 834–845

[154] Yan, L and Spitznagel, EL (2005). Meta-analysis of soy food and risk of prostate cancer in men. *International journal of Cancer*, 117, 667–669

[155] Trock, BJ; Hilakivi-Clarke, L; Clarke, R (2006) Meta-analysis of soy intake and breast cancer risk. *Journal of The National Cancer Institute*, 98, 459–471

[156] Wu, AH; Wan, P; Hankin, J; Tseng, CC; Yu, MC; Pike, MC (2002). Adolescent and adult soy intake and risk of breast cancer in Asian-Americans. *Carcinogenesis,* 23, 1491–1496

[157] Sarkar, FH; Li, YW (2004). Cell signaling pathways altered by natural chemopreventive agents. *Mutation Research*, 555, 53–64

[158] Pettersson, K and Gustafsson JA (2001). Role of estrogen receptor beta in estrogen action. *Annual Review of Physiology*, 63, 165–192

[159] Lindberg, MK; Moverare, S; Skrtic, S; Gao, H; Dahlman-Wright, K; Gustafsson, JA; Ohlsson, C (2003). Estrogen receptor (ER) β reduces ER α-regulated gene transcription, supporting a 'Ying Yang' relationship between ERα and ERβ in mice. *Molecular Endocrinology*, 17, 203–208

[160] Leygue, E; Dotzlaw, H; Watson, PH; Murphy, LC (1998). Altered estrogen receptor α and β messenger RNA expression during human breast tumorigenesis. *Cancer Research*, 58, 3197–3201

[161] Power, KA and Thompson, LU (2003). Ligand-induced regulation of ERα and ERβ is indicative of human breast cancer cell proliferation. *Breast Cancer Research Treatment*, 81, 209–221

[162] Cappelletti, V; Miodini, P; Di Fronzo, G; Daidone, MG (2006). Modulation of estrogen receptor-beta isoforms by phytoestrogens in breast cancer cells. *International Journal of Oncology*, 28, 1185–1191

[163] Qin, W; Zhu, W; Shi, H; Hewett, JE; Ruhlen, RL; MacDonald, RS; Rottinghaus, GE; Chen, YC; Sauter, ER (2009). Soy isoflavones have an antiestrogenic effect and alter mammary promoter hypermethylation in healthy premenopausal women. *Nutrition and Cancer,* 61, 238–244

[164] Majid, S; Dar, AA; Shahryari, V; Hirata, H; Ahmad, A; Saini, S; Tanaka, Y; Dahiya, V; Dahiya, R (2010). Genistein reverses hypermethylation and induces active histone modifications in tumor suppressor gene B-Cell translocation gene 3 in prostate cancer. *Cancer*, 116, 66-76

[165] Li, Y; Vandenboom, TG; Kong, D; Wang, Z; Ali, S; Philip, PA; Sakar, FH (2009). Up-regulation of miR-200 and let-7 by Natural Agents Leads to the Reversal of Epithelial-toMesenchymal Transition in Gemcitabine-Resistant Pancreatic Cancer Cells. *Cancer Research*, 69, 6704- 6712

[166] Steiner, C; Arnould, S; Scalbert, A; Manach, C (2008). Isoflavones and the prevention of breast and prostate cancer: new perspectives opened by nutrigenomics. *British Journal of Nutrition*, 99, ES78–ES108

[167] Albertazzi, P; Pansini, F; Bonaccorsi, G, Zanotti, L; Forini, E; De Aloysio, D. (1998). The effect of dietary soy supplementation on hot flushes. *Obstetrics & Gynecology*, 1, 6-10

[168] Chiechi, LM; Putignano, G; Guerra, V; Schiavelli, MP; Cisternino, AM; Carriero, C. (2003). The effect of a soy rich diet on the vaginal epithelium in postmenopause: a randomized double blind trial. *Maturitas*, 45, 241-246

[169] Albert, A; Altabre, C; Baró, F; Buendía, E; Cabero, A; Cancelo, MJ.; Castelo-Branco, C; Chantre, P; Duran, M; Haya, J; Imbert, P; Juliá, D; Lanchares, JL; Llaneza; Manubens, PM; Miñano, A; Quereda, F; Ribes, C; Vázquez, F. (2002). Efficacy and safety of a phytoestrogen preparation derived from Glycine max (L.) Merr in climacteric symptomatology: A multicentric, open, prospective and non-randomized trial. *Phytomedicine*, 9, 85-92

[170] Levis, S; Strickman-Stein, N; Ganjei-Azar, P; Xu, P; Doerge,DR; Jeffrey Krischer, J. (2011). Soy Isoflavones in the Prevention of Menopausal Bone Loss and Menopausal Symptoms a Randomized, Double-blind Trial. *Archives of Internal Medicine*, 171, 1363-1369

[171] Setchell, KDR and Cassidy, A (1999). Dietary isoflavones: Biological Effects and Relevance to Human Health. *Journal of Nutrition*, 129: 758S–767S

[172] Adlercreutz, H; Hamalainen, E; Gorbach, S; Goldin, B (1992). Dietary phyto-oestrogens and the menopause in Japan. *Lancet*, 339, 1233

[173] Huntley, AL and Ernst, E (2004). Soy for the treatment of perimenopausal symptoms—a systematic review. *Maturitas*, 47, 1–9

[174] Setchell KD; Brown NM; Lydeking-Olsen E (2002). The clinical importance of the metabolite equol: a clue to the effectiveness of soy and its isoflavones. *Journal of Nutrition*, 132, 3577–84

[175] Rowland IR; Wiseman H; Sanders TA; Adlercreutz H; Bowey EA (2000). Interindividual variation in metabolism of soy isoflavones and lignans: influence of habitual diet on equol production by the gut microflora. *Nutrition and Cancer*, 6, 27–32

[176] Song KB; Atkinson C; Frankenfeld CL; Jokela T; Wähälä K; Thomas WK; Lampe, JW (2006). Prevalence of Daidzein-Metabolizing Phenotypes Differs between Caucasian and Korean American Women and Girls. *Journal of Nutrition*, 136, 1347–51

[177] Freedman, RR (1998). Biochemical, metabolic, and vascular mechanisms in menopausal hot flashes. *Fertility and Sterility*, 70, 332–337

[178] Kronenberg, F; Cote, LJ; Linkie, DM; Dyrenfurth, I; Downey, JA (1984). Menopausal hot flashes: thermoregulatory, cardiovascular, and circulating catecholamine and LH changes. *Maturitas*, 6, 31–43

[179] Miquel, J; Ramírez-Bosca, A; Ramírez-Bosca, JV; Diaz Alperi, J (2006). Menopause: A review on the role of oxygen stress and favorable effects of dietary antioxidants. *Archives of Gerontology and Geriatrics*, 42, 289–306

[180] Nelson, HD; Vesco, KK; Haney, E; Fu, R; Nedrow, A; Miller, J; Nicolaidis, C; Walker, M; Humphrey, L (2006). Nonhormonal therapies for menopausal hot flashes: systematic review and meta-analysis. *Journal of American Medical Association*, 295, 2057-71

[181] Boulet, MJ; Oddens, BJ; Lehert, P; Velmer, HM; Visser, A (1994). Climacteric and menopause in seven south-east Asian countries. *Maturitas*, 19, 157–76

[182] Brezinski, A; Adlercreutz, H; Shaoul, R (1997). Short term effects of phytoestrogen-rich diet on postmenopausal women. *Menopause*, 4, 89–94

[183] Murkies AL; Wilcox G; Davis SR (1998). Clinical review 92: phytoestrogens. *Journal of Clinical Endocrinology and Metabolism*, 83, 297–303

[184] Kurzer, MS (2008). Soy consumption for reduction of menopausal symptoms. *Inflammopharmacology*, 16, 227–229

[185] Shen, W and Stearns, V (2009). Treatment strategies for hot flushes. *Expert Opinion on Pharmacotheraphy*. 10, 1133–1144

[186] Williamson-Hughes PS; Flickinger BD; Messina MJ; Empie MW (2006). Isoflavone supplements containing predominantly genistein reduce hot flash symptoms: a critical review of published studies. *Menopause*, 13, 831–839

[187] Jou, HJ; Wu, SC; Chang, FW; Ling, PY; Chu, KS; Wu, WH. (2008). Effect of intestinal production of equol on menopausal symptoms in women treated with soy isoflavones. *International Journal of Ginecology and Obstetrics*, 102, 44-49.

[188] Eden, J.A. (2012). Phytoestrogens for menopausal symptoms: A review. *Maturitas*, 72, 157-159

[189] Wilcox G; Wahlqvist ML; Burger HG; Medley G (1990). Oestrogenic effects of plant foods in postmenopausal women. *British Medical Journal*, 301, 905-906

[190] Baird, DD; Umbach, DM; Lansdell, L; Hughes, CL; Setchell, KD; Weinberg, CR; Haney, AF; Wilcox, AJ; Mclachlan, JA (1995). Dietary intervention study to assess the estrogenicity of dietary soy among postmenopausal women. *The Journal of Clinical Endocrinology and Metabolism*, 80, 1685-1690

[191] Brouns, F and Vermeer, C (2000). Functional foods for reducing the risks of osteoporosis. *Trends in Food Science and Technology*, 11, 22-33

[192] Duncan, AM; Phipps, WR; Kurzer, MS (2003). Phyto-oestrogens. *Best Practice and Research Clinical Endocrinology and Metabolism*, 17, 253-271

[193] AESAN (2007). Informe del Comité Científico de la Agencia Española de Seguridad Alimentaria y Nutrición (AESAN) en relación con las consecuencias asociadas al consumo de isoflavonas. *Revista Comité Científico*, 5, 77-94

[194] Vidal, V; Kindblom, L; Ohlsson, C (1999). Expression and localization of estrogen receptor b in murine and humane bone. *Journal of Bone and Mineral Research*, 14, 923-929

[195] Chiechi, LM; Secreto, G; D'Amore, M; Fanelli, M; Venturelli, E; Cantatore, F; Valerio, T; Laselva, G; Loizzi, P (2002). Efficacy of a soy rich diet in preventing postmenopausal osteoporosis: the Menfis randomized trial. *Maturitas*, 42, 295-300

[196] Crisafulli, A; Altavilla, D; Squadrito, G (2004). Effects of the phytoestrogen genistein on the circulating soluble receptor activator of nuclear factor kappa ligand-oestoprotegerin system in early postmenopausal women. *The Journal of Clinical Endocrinology and. Metabolism*, 89, 188-192

[197] Messina, M (2000). Soyfoods and soybean phyto-oestrogens (isoflavones) as posible alternatives to hormone replacement therapy (HRT). *European Journal of Cancer*, 36, S71-S77

[198] Arjmandi, BH; Alekel, L; Hollis, BW; Amin, D; Stacewicz-Sapuntzakis, M; Guo, P; Kukreja, SC (1996). Dietary soybean protein prevents bone loss in an ovariectomized rat model of osteoporosis. *Journal of Nutrition*, 126, 161-7

[199] Fanti, O; Faugere, MC; Gang, Z; Schmidt, J; Cohen, D; Malluche, HH (1998). Systemic administration of genistein partially prevents bone loss in ovariectomized rats in a nonestrogen-like mechanism. *American Journal of Clinical Nutrition*, 68,1517S

[200] Jayo, MJ; Anthony, MS; Register, TC; Rankin, SE; Vest, T; Clarkson, TB (1997). Dietary soy isoflavones and bone loss in ovariectomized monkeys. *FASEB Journal*, 11, S228

[201] Potter, SN; Baum, JA; Teng, H; Stillman, RJ; Shay, NF; Erdman, JW (1998). Soy protein and isoflavones: their effects on blood lipids and bone density in postmenopausal women. *American Journal of Clinical Nutrition*, 68, 1375S–9S

[202] Horiuchi, T ; Onouchi, T ; Takahashi, M ; Ito, H ; Orimo, H (2000). Effect of soy protein on bone metabolism in postmenopausal Japanese women. *Osteoporosis International,* 11, 721–724

[203] Kritz-Silverstein D and Goodman-Gruen DL (2002). Usual dietary isoflavone intake, bone mineral density, and bone metabolism in post-menopausal women. *Journal of Women's Health and Gender-Based Medicine*, 11, 69–78

[204] Tsuchida, K; Mizushima, S; Toba, M; Soda, K. (1999). Dietary soybeans intake and bone mineral density among 995 middle-aged women in Yokohama. *Journal of Epidemiology*, 9, 14–19

[205] Kardinaal, AF; Morton, MS; Bruggemann-Rotgans, IE; van Beresteijn, EC (1998). Phyto-oestrogen excretion and rate of bone loss in post-menopausal women. *European Journal of Clinical Nutrition*, 52, 850–855

[206] Nagata, C; Shimizu, H; Takami, R; Hayashi, M; Takeda, N; Yasuda, K (2002). Soy product intake and serum isoflavonoid and estradiol concentrations in relation to bone mineral density in post-menopausal Japanese women. *Osteoporosis International*, 13, 200–204

[207] Cassidy, A; Albertazzi, P; Lise Nielsen, I; Hall, W; Williamson, G; Tetens, I; Atkins, S; Cross, H; Manios, Y; Wolk, A; Steiner, C; Branca, F. (2006). Critical review of health effects of soyabean phytoestrogens in post-menopausal women. *Proceedings of the Nutrition Society*, 65, 1, 76-92

[208] Song, YJ; Paik YH; Joung, H (2008). Soybean and soy isoflavone intake indicate a positive change in bone mineral density for 2 years in young Korean women. *Nutrition Research*, 28, 25–30

[209] Taku, K; Melby, MK; Nishi, N; Omori, T; Kurzer, MS (2011). Soy isoflavones for osteoporosis: An evidence-based approach. *Maturitas* 70, 333–338

[210] Dalais, FS; Ebeling, PR; Kotsopoulos, D; McGrath, BP; Teede, HJ (2003). The effects of soy protein containing isoflavones on lipids and indices of bone resorption in Postmenopausal women. *Clinical Endocrinology*, 58, 704-709

[211] Gallagher, JC; Satpathy, R; Rafferty, K; Haynatzka, V (2004). The effect of soy protein isolate on bone metabolism. *Menopause*, 11, 290-298.

[212] Liu, J; Ho, SC; Yi-xiang Su, Y; Wei-qing Chen, W; Zhang, C; Chen, Y (2009). Effect of long-term intervention of soy isoflavones on bone mineral density in women: A meta-analysis of randomized controlled trials. *Bone*, 44, 948-953.

In: Legumes
Editors: Hiroto Satou and Ren Nakamura
ISBN: 978-1-62808-280-7

Chapter 5

FERMENTATION OF LESSER KNOWN LEGUMES

E. O. Farinde* and S. B. Fasoyiro
Institute of Agricultural Research and Training, P.M.B. 5029, Ibadan, Nigeria

ABSTRACT

Legumes belong to the family *Leguminosae*. Legumes are major sources of plant protein and oil. Some of the major and well known legumes are cowpea, peas, clover, lentils, lupins, mesquite, carob, peanut, locust bean, groundnut, wisteria, and kenturkey coffee tree. Some minor and lesser known legumes are pigeon pea, jack bean, sword bean, gram pea, moth bean, mungo bean, winged bean, velvet bean, kidney bean, lima bean, African yam bean, black gram and bambara groundnut. The lesser legumes have been reported to be less accepted either due to their long cooking time or their anti-nutrient content. This had been improved through the use of fermentation technologies, hence the theme fermentation of lesser legumes. This chapter focused on the importance of legumes, some examples of lesser legumes, nutrient composition and anti nutrient content of some of the lesser legumes , methods of fermentation of some lesser legumes with emphasis on lima bean, pigeon pea and African yam bean. Other areas discussed are the effects of fermentation on the anti nutrient content of these lesser legumes and the micro organisms involved in their fermentation process. Cooking is a unique and compulsory precursor of fermentation to processing these lesser legumes. Fermentation improved the nutrient composition and reduced the anti-nutrient contents of the lesser legumes reviewed. Protein, fat, and ash content of lima bean were increased from 16.81%, 1.30% and 2.41% in the raw lima bean to 17.52%, 2.17% and 2.68% respectively at 72 h of fermentation. Minerals (Mg, P and K) also increased with fermentation time in lima bean. Mg increased from 8,100 ppm in the raw lima bean to 8200 ppm at 72 h of fermentation , P and K also increased from 8,400 ppm, 2,100 ppm in the raw lima bean to 12,800 ppm, 6,900 ppm respectively at 72 h of fermentation. . Tannin, phytate, cyanide and trypsin inhibitors were reduced from 11.0 mg/kg, 17.0 mg/kg, 1.87 mg/kg and 6.32 mg/kg in the raw lima bean to 6.01 mg/kg, 5.52 mg/kg, 0.80 mg/kg and 1.0 mg/kg respectively at 72 h of fermentation. In fermented pigeon pea seeds, phosphorus content was increased to 0.51 mg/100g in comparison to 0.44 mg/100g in the raw seeds. There was no significant difference in the sodium, potassium, calcium and

* Corresponding author E-mail: osekinat@yahoo.co.uk.

magnesium contents of the raw pigeon pea seeds and the fermented seeds. Tannin and phytate in raw seeds were significantly reduced from 4.23 mg/100g and 2.20 mg/100g to 1.19 mg/100g and 1.65 mg/kg by fermentation. Fermentation reduced the flatulence potential of African yam bean, α-galactoside was reduced to the range of 22 – 39 %. The microorganisms involved in the fermentation process were found to be mainly *Bacillus species* and were common to almost all the lesser legumes reviewed. They include *Bacillus subtilis, B. plumilus, B. licheniformis,* and *B. megatarium. Bacillus species* are able to break down protein, carbohydrate and fat, which are the major constituents of these legumes. Fermented lesser legumes have their nutritional quality and acceptability improved over the starting material.

1. Introduction

Legume is a plant belonging to the family *Fabeceae* or *Leguminosae*. Legume is synonymous with pod and fruit. The seeds are inside the pod. Developed countries produce grain legumes yielding an average of between 500 to 2000 kg/ha. Developing countries have their production yield much lower but there is improvement in the growth potential of legumes production in the developing countries. Grain legumes are cultivated mainly for their seeds. They are called Pulses. They comprise of different types of beans. Their seeds are usually consumed in form of matured seeds or immature pods. The matured seeds are stored as dried seeds prior to processing for consumption. The immature pods are harvested and processed for consumption as green vegetables when the pods are still fresh and the fibre still soft. Oil can also be extracted from the dried matured seeds.

2. Importance of Legumes

Legumes fix atmospheric nitrogen through nitrogen cycle with the action of nitrogen fixing bacteria (*Rhyzobium sp*) and cyanobacteria in the soil. The symbiotic relationship between the legume and the nitrogen fixing bacteria result in conversion of atmospheric nitrogen to ammonium and nitrates in the soil and these nitrogenous compounds are used in forming proteins which the bacteria share with the host plant (the legume).

Legumes are very important in human diet because of their high protein content compared to other food groups such as cereals and tubers. They serve as main sources of both protein and calories in many of the tropical and sub-tropical areas. Dry grain legumes and products developed from legumes have been established as the richest sources of food protein derived from plants [1]. Grain legumes contain essential amino acids which include lysine, leucine, isoleucine, tryptophan, methionine, arginine, aspartic acid, thronine, cystine, valine, phenylalanine and hystidine. Legumes are very rich in lysine but poor in methionine. Because cereals are very rich in methionine, legumes serve to compensate lysine that is lacking in cereals.

Some legumes such as *leucaena* and *sesbania* species are grown and used for fallow to improve the fertility of the soil.

Forage legumes are grown in pasture and used as feeds for livestock. They include Alfallfa, Clover, Vetch, Albizia.

3. Major and Minor Legumes

Legumes are referred to as major or minor depending on their popularity, their discovery and the extent of their utilization.

Some of the major and well known legumes are cowpea, peas, clover, lentils, lupins, mesquite, carob, peanut, locust bean, groundnut, wisteria, and kenturkey coffee tree. These major legumes are popular and highly utilized.

Lesser legumes or minor legumes are legumes of minor current economic significance which are not yet very popularly utilized. Some of them have their seeds, pods and tubers edible, while some of them have their pods and seeds useful for forage. Some of them have their immature pods useful for green vegetables and their leaves for salad or spinach preparation [2]. These minor legumes are well spread over the sub tropics areas. Their production/yield potential is now increasing due to the awareness of their economic and nutritional benefit. Most of them are less accepted for consumption because of their long processing procedure and their anti-nutritional factor. A lot of research has been carried out to remove these barriers from these important lesser known legumes. Processing such as soaking, roasting, germination, cooking, fermentation, were found to reduce or totally remove the anti-nutritional factors in these minor legumes thus bringing out the quality in them.

4. Some Important Lesser Legumes

Some important minor legumes which although under- utilized, play vital role in supplying protein in diets of people in most developing countries particularly Africa, include pigeon pea, jack bean, sword bean, gram pea, moth bean, mungo bean, winged bean, velvet bean, kidney bean, lima bean, African yam bean, black gram and bambara groundnut. The botanical name and the common names of some lesser legumes are shown in Table 1.

Table 1. The botanical name and the common names of some lesser legumes

Botanical name	Common names
Cajanus cajan	Pigeon pea, red gram
Canavalia ensiformis	Jack bean/horse bean
Canavalia gladiata	Sword bean
Cicerarietinum	Bengal gram, chicken pea
Dolichus biflorus	Horse bean
Dolichos lablab	Field bean
Hylablab purpureus	Hyacinth bean
Mucuna pruriens	Velvet bean
Phaseolus aconitifolius	Moth bean
Phaseolus aureus	Mung bean, green gram, golden gram
Phaseolus mungo	Mungo bean, black gram
Phaseolus vulgaris	Kidney bean, common bean, pink bean
Phaseolus lunatus	Limabean, Madagascar bean, butter bean
Psophocarpus tetragonolobus	Winged bean
Sphenostylis stenocarpa	African yam bean
Vigna Mungo	Black gram
Vigna subtereranea	Bambara groundnut, Madagascar groundnut

Bambara Groundnut

Bambara Groundnut (*Vigna, subteraneae)* is regarded as the third most important crop after groundnuts and cowpea in Africa. Other names include jugo beans (South Africa), Kwaruru (Hausa, Nigeria), Okpa (Ibo, Nigeria), Epa-Roro (Yoruba, Nigeria) and Nyimo beans (Zimbabwe). It originated in Central Mali [3]. Bambara groundnut can be consumed as matured dry seeds or as immature pods. It is a nutritious food, containing over 22% protein [4]. Due to their protein content, they are used occasionally as substitutes for meat or fish [5]. It is also high in carbohydrate, about 63% but low in fat. Bambara groundnut contains all the essential amino acids such as lysine, threonine, tryptophan, leucine, isoleucine, phenylalanine, arginine, hystidine, methinine, and valine. It has agronomic advantages of been drought tolerant and soil fertility improver. The seeds are of various sizes and colors.

Figure 1. Bambara groundnut.

Pigeon Pea

Pigeon pea (*Cajanus cajan*) is an important grain legume crop of the agricultural zone in the semi-arid and tropics. It is cultivated in more than twenty five sub tropics and tropical countries either as a sole crop or intermixed with cereals and legumes. It is a perennial crop that grows up to five years. It is drought and high temperature tolerant than most other crops [6]. It can serve both as food crop or as forage or cover crop. As a food crop it can be used as dried pea, flour or as green vegetable.

Pigeon pea is an affordable and available under-utilized legume which grows well in Nigeria and can be used for food security especially when used to address malnutrition. Pigeon pea is a source of plant protein that can contribute up to 20-25% protein content into human diet. Pigeon peas are nutritionally important as they contain they contain amino acids profile similar to that of soy bean [7], they contain lysine and tryptophan which could be utilized in complementing cereals and other staples. In combination with cereal, pigeon pea makes a well balanced food. It is also a good source of minerals such as calcium and magnesium.

Figure 2. Pigeon pea.

Kidney Bean

Kidney bean *(Phasolus vulgaris)* also known as common bean, pink bean, salad bean, snap bean, dry bean, navy bean are consumed either as immature vegetables or as mature dry beans. The beans are usually kidney-shaped, up to 1.5 cm long. They are of different colors and are often mottled in two or more colors. Kidney beans are rich in soluble fibre. They supply high quality protein which provides a healthy alternative to meat or other animal protein [8].

Figure 3. Kidney bean (brown).

Lima Bean

Lima beans are of two types, the large white seeded lima bean (*Phaseolus limensis L*) and the sieva or smaller lima bean (*Phaseolus lunatus L*). The larger white seeded type originated in Peru while the small seeded lima bean originated in Mexico. USA is the world leading producer of lima bean followed by Madagascar and Peru. In Nigeria, lima bean is cultivated and well adapted to humid rain forest of southern Nigeria. It is usually intercropped with cassava, yam and pepper. It is mainly produced for household consumption as only about 35% of the grain is sold [9]. Lima bean is one of the lesser legumes that are not well utilized in Nigeria [9]. It's hard to cook phenomenon makes it less acceptable for consumption as other beans or legumes such as cowpea. In Africa, about 120,000 to 200,000 hectare is devoted to lima bean cultivation in the sub-humid and humid areas [10]. Currently, the production of lima bean in Africa is only sufficient for local consumption, more intensive management practice will be needed to improve the production of lima bean in order to change the yield in Africa [11]. In Western Nigeria, lima bean yield of 3000 kg/ha have been reported under experimental condition showing that there is good yield potential of lima bean in the humid tropics [12]. An intensive lima bean improvement programme has been carried out at IITA in Nigeria. The experimental yield exceeding 5000kg dry seeds ha^{-1} was obtained from the vining types under optimal condition [13]. Lima bean is nutritious. It is a very good source of protein, dietary fibre, potassium, iron, copper, phosphorous, magnesium and thiamin [14].

Figure 4a. Lima bean (*Phaseolus lunatus*).

Figure 4b. (Phaseolus limensis).

African Yam Bean

African bean (*Sphaenostylis stenocarpa*) is grown in West African countries. In Nigeria it is found commonly cultivated in the South West. African yam bean can be consumed as seed or as tuber. The seeds are of various sizes and color.

The seed grains and the tubers are the two major organs of immense economic importance as food for Africans. This indigenous crop has huge potential for food security in Africa. In West Africa, the seeds are preferred to the tubers but the tubers are preffered in East and Central Africa [15]. The crop replaces cowpea in some parts of southwestern Nigeria, the crop is well balanced in essential amino acids and has a higher amino acid content than pigeon pea, cowpea, and bambara groundnut [16].

Figure 5a. African yam bean (white).

Figure 5b. African yam bean (brown).

The protein content in AYB grains ranged between 21 and 29% [17]. Another positive contribution of the crop to food security is the identification of the presence of lectin in the seeds, which could be a potent biological control for most leguminous pests.

5. Nutritional Composition of Some Lesser Legumes

The nutritional composition of some lesser known legumes is shown in Table 2.

They are all rich in protein and carbohydrate, they contain less fat when compared with the fat content of most of the major legumes. Their calorie content is within the range of 300 and 400 kcal which is relatively high enough as energy required in any daily intake diet. Kidney bean and bambara groundnut are richer in calcium when compared to the calcium content of lima bean, pigeon pea and African yam bean.

6. Anti-Nutritional Factors in Some Lesser Known Legumes

Generally, the lesser legumes contain some anti nutrients. The anti nutritional factors found common to most of these legumes include trypsin and chymotrypsin inhibitor, heamaglutinins, tannin, polyphenols, phytate or phytic acid, cyanogenic glucoside, saponin.

Tannin and polyphenolic substances are more concentrated in the seed coat of these legumes. They interfere with the digestibility of the legumes, Haemaglutin affect protein utilization.

Table 2. The nutritional composition of some lesser known legumes in Nigeria

Legume	Calorie (kcal)	CHO (%)	Protein (%)	Fat (%)	Ash (%)	Fibre (%)	K (mg/100g)	P (mg/100g)	Ca (mg/100g)	Mg (mg/100g)
Lima bean (*Phaseolus lunatus*)	340	2.4	21.7	2.4	4.1	5.3	2.10	8.40	0.05	8.10
Pigeon pea (*Cajanus cajans*)	315	56.8	22.6	1.5	4.4	2.7	1.8	3.48	1.0	1.45
African yam bean (*Sphenostylis stenocarpa*)	320	59.8	22.2	4.5	4.1	5.2	1.8	2.6	0.02	0.9
Bambara groundnut (*Vigna subterranea*)	390	62.4	22.0	8.6	4.2	1.5	2.6	1.6	1.07	2.9
Kidney bean (*Phaseolus vulgaris*)	350	61.0	22.6	1.2	4.5	5.2	0.8	2.0	1.42	0.4

Phytohaemaglutinin agglutinate red blood cells, Lecitin binds intestinal mucosa cells causing disruption in the small intestine. Trypsin inhibitor affects the digestion of protein (trypsin), protease inhibitor inhibit proteolytic activities of the enzyme protease resulting in slow growth. The antinutritional factors contained in lima bean include trypsin and chymotrypsin inhibitors [18], haemagglutinins and cyanogenic glucosides [19], polyphenols and phytic acid [20] and various oligosaccharides which cause flatulence when lima beans are consumed. Pigeon pea contains tannin, phytate and trypsin inhibitor [21, 22].

7. FERMENTATION

Fermentation is defined as a process whereby food materials particularly cereals and legumes are subjected to the actions of microorganisms or enzymes for biochemical changes to take place to modify the food and produce desirable attributes such as flavor and textural changes in the food [23]. In a general sense, fermentation is the conversion of a carbohydrate such as sugar into an acid or an alcohol. More specifically, fermentation refers to the use of yeast to change sugar into alcohol or the use of bacteria to create lactic acid in certain foods. Fermentation occurs naturally in many different foods under the right conditions.

Food fermentation processes often result in changes in flavor relative to the starting material/ingredients. It involves very complex ecosystem with active enzyme activities from the raw material/ingredients interacting with the metabolic activities of the fermenting organisms. Fermentation is an age long method of processing cereals and legumes It modifies some physical characteristics of cereals and legumes, increases level of some nutrients digestibility and bioavailability [24], decreases level of antinutrients, increases nutrient density [25, 26] and impacts some antimicrobial properties [27, 28]. Fermentation of grains and oil seeds results in increased nutritional value and wholesomeness over the starting material [29].

Food fermentation can be broadly classified into solid state (without soaking) and submerged (with soaking in water) fermentation.

a) Solid state fermentation

Solid state fermentation (SSF) is a process where microbial growth and product formation occur on the surface of the solid materials. This process occur when there is no free water and moisture is absorbed to the solid matrix [30]. Solid state fermentation has a lot of advantages over the submerged fermentation. These include lower cost, improved product recovery and characteristics, higher product yield and reduced energy requirement [31]. Examples of fermented foods produced through solid state fermentation are gari, natto, dawadawa, tempeh.

b) Submerged fermentation

Submerged fermentation (SmF) on the other hand is the process in which microorganisms degrade carbohydrate in liquid medium with plenty of available free water [32]. Examples of fermented foods produced through submerged fermentation are fufu, lafun, yoghurt, beverages such as wine and beer.

8. Importance of Fermentation of Lesser Legumes

These lesser legumes often contain undesirable substances called anti nutrients. Example is trypsin inhibitor. The anti - nutrients are usually reduced or removed through processing such as soaking, boiling, roasting and fermenting.

Fermentation serves as preservative method by creating an undesirable condition for the spoilage micro organisms through increase in the acidity. There are several reasons why fermentation of legumes is necessary.

1. Legumes often contain undesirable substances, such as the trypsin inhibitor and phytic acid which interfere with digestion of useful nutrients such as protein and minerals. These undesirable substances are reduced or removed through processing such as soaking, cooking, germinating and roasting.
2. Moist legume products spoil readily but when fermented and then refrigerated, keep better and longer.
3. All the legume fermentations involve the action of proteolytic and lipolytic enzymes which result in the final product been more digestible.
4. Fermentation improves the quality attributes of the product over the starting material thus increases the consumer acceptability. Invariably the final product has a changed flavor and even texture and taste.
5. Fermentation of legumes allow for better tolerance of certain products for example a lot of people cannot tolerate soy milk because of the beany flavor in the milk, but when the milk is fermented, the beany flavor is removed and as such becomes more acceptable to the consumer.
6. In many instances the microorganisms increase nutrients such as vitamins, including riboflavin and vitamin B12 in some of the bacterial fermentations.
7. Fermentation may reduce energy requirements for cooking as most of the lesser legumes have the hard to cook phenomenon. Thus, a short cooking may be the major energy input required while the microorganisms do the rest of the work by using energy from the substrate
8. Fermentation process results in increase in the acidity of food thereby reducing the incidence of spoilage microorganisms.

9. Fermented Legumes Products

Some oriental fermented legume foods include *kinema* [33], *natto* [34], *meju* [35].

They are legumes fermented by *Bacillus* species and they appear similar to soy *daddawa* except that *kinema* is made from crushed soybean while *natto* and *meju* are made from whole soybean. Others include *Tofu* or soybean cheese fondly consumed by the Chinese. *Miso* and *tempeh* are consumed by the Indonesians [35].

Indigenous African fermented legumes include iru from African locust bean (*Parkia biglobosa)* [36, 37, 38], soybean (*Glycine max)* [39, 40, 41] and bambara groundnut (*Vigna subterranean*) [42].

10. Some Fermented Lesser Known Legumes

Some fermented lesser legumes are *Iru* from lima bean [43]. *Owoh* from African yam bean [44]. *Iru* from pigeon pea [45, 46]. Pigeon pea flour from pigeon pea [22]. Tempeh from African yam bean [47].

11. Methods of Fermentation of Some Lesser Legumes (Lima Bean, Pigeon Pea and African Yam Bean)

11.1. Fermentation of Lima Bean to Produce *Iru* (Condiment)

Lima bean can be fermented to produce *iru* analogue as follows:

Sorting of lima bean seeds to remove dirts and broken seeds, roasting of the limabean seeds in an open frying pan for 10 minutes, dehulling to remove seed coat, cooking of the dehulled beans with 1000mls of water for 40 minutes, draining of water from the cooked beans, pouring of the cooked beans into calabash lined with clean banana leaves, placing of the calabash and its content into an incubator, at $30^{o}C \pm 5^{0}C$ for 72 hours during which the characteristic *iru* flavor develops. The lima bean *iru* could be stored in the refrigerator or dried and grounded into powder to extend it's shelf- life. [43]. The flow chart of lima bean *iru* processing is shown in Figure 6.

11.2. Fermentation of Lima Bean to Produce Good Quality Flour

Good quality flour from fermented lima bean can be produced as follows:

Soaking the lima bean seeds in water for 12 h, dehulling of the beans to remove testa, followed by wet milling of the dehulled beans into paste, drying of the paste at $60^{o}C$ till it is well dried, and grinding into flour. This will be followed by sieving the flour using a 40 mm mesh screen, inoculating the flour with *Rhizopus oligosporous* and fermenting the flour for 6 days at ambient and at pH range of 4.10 to 6.20 [48]. Next is drying of the fermented meal in an oven at $60^{o}C$ and milling of the dried meal into flour.

11.3. Fermentation of Pigeon Pea to Produce Good Quality Flour

Fermentation of pigeon pea to produce pigeon pea flour was described by [22] as follows: Sorting of the pigeon pea seeds, weighing, soaking in tepid water for 72 h, wet milling and sieving under water using muslin cloth. , washing and re-sieving, allowing the sediment to settle and fermenting for 24 h, discarding of the supernatant and draining the sediment off water by pressing out the water to get a solid sediment. Oven-drying at 60°C and milling into flour, sieving and storing under appropriate condition till when needed.

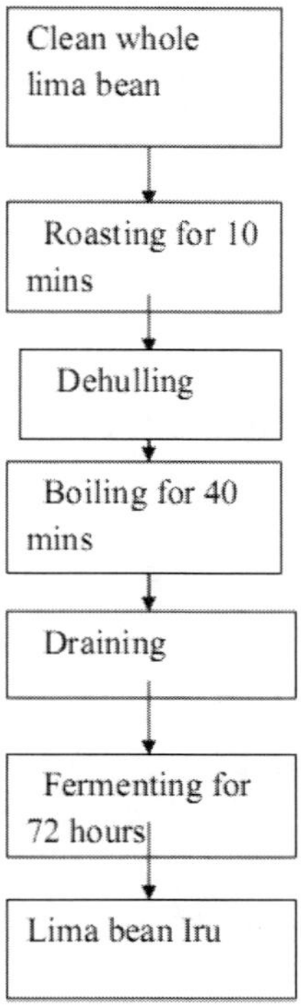

Figure 6. Flow chart of fermentation of lima bean to produce *iru*.

11.4 Fermentation of Pigeon Pea to Produce *Iru* (Condiment)

Fermentation of pigeon pea to produce *Iru* condiment is as follows: Sorting of the pigeon pea seeds, washing and roasting for 10 min, dehulling manually or with mortar and pestle, cooking for 20 – 45 min depending on the hardness of the seeds, draining and wrapping with banana leaves with or without wood ash sprinkled on it and allowing to ferment for two to three days at 37°C [44, 46].

11.5. Fermentation of African Yam Bean to Produce *Tempeh*

Fermentation of African yam bean to produce tempeh was described by [49] and [47] as follows: Sorting of the African yambean seeds, dehulling, soaking for 24 hours, cooking for 30 min in water, inoculating with spores suspension of *Rhizopus oligosporous* at 30°C for 48 h, mixing and packing into perforated plastic bags. Alternatively, 1% citric acid could be used instead of water.

11.6. Fermentation of African Bean to Produce *Otiru* (Condiment)

Fermentation of African yam bean to produce *otiru* (a condiment) was described by [50] as follows: Sorting of the raw seeds, washing in clean water, soaking in water for 12 h, dehulling, washing under water to remove the hulls, boiling for 3 h, draining, wrapping in clean plantain leaves in a calabash and fermenting at ambient temperature for 120 hours.

12. Microorganisms Involved in Fermentation of Some Lesser Legumes (Lima Bean, African Yam Bean, and Pigeon Pea)

Microorganisms involved in the fermentation of lima bean to produce *iru,* fermentation of lima bean to produce lima bean flour, fermentation of pigeon pea to produce pigeon pea flour, fermentation of African yam bean to produce tempeh and fermentation of African yam bean to produce otiru are shown in Table3.

Table 3. Microorganisms involved in the fermentation of some lesser legumes

Product	Implicating microorganism	Reference
Lima bean *iru*	*Bacillus subtilis B.licheniformis, B. pumilus, Corynebacterium sp*	Farinde et al., 2011
Lima bean flour	*Rhizopus oligosporous*	Onoja et al., 2001
Pigeon pea flour	*Bacillus megatarium, B.subtilis, Staphylococcus sp. and Saccharomyces cerevisiae*	Oyarekua, 2011
Tempeh	*Rhizopus oligosporous*	Akinyele and Akinlosoyu, 1991; Azeke, 2007
Product	Implicating microorganism	Reference
African yam bean *otiru*	*Bacillus coagulans, Lactobacillus jensenii, Aerococcus viridians Pediococcus cerevisiae, Saccharomyces cerevisiae Candida mycoderma*	Jeff- Agboola, 2007

14. Effect of Fermentation on the Nutrient Composition of Lima Bean, Pigeon Pea and African Yam Bean

14.1. Effect of Fermentation on the Nutrient Composition of Lima Bean

Fermentation was found to improve the nutrients content in fermenting lima bean [43]. When lima bean was fermented to *iru* analogue, processing such as dehulling and cooking reduced the nutrient content of lima bean, fermentation however improved some of the nutrients such as protein fat and ash while it decreased crude fibre and carbohydrate. Processing decreased the protein content of lima bean from 21.7% in the raw to 16.81%. Fermentation increased this to 17.5%. Fat and Ash content also decreased with processing from 2.41% and 4.1% in the raw lima bean to 1.30% and 2.41% respectively, fermentation improved this by increasing the fat content to 2.17% and the ash to 2.68%.

The crude fibre and the carbohydrate decreased with fermentation time. Crude fibre decreased from 5.38% in the raw lima to 1.48% after 72 h of fermentation. Carbohydrate decreased from 52.42% in the raw lima bean to 43.73% after 72 h of fermentation. Magnesium content increased from 8.10 in the raw to 11.4 mg/110 after 72 hr of fermentation. Phosphorous and potassium increased with fermentation from 8.40 and 2.10 mg/100g in the raw to 7.48 and 2.60 mg/100g at 72 h of fermentation respectively. Calcium content reduced with fermentation.

14.2. Effect of Fermentation on the Nutrient Composition of Pigeon Pea

The effect of fermentation on the proximate and mineral composition of pigeon pea were studied [45]. Protein content in the fermented seeds was higher 24.12% in comparison with the 20.02% in the raw seeds The carbohydrate content decreased 53.93% when compared with 57.22% in the raw seeds, No significant difference ($p < 0.05$) was observed between the fat and fibre content of the raw and fermented seeds. For the mineral content, phosphorous content increased from 0.444 mg/100g in the raw seed to 0.541 mg/100g in the fermented pigeon pea seeds. No significant change was observed in the calcium, sodium, potassium and magnesium content of the raw and fermented seeds.

14.3. Effect of Fermentation on the Nutrient Composition of African Yam Bean Bean

When African yam bean seeds were fermented for 24 h, protein content increased from 20.05 % in the raw seeds to 21.06 % in the fermented seeds. Carbohydrate content also increased from 71.31 % in the raw seeds to 72.05 % in the fermented seeds. Fat content also increased from 0.84% in the raw seeds to 1.56 % in the fermented seeds. However, the ash content decreased from 4.46 % in the raw seeds to 3.13 % in the fermented seeds and fibre content also decreased from 2.51 in the raw seeds to 2.20 % in the fermented African yam bean seeds [51].

15. Effect of Fermentation on the Anti Nutrient Content of Lima Bean, Pigeon Pea and African Yam Bean

15.1. Effect of Fermentation on the Anti Nutrient Content of Lima Bean

According to [43], all the anti-nutrients analyzed (tannin, phytate, cyanide, trypsin inhibitor) during the fermentation of lima bean to produce *iru* analogue decreased with fermentation. Tannin content decreased significantly ($p < 0.05$) from 23.55 in the raw lima beans to 2.0 mg/kg at 72 of fermentation. Fermentation significantly decreased phytate content ($p < 0.05$) from 31.56 in the raw lima beans to 7.0 mg/kg at 72 *h* of fermentation. Cyanide content decreased from 5.20 in the raw lima beans to 0.25 mg/kg after fermenting for 72 h while trypsin inhibitor decreased from 28.36 in the raw lima beans to 1.76 mg/kg at 72 h of fermentation. Trypsin Inhibitor content also decreased significantly ($p < 0.05$) with increase in fermentation time in lima bean.

15.2. Effect of Fermentation on the Anti Nutrient Content of Pigeon Pea to Produce Condiment

The tannin and phytate content of raw and fermented pigeon pea were studied by [45]. Generally tannin and phytate content reduced in the fermented pigeon pea seeds in comparison with the raw seeds. Tannin content of the raw seeds was significantly ($p<0.05$) reduced from 4.23 mg/100g to 1.19 mg/100g while phytate content reduced significantly ($p < 0.05$) from 2.20 mg/100g to 1.65 mg.100g by fermentation.

15.3. Effect of Fermentation on the Anti Nutrient Content of Pigeon Pea to Produce Flour

Fermentation decreased all the anti nutrients analyzed (tannin, phytate, cyanide and trypsin inhibitor) when pigeon pea was processed into flour [22]. Tannin decreased from 2.0 in the raw pigeon pea seeds to 1.20 mg/100g in the fermented pigeon pea. Phytate decreased from 0.013 in the raw bean to 0.006 mg/100g in the fermented bean.

Cyanide also decreased from 3.05 in the raw bean to 2.48 mg/100g in the fermented pigeon pea.

Fermentation also decreased trypsin inhibitor content from 8.62 in the raw pigeon pea to 4.18 TIU in the fermented pigeon pea [22].

15.4. Effect of Fermentation on the Anti Nutrient Content of African Yam Bean

During tempeh production, fermentation with *Lactobacillus plantarum* was found to reduce the flatulence causing α-galactose, cyanogenic glucoside, tannin and phytic acid [52].

Lactic acid fermentation was more effective than the traditional cooking in reducing the flatulence potential (oligosaccharides).

Similarly, fermentation with *Rhizopus oligosporous* resulted in total removal of α-galactose, and phytic acid, 50% reduction in cyanogenic glucoside and 20% reduction in tannin [52].

CONCLUSION

Fermentation is a long age processing technique which is still relevant till today both in traditional and modern food processing. Fermentation is a cheaper means of improving the utilization of under- utilized legumes especially in Africa among the rural poor for improved protein intake. However in the traditional setting, fermentation needs to be upgraded for convenience and development of hygienic safe products.

REFERENCES

[1] Deshpande, S. S. (1992). Food legumes in human nutrition. A personal perspective. *Crit. Rev. Food Sci. Nutr.*, 32: 333-363.

[2] Smart, J. (1990). *Evolution and genetic resources* Pg 294-309.

[3] Nwanna L. C., Enujiugha V. N., Oseni, A. O. and Nwanna, E. E. (2005). Possible effects of fungal fermentation on Bambara groundnuts (*Vigna subterranean* (L.) Verde) as a feedstuff resource. *J. Food Tech.*, 3(4): 572-575.

[4] Minka, S. R. and Bruneteau, M. (2000). Partial chemical composition of Bambara pea (*Vigna subterranean* (L.) *Verde). Food Chemistry*. 68: 273-276.

[5] Campbell–Platt, G. (1980). African locus bean (*Parkia* spp.) and its West – African fermented food product, '*dawadawa*'. *Ecology Food Nutrition* 9, 123-132.

[6] Sighn, U., Jambunathan, R., Saxena, K. B. and Subrahmanyam, N. (1990). Nutritional quality evaluation of newly developed high protein genotype of pigeon pea (Cajanus cajan L.). *J. Sci. Food Agric*. 50: 201 – 209.

[7] Apata, D. F. and Ologhobo, A.D. (1994). Biochemical evaluation of some Nigerian legume seeds. *Food Chem*. 49: 333 – 338.

[8] Akande, K. E., Abubakar, M.M., Adegbola, T. A., bogoro, S. E. and Doma, U. D. (2010). Chemical evaluation of the nutritive Quality of pigeon pea (Cajans cajaan L.). *Int. Journal of Poultry Science*, 9 (1): 63 - 65.

[9] Saka, J. O., Ajibade, S. R., Adeniyan, O. N., Olowoyo, R. B. and Ogunmodede, B. A. (2004). Survey of under- utilized grain legume production system in South West Africa zone of Nigeria. *Journal of Agric. and Food Information* 6 (2): 93 – 108.

[10] Rachie, K. O. and Silvester, P. (1977). *Grain legumes. Food Crops of the Lowland Tropics.* (leakey, C.L.A. and wills, J. B. eds.) pp 41 – 74. Oxford: Oxford Univrsity Press.

[11] Lyman, S. M., Baudoin, J. P. and Hildago, R. (1985). Limabean (*Phaseolus lunatus*) In: *Grain Legume Crops*. Edited by Summerfield, R. J. and Roberts, E. H. Williams Collins sons and Co Ltd London, United Kingdom.

[12] Kay, D. E (1979). Food Legumes. *Crops and Product Digest* No. 3. Tropical Products Institute, Ministry of Oversea Development, London, pp. 225 -243.

[13] IITA (1979). Annual Report for 1978. Ibadan, Nigeria: *International Institute of Tropical Agriculture*, 44 - 46.

[14] WHFoods (2010). Lima beans http://www.whfooods.com/genpage.php? tname=food spice$dbid=59. Date retrived: 14-06-11.

[15] Potter, D. 1992. Economic botany of *Sphenostylis* (Leguminosae). *Economic Botany*, 46: 262-275.

[16] Okpara, D.A. and C.P.E. Omaliko. 1995. Effects of staking, nitrogen and phosphorus fertilizer rates on yield and yield components of African yam bean (*Sphenostylis stenocarpa*). *Ghana Journal of Agricultural Science* 28:23–28.

[17] Uguru, M.I. and S.O. Madukaife. 2001. Studies on the variability in agronomic and nutritive characteristics of African yam bean (*Sphenostylis stenocarpa* Hochst ex. A. Rich. Harms). *Plant Production and Research Journal* 6:10-19.

[18] Ologhobo, D. A. and Fetuga, B. L. (1983a). Trypsin inhibitor activity in some lima bean (*Phaseolus lunatus)* varieties as affected by different processing methods. *Nutr. Rep. Int*. 27: 41 -47.

[19] Ologhobo, D. A., Fetuga, B. L. and Tewe, O. O. (1984). The cyanogenic glucoside content of raw and processed lima bean varieties. *Food Chem.* 13: 117 – 128.

[20] Ologhobo, D. A. and Fetuga, B. L. (1982). Polyphenols, phytic acid and other phosphorous compounds of lima bean (Phaseolus lunatus). *Nutr. Rep. Int*., 26: 605 – 611.

[21] Fasoyiro, S. B., Ajibade, S. R., Omole, A. J. Adeniyan, O. N and Farinde, E.O. (2006). Proximate, minerals and antinutritional factors of some under-utilized grain legumes in the South West Nigeria. *Journal of Nutrition and food Science*, 36 (1) : 18-23.

[22] Oyarekua, M. A. (2011). Biochemical and microbial changes during production of fermented pigeon pea (Cajanus cajan) flour. *African Journal of Food Science and Technology*. 2 (10): 223 - 231

[23] Campbell-Platt, G. (1987). Fermented Foods of the World. Dictionary and guide. Butterworths, London, UK.

[24] WHO, (1998). Complementary Feeding of young Children in Developing Countries. WHO, Geneva, pp. 130 -131.

[25] Tonikins, A. Alnwick, D. and Haggerty, P. (1987). Fermented Foods for Improving young Child Feeding in Eastern and Southern Africa. In: *Improving Young Child Feeding in Eastern and Southern Africa,* Alnwick, D., S. Moses and O. G. Schmidt (Eds.). International Development Research Centre, Ottawa, pp. 198 - 208.

[26] Nnam, N. M. (1999). Chemical and organoleptic evaluation of complementary foods formulated from maize, cowpea and plantain flours. *Nig. J. Nutr. Sci*., 20: 35-40.

[27] Mensah, P P. A.., Tomkins, A. M., Drasar, B. S. and T. J. Harrison, (1990). Fermentation of cereals for reduction of bacterial contamination of weaning food in Ghana. *Lancet,* 336: 140-143.

[28] Mensah, P.P.A., Tomkins, A.M., Draser, B.S. and Harrison, T.J. (1991). Antimicrobial effect of fermented Ghanian maize-dough. *J. Applied Bacteriol.* 70: 203-210.

[29] Quinn, M. R and Beuchart, L. R. (1975), Funtional property changes resulting from fungal fermentation of peanut flour. *J. Food science*; 40: 475-478.

[30] Pandey, A., Soccol, C. R. and Mitchell, D. (2000). New development in solid state fermentation: Bioprocesses and Products. *Product and Biochemistry*. 35:1153-1169.

[31] Ray, R. C, Shetty, S. and Ward O. P. (2008). Solid state fermentation and value-added utilization of horticultural processing wastes. *Microbial Biotechnology in Horticulture*, 1: 231-272.

[32] Ray R. C and Ward O. P. (2006). Post - harvest microbial biotechnology of tropical root and tuber crops. In: *Microbial Biotechnology in Horticulture*, Vol 1 (edited by R.C Ray and O.P Ward) pp. 349-356.

[33] Sarkar, P. K., Jones, L. J., Craven, G. S. , Somerset, S. M. and Palmer, C. (1997). Amino acid profile of kinema, a soybean fermented food. *Food Microbiol.* 59: 69 – 75.

[34] Ohta , T. (1986). Natto: In Legume –Based Fermented Foods (Eds N.R. Reddy, M.D. Pierson and D.K. Salunkhe) pp. 85-89. Boca Raton, Florida, USA.

[35] Kim, D., Lim, D., Bai, S. and Chun, S. (1997). Fermentation characteristics of whole soybean meju model system inoculated with four *Bacillus* Strain. *Korean Journal of Food Science and Tech.* 29: 1006 – 1015.

[36] Frazier, W. C. and Westhoff, D. C. (2005). Food Microbiology. Fourth Edition.Tata McGray-Hill. 7, West Patel Nagar, New Delhi.

[37] Odunfa, S. A. (1981a). A note on the microorganisms associated with the fermentation of African locust bean *(Parkia filicoidea)* during *iru* production. *J. Plant Foods* 3: 245-250.

[38] Antai, S. P. and Ibrahim, M. H. (1986). Microorganisms associated with African locust bean *(Parkia filicoidea Welw)* fermentation for dawadawa production. *J. Appl. Bacteriol.* 61:145-148.

[39] Popoola, T. O. S. and Akueshi, E. O. (1985). Microorganisms associated with fermentation of soybean for the production of *daddawa* (a condiment). *Nigerian Food Journal* 2: 194-196.

[40] Ogbadu, C.O. and Okague, R.N. (1988). Bacterial fermentation of soybeans for *daddawa* production. *J. Appl. Bacteriol.* 65: 353-356.

[41] Omafuvbe, B. O., Shonukan, O. O. and Abiose, S. H. (2000). Miocrobiological and biochemical changes in the traditional fermentation of soybean for *soy dadawa,* Nigerian fermented condiment. *Food Microbiol.*, 17: 469-474.

[42] Barimalaa, I. S., Achinewhu, S. C., Yibatima, I. and Amadi, E. N. (1989). Studies on the solid substrate fermentation of *bambara* groundnut (*Vigna subterranean* (L) Verdc). *Journal of Sci. Food Agric.* 66:443-446.

[43] Farinde, E.O., Adeniran, H. A. and Abiose, S.H. (2011). Proximate composition, mineral content and sensory assessment of an *iru* analogue produced from fermented lima bean (Phaseolus lunatus). *Ife journal of Technology.* 20 (2): 1 – 6.

[44] Ogbonna, D. N., Sokari, T.G. and Achinewhu, S. G. (2001). Development of an *owoh* – type product from African yam beans *(Stenostylis stenocarpa)* (Hoechs ex. A. Rich Harms) by solid substrate fermentation. *Plant Foods. Hum. Nutrition.* 56: 183 -184.

[45] Fasoyiro S.B, Obatolu V.A, Ashaye O.A, Adeojo E.A and Ogunleti D.O (2009a). Chemical and sensory qualities of pigeon pea (*Cajanus cajan*) developed into a local spice *dawadawa* . *Nigerian Food Journal* 27:150-159.

[46] Fasoyiro S.B, Taiwo L.B, Osuntade . Arowora K.A, Ejoh S.I (2009b). Development of fermented pigeon pea (*Cajanus cajan*) dawadawa spice into shelf-stable products. *Moor Journal of Agricultural Research* 10:80:88.

[47] Azeke, M. A. (2007). Nutritional value of African yam bean: Improvement by solid substrate fermentation using *tempeh* fungus, Rhizopus oligosporous. *J. Food Science.* 87: 279 – 304.

[48] Onoja, U. S., Dibua, U. M. E., Eze, J. I. (2011). *Rhizopus Oligosporus* Fermentation of Lima Bean (*Phaseolus Lunatus*) Flour: Effect on Nutritional, Anti-nutritional and Toxic Factor. JHER. 14: 244-254.

[49] Akinyele, I. O. and Akinlosotu, A. (1991). *Food Chem.* 41, 43-53.

[50] Jeff-Agboola, Y. A. (2007). Microorganisms associated with fermentation of African yam bean (*Sphaenostylis stenocarpa Harms*) seeds for the production of *otiru. Research journal of Microboiology*, 2: 816 – 823.

[51] Onaja, U. S and Obizoba, I. C. (2009). Nutrient composition and organoleptic attributes of gruel based of fermented cereal, legume, root and tuber flour. *Journal of Tropical Agriculture, Food, Environment and Extension.* 8:162 – 168.

[52] Azeke, M. A. (2003). Characterization and improvement of the nutritional value of African yam bean (*Sphaenostylis stenocarpa*) by non - traditional processing methods. Thesis, Agricultural Faculty, University of Bonn, with Federal centre of Cereal, Potato and Lipid Research, Detmold.

In: Legumes
Editors: Hiroto Satou and Ren Nakamura

ISBN: 978-1-62808-280-7

Chapter 6

DARK AND BRIGHT FACETS OF THE NUTRITIONAL VALUE OF GRASS PEA (*LATHYRUS SATIVUS L.*) SEEDS

Angela R. Piergiovanni*[*], *Venturino Bisignano*, *Giambattista Polignano and Lucia Lioi
Istituto di Genetica Vegetale (IGV-CNR), Bari, Italy

ABSTRACT

Grass pea (*Lathyrus sativus* L.) is an annual food, feed and fodder crop cultivated in temperate regions of Asia, Europe, America and East Africa. This pulse has several advantageous traits such as: high yielding, nitrogen fixation, extensive tolerance of drought and water-logging, adaptation to a wide range of soil types, resistance to insects and pests. Grass pea is a popular source of vegetable proteins in subsistence farming systems, mainly in some developing countries. However, an extensive use for food and manufacturing feedstock is limited by the presence in seeds of appreciable quantities of β-N-oxalyl-L-α,β-diaminoproprionic acid (β-ODAP), in addition to other antinutritional factors typical of pulses (i.e.: digestive enzyme inhibitors, lectins, phytic acid, etc.). ODAP is a non protein amino acid associated with neurolathyrism, a neurological disorder that manifests in humans and animals when grass pea seeds became a staple food for long periods. As a consequence of the reputed grain toxicity, a significant drop of grass pea cultivation took place in the second half of 20th century at worldwide level. Presently, it is considered a model for sustainable agriculture, so efforts are underway for the reintroduction in North and Latin America, Australia and Southern Europe. The aim of these initiatives is to counter the effect of climate change on agricultural productivity and to provide an alternative crop in the soils overexploited by continuous cereal cultivation. The purpose of this review is to give an overview of the most recent literature dealing with nutritional and nutraceutical quality of grass pea seeds.

[*] Corresponding author: Angela Rosa Piergiovanni, CNR-Istituto di Genetica Vegetale (IGV), Via Amendola 165/a, 70126 Bari, Italy. Phone: +39 080 5583400; Fax: +39 080 5587566. E-mail: angelarosa.piergiovanni@igv.cnr.it.

INTRODUCTION

Grass pea or chickling pea (*Lathyrus sativus* L.) is an annual cool-season pulse belonging to *Lathyrus* genus. It was first domesticated some 7-8,000 years ago, but the centre of origin is still debated. The combination of archaeobotanical and phytogeographycal evidences gathered in the last decades suggests that the grass pea cultivation started in the early Neolithic period (ca 6th millennium BC) in the Balkan peninsula (Kislev 1989). Grass pea survives and yields relatively well on low-fertility soils, is a hardy, drought tolerant subsistence crop and does not get affected by water-logging situations (Campbell et al. 1994; Dadi et al. 2003). As other pulses, it adds nitrogen to the soil through N symbiotic fixation, thereby conferring protein benefits for the subsequent non-legume crop and reducing pest outbreaks through intercropping with other food crops (Peoples et al. 2008). Being a very adaptable species, it is cultivated in different areas of the world, and at altitudes ranging from 0 to more than 1500 m above sea level (Tadesse & Bekele 2003). Globally, area under grass pea is estimated at 1.5 million ha with annual production of 1.2 million tons (Kumar et al. 2011). This pulse has great economic and ecological significance in some regions of South Asia, and highlands of Ethiopia, where represents the last resort in times of famine due to drought, floods or conflicts (Campbell et al. 1994; Dadi et al. 2003; Dixit et al. 2008). The grass pea cultivation is practiced to a much lesser extent, in southern and central Europe, west Asia and South America (Tavoletti et al. 2005; De la Rosa & Martin 2001; Milczak et al. 2001; Mera et al. 2000). Grass pea has a long history of cultivation in Italy, but presently occupies a niche role in human nutrition, being cultivated by organic farmers which continue small scale production for household consumption or local market request (Tavoletti et al. 2005; Piergiovanni et al. 2011).

It is well known that nutritional value of pulse seeds is related to the content of biologically available nutrients, as well as to the efficacy of processing methods on removing or inactivating the anti-nutritional compounds. The equilibrium between these two components assumes a high relevance for grass pea that presents a fascinating paradox being at the same time a lifesaver, and a harmful food. Several studies (Hambury et al. 2000; Urga et al. 2005; Piergiovanni et al. 2011) showed that grass pea seeds have a protein content (240-343 g kg^{-1} dm) comparable with pea (*Pisum sativum* L.), lentil (*Lens culinaris* Medik.) or faba bean (*Vicia faba* L.) seeds, but the nutritional potential is limited by the combined action of several antinutritional compounds such as digestive enzyme inhibitors, lectins, tannins, polyphenols and phytate (Enneking 2011). Moreover, seeds are deficient in methionine, lipids, and C, D and B1 vitamins. However, the major nutritional trouble is the presence of significant amounts of β-N-oxalyl-L-α,β-diaminopropionic acid (β-ODAP), a non protein amino acid responsible of neurolathyrism (Rao et al. 1964). This is a spastic para-paresis caused by hyper-excitation of motorneurons that produces spastic movements of the legs in up to 6% of the population, mainly young males (Kumar et al. 2011). The link between excessive consumption of grass pea and neurolathyrism is known since prehistoric times; it has been described by the Greek physician Hippocrates, as well as in ancient Indian documents, but the conditions for disease development are still debated (Tekle Haimanot et al. 1993, Enneking 2011). Despite these concerns, million of peoples living in the poorest areas of the world consume it on a regular basis with no apparent ill effects, except when the consumption is excessive. Most of consumers know that lathyrism is caused by grass pea but cannot afford to buy other pulses (Girma et al. 2011). Based on these considerations, the decision to ban grass pea requires a careful re-examination of its true nutritional value.

Historically, the main objectives of grass pea breeding have been the development of high-yielding varieties ODAP-free, or with a content safe for human consumption (less than 0.2% is currently indicated), and the improvement of agronomic performances. This review will give an overview of the more recent studies dealing with nutritional, toxic, and nutraceutical characteristics of grass pea seeds.

Grass Pea Neurotoxicity

It is known that if non protein amino acids are taken up by vertebrates, insects or microorganisms they may interfere with several processes. If accepted in protein biosynthesis in place of the normal amino acid, the non protein ones can lead to defective proteins, may competitively inhibit uptake systems for amino acids, could generate some pathologies and so on. Naturally occurring ODAP exists in two isomeric forms named α and β, with α-isomer being less toxic (De Bruyn et al. 1994). The β-isomer accumulates in all grass pea tissues at all development stage, though the highest content has been observed in the 10-day old seedlings (Jiao et al. 2006). The concentration of the β-isomer is about 95% of the total ODAP present in dry seeds, and its isomerisation to the α-isomer depends on temperature, cooking time and pH. The equilibrium concentration of α and β-isomer is in the ratio of 30:70 at room temperature and 40:60 at 55-60°C (Kumar et al. 2011). Thus, thermal isomerisation occurring during food processing can be used, to some extent, for seed detoxification. Most studies following the β-ODAP discovery in 1960's as a neurotoxic amino acid, have primarily focused on the mechanism of toxicity (Rao 1978; Mehta et al. 1983). The effects of feeding grass pea are species-specific. While animals seem to eliminate the bulk of ingested ODAP via the kidneys and urine, human consumers appear to be metabolising more than 99% of ingested toxin (Yan et al. 2006; Rao 2011). Breakdown of ODAP in the gastrointestinal tract is one possible mechanism (Enneking 2011). Since the gastrointestinal tract has an important role in toxin uptake, its role in neurolathyrism needs to be examined in deep. Epidemiological studies evidenced that less than 4% of grass pea consumers, especially young man, are affected by neurolathyrism (Rao 2011). For this reason, a wide scale screening of habitual consumers would help to establish if there are genetic variations for the capability to metabolise the toxin. On the other hand, there are evidences that the malnutrition plays an important role in the development of neurolathyrism. Recent studies suggested that nutritional deficiencies and oxidative stress could predispose to the development of pathology (Yan et al. 2006). A working hypothesis is that a lack of reduced thiols makes consumers more susceptible to neurolathyrism. A reduced methionine flux into the central nervous system in consequence of prolonged consumption of grass pea as staple, makes these cells more sensitive to the additional oxidative stress produced by over-excitation of the motor neurons by β-ODAP. Fikre et al. (2010), and Enneking (2011) suggested that consumption of methionine-rich foods has a protective action against the neuro-excitatory action of β-ODAP. This kind of foods can counterbalance the significant and persistent decrease of methionine concentration in plasma observed in grass pea consumers.

Table 1. Variation range of grass pea seed components as deduced from the literature. The values are expressed on dry matter basis

Component	Range	Source
Protein (g kg^{-1})	240-343	Piergiovanni et al. 2011; Hanbury et al. 2000
Starch (g kg^{-1})	412-523	Hanbury et al. 2000
β-ODAP (g kg^{-1})	0.1-9.2	Kumar et al. 2011
Ash (g kg^{-1})	20-39	Hanbury et al. 2000
Fat (g kg^{-1})	7-17	Hanbury et al. 2000; Yan et al. 2006
Crude fibre (g kg^{-1})	35.6-100	Hanbury et al. 2000; Urga et al. 2005
Phytate (g kg^{-1})	4.9-10.9	Srivastava & Khokhar 1996; Hanbury et al. 2000
Polyphenols (g kg^{-1})	6.0-6.9	Srivastava & Khokhar 1996
Trypsin inhibitors (TIU mg^{-1} dm)	11.6-33.1	Piergiovanni et al. 2011; Srivastava & Khokhar 1996
Amylase inhibitor activity (AIU)	0.8-1.2	Srivastava & Khokhar 1996
Condensed tannins (catechin equivalent %)	0.0-0.77	Hanbury et al. 2000
Hemoagglutinating activity	+	De Falco et al. 2007
Calcium (g kg^{-1})	0.7-2.8	Hanbury et al. 2000; Urga et al. 2005
Phosphorus (g kg^{-1})	2.3-4.4	Hanbury et al. 2000; Urga et al. 2005
Lysine (g 16 g N^{-1})	4.08-9.65	Hanbury et al. 2000
Threonine (g 16 g N^{-1})	2.59-5.15	Hanbury et al. 2000
Methionine (g 16 g N^{-1})	0.24-1.0	Hanbury et al. 2000
Cysteine (g 16 g N^{-1})	1.3-1.53	Hanbury et al. 2000
in vitro digestibility (%)	79.9-93.4	Hanbury et al. 2000; Piergiovanni unpublished

AIU- Amylase inhibitor unit calculated in terms of mg pancreatic amylase inhibited per g of sample.
TIU – Trypsin inhibitor unit.

NUTRITIONAL AND TOXIC FEATURES

Based on the high protein content of seeds, the grass pea can be rightly considered as one of the most promising sources of vegetable food, for wide and expanding populations of marginal areas of Africa and Asia. In Europe, where diversification of cropping systems and more sustainable agriculture are the main concerns, grass pea could represent a good alternative to imported soya, as well as, to the use of animal proteins in animal feed. Data available in the literature show that the content of basic grass pea nutrients as well as the *in vitro* digestibility, are similar to those of the most widely consumed legumes (Hanbury et al. 2000; Yan et al. 2006). On the whole, the amino acid profile of proteins, the ether extract content and the fatty acid composition, and macro- and micro-elements (Table 1), are comparable to those of many grain legumes. Hemoagglutinating activity is almost ten times less than that of common bean, and show a low variation among cultivars (De Falco et al. 2007). Seed processing before consumption has a relevant role in the improvement of nutritional value. Soaking followed by water decantation, seed dehulling, thermal processing, extrusion and fermentation reduce considerably levels and activities of the main antinutritional compounds (Srivastava & Khokhar 1996; Stodolak & Starzynska-Janiszewska 2008). Phenolic compounds are mainly leached during soaking and conventional cooking. Thermal denaturation is the most important strategy in reducing both trypsin inhibitors and inositol phosphates (Stodolak & Starzynska-Janiszewska 2008). Conventional cooking improves protein bioavailability by over 45% as compared to raw seeds, while the increase is over 21% after seed fermentation.

After discovering the neurotoxic effect of β-ODAP, major efforts have done in order to identify the most efficacious post-harvest processing protocols to reduce the toxicity at a safety level for consumption. Water solubility and thermal instability of β-ODAP can be useful tools in seed detoxification (Srivastava & Khokhar 1996). Various methods have been evaluated, but no one is really effective in the complete toxin removal. Contrasting results on the detoxification extent have been reported in the literature. A reduction close to 60% can be obtained adding alkaline substances in cooking water (Kumar et al. 2011). Steeping seeds in a large volume of cold water for 12h or in hot water for 3h removes from 30 to 40% of the neurotoxin. The loss of β-ODAP may reach 90%, cooking seeds soaked in tap water, in an open pan or under pressure (Srivastava & Khokhar 1996). Roasting, dry autoclaving, extrusion or fermentation are less efficient operations, producing a β-ODAP reduction from 40 % to 70% (Grela et al. 2001; Misra et al. 2009; Kumar et al. 2011).

Nutraceutical Features

There can be no doubt that antinutritional compounds contained in seed pulses have a biological function. They have certainly a role in seedling physiology, can be important for plant fitness, or represent defence factors towards animals, microorganisms, and virus. In the last years, there are increasing evidences that some of these compounds could have nutraceutical properties. However, if grass pea could be a rich source of substances promoting human health has not been fully explored.

Like other legumes, the trypsin inhibitors (TI) are the main antinutritional compounds of grass pea seeds. The presence of these compounds in pulse seeds has been related to different functions such as storage of sulphur amino acids, regulation of endogenous proteolytic enzymes, and involvement in the resistance to seed storage pest (Molsov & Valueva 2005). The TIs have a low effect on the nutritional value of cooked seeds since, being proteins, are heat-labile. However, their influence on the nutritional value cannot be disregarded when seeds are used as feed for monogastric animals, whose growth rate is depressed by high TI levels. Studies on *in vitro* and *in vivo* model systems and human trials have shown that TI could have anti-carcinogenic and anti-inflammatory properties (Clemente et al. 2013). This ability is also supported by the observation that populations consuming relatively high amounts of legumes show an inferior incidence of some cancer types. This arouses considerable interest by nutritionists for potential use of TI in cancer therapy. The TI amount in grass pea seeds varies considerably according to cultivar (10 - 44 TIU mg^{-1}), suggesting potential interest for breeding purposes (Aletor et al. 1994; Piergiovanni et al. 2011). Overall, the range of variation is comparable to that of common bean, that is recognised as the most widely grown legume species (Piergiovanni & Pignone 2003). A recent study on fourteen Italian landraces, cultivated for three growing seasons in two different environments, evidenced a threefold variation of TI (11.6 - 33.1 TIU $mg^{-1}dm$) between the extreme values (Piergiovanni et al. 2011). Moreover, this study showed that the TI amount of grass pea, like in other pulses, varies in somewhat among years and growing locations (Table 2). The amount of rainfall during specific vegetative stages seemed able to affect TI storage, being recorded an increase of inhibitors in correspondence of drought during seed development (Piergiovanni et al. 2011).

Table 2. TI variation recorded for fourteen grass pea landraces cultivated for three years in two locations of Southern Italy

Year of cultivation	Growing location	
	A	B
I	11.6-23.9	18.2-33.1
II	15.9-20.2	11.6-15.3
III	12.0-16.4	14.7-19.4

Table 3. Mean values, ranges, coefficients of variation observed in 117 accessions belonging to the grass pea collection held at IGV-CNR, Bari (Italy) gene bank

Trait	Mean	Min	Max	CV (%)
Protein (g kg^{-1})	266	230	311	1.24
Ash (g kg^{-1})	33	21	40	0.27
ODAP (g kg^{-1})	3.3	1.3	6.1	25.07
100 seeds weight (g)	21.2	6.4	72.0	10.17
Yield (t ha^{-1})	2.72	1.61	3.28	12.4

Legumes contain a higher concentration and a more diverse range of non protein amino acids than any other plant species. These compounds constitute the major factor limiting exploitation of alternative grain legumes being mainly concentrated in seeds. As concerns *Lathyrus* species, it is known that all those synthesising β-ODAP also accumulate L-homoarginine (Har). Together these amino acids make up about 90% of ninhydrin-reacting compounds in ethanol extracts (Lambein et al. 1992). Har has multiple and contrasting roles in human and animal physiology. It is converted into lysine by mammalian liver, is a modulator of the biosynthesis of nitric oxide which, in turn, reduces the excitation of neuronal receptors, and could modulate the toxicity of β-ODAP (Shamin et al. 2002), but is also present in gene activator-repressor histones involved in most cancer types. The variation of Har amount in grass pea seeds is still under-investigated. The episodic studies available in the literature, referring to a few genotypes grown in disparate environments for a single year, report contrasting results on the amount in seeds as well as on the variation within the grass pea germplasm. A recent study investigated the Har variation in seven Italian grass pea landraces in relation to year of cultivation and growing location (Piergiovanni & Damascelli

2011). Although this study cannot allow drawing final conclusions, some working hypotheses can be devised. A statistically significant positive correlation ($P < 0.05$) between the amount of β-ODAP and Har was recorded (Figure 1). This might be highly desirable if Har could really modulate the β-ODAP toxicity. Moreover, the effects on Har amount of both season and growing location were found statistically significant (Figure 2). Similarly to other secondary metabolites, also Har storage could be affected by the amount of rainfall during seed development. Lambein et al. (1990) postulated that β-ODAP may have a role in drought tolerance, since its level can double in plants suffering water stress. Our studies (Piergiovanni et al. 2011; Piergiovanni & Damascelli 2011) confirm that drought increases the toxin accumulation in seeds, but add, as new information, that also the TI and Har amounts increase in water stressed plants.

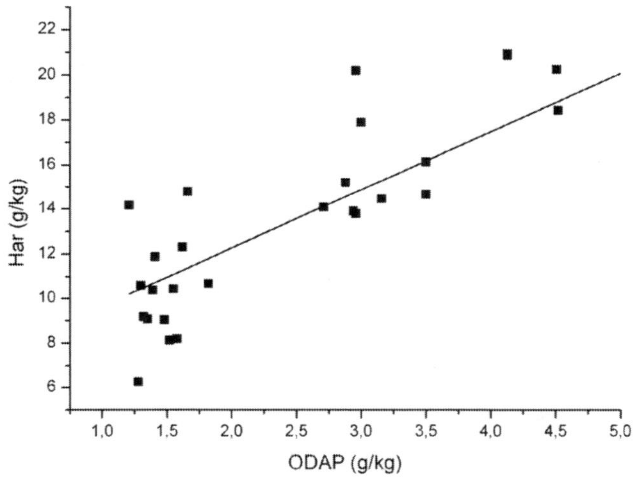

Figure 1. Representation of relationship between β-ODAP and Har content.

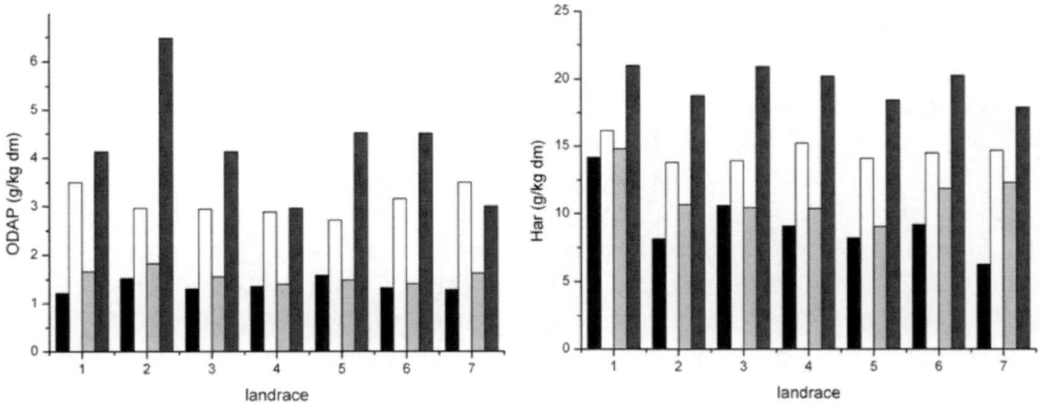

Figure 2. Effect of genotype, year of cultivation and growing location on β-ODAP and Har (left and right diagram, respectively) content recorded for seven grass pea landraces. The code color of bars is: location A - 1th and 2th year, black and white bar, respectively; location B - 1th and 2th year, grey and dark grey bar, respectively.

Polyphenols are a large group of secondary metabolites with functional value and include different subclasses such as flavones, isoflavones and antocyanines. Traditionally, polyphenols have been considered antinutritionals because of their chelating properties that reduce mineral absorption. In recent years polyphenols are receiving more attention for their health promoting effects, mainly related with antioxidant activity, in the treatment of Parkinson and Alzheimer diseases. Experimental evidences showed that they are more abundant in hull than in cotyledons, so a higher content was detected in small-seeded varieties. Pastor–Cavada et al. (2009) reported for grass pea an average value of 4.3 mg g^{-1} , that is lower than those recorded in soybean and chickpea, but higher to that of common bean.

GERMPLASM EXPLOITATION

At the same time of toxicological studies, research efforts have been directed to improve grass pea through genetic and agronomic manipulations (Kumar et al. 2011). Breeders were trying to select grass pea varieties with no or very low β-ODAP content, agronomists were working to find agronomic practices to reduce the storage of this neurotoxin in seeds, and toxicologists were trying to develop post harvesting detoxification processes. Four decades has elapsed since the discovery of β-ODAP as the neurotoxic constituent of grass pea seeds, but, still to the present, breeders could not succeed in release toxin free varieties. Like other species, the germplasm collections of grass pea represent the reservoir of genetic material that breeders can use for schedule any breeding activity. Major collections are present in India, Bangladesh, Canada, Turkey, and Ethiopia with 3754, 2130, 953, 600, and 368 accessions, respectively. The International Centre for Agricultural Research in the Dry Areas (ICARDA, Syria) has assembled a large collection (3001 accessions), with a view firstly to genetic resource conservation and secondly to germplasm exploitation (Robertson et al. 1996). Some European countries have small collections of their own germplasm. Evaluations of these collections have been sporadically carried out, and only for the traits of breeders' interest, such as plant morphology, biomass, seed yield, and seed weight. However, the primary objective of germplasm screenings was the variation of the ODAP content. These efforts failed to find toxin-free genotypes, but indicated a wide range (0.2-7.6 g kg^{-1}) of toxin variation (Kumar et al. 2011). Ethiopian germplasm was the most variable, while accessions from near East have generally a lower content, as compared to those from Indian subcontinent and Ethiopia (Yan et al. 2006). In Italy there are only small collections scattered in national and regional institutions. The Institute of Plant Genetics (IGV-CNR, Bari) maintains a collection including 332 accessions belonging to grass pea and two allied species *L. cicera* L. and *L. ochrus* L. (DC.). The ninety-seven Italian accessions are mostly landraces from Sicily, Apulia, and Basilicata regions (southern Italy). A survey of this collection started in 1999 recording 18 morpho-agronomic traits as well as protein and β-ODAP contents (Granati et al. 2003). The Principal Component Analysis of collected data showed the independent variation of β-ODAP, which suggests that it should be possible to manipulate grain quality traits independently. Further studies on a *Lathyrus* core collection allowed to identify a few lines with useful morpho-agronomic traits (i.e. good yield potential, large and light-coloured seeds), relatively low β-ODAP content (0.24-0.64%), and significant stability of performances over different growing seasons (Polignano et al. 2009).

FUTURE DIRECTIONS

The development of grass pea safe varieties requires new information generated through multidisciplinary researches, being the β-ODAP content a polygenic trait with a complex inheritance (Kumar et al. 2011). In this frame, there is the need to develop a large number of molecular markers, which could be used in different molecular breeding strategies.

Several DNA-based studies on grass pea diversity and relationships with related species, have been so far carried out. Internal Transcribed Spacer (ITS) of nuclear ribosomal DNA, as well as specific regions of chloroplast DNA have been used to estimate the phylogeny of some *Lathyrus* species (Asmussen & Liston 1998; Kenicer et al. 2005). At the same time, some molecular markers such as ISSR (Inter Simple Sequence Repeat), RFLP (Restriction Fragment Length Polymorphism), RAPD (Random Amplified Polymorphic DNA), and STS (Sequence-Tagged-Site) have been successfully used to evaluate interspecific relationships within the *Lathyrus* genus (Croft et al. 1999; Belaïd et al. 2006; Chtouron-Ghorbel et al. 2001; Skiba et al. 2003; 2004).

In particular, AFLPs (Amplified Fragment Length Polymorphisms), a class of molecular markers detecting polymorphism at a great number of *loci*, and requiring no prior sequence knowledge, have proved to be useful for assessing genetic relationships among grass pea landraces (Badr et al. 2002; Tavoletti & Iommarini 2007; Tavoletti et al. 2011).

In a recent study, biochemical (seed storage protein patterns) and molecular (SSR and AFLP) markers were tested to evaluate the levels of genetic variation within and among grass pea landraces mainly collected in southern Italy, so to estimate genetic relationships and evaluate agronomic performances in different southern marginal areas (Lioi et al. 2011). This study supported the general utility of the molecular marker approach for genome analysis, even though a quite low level of polymorphism was observed in the analyzed material. For AFLP analysis, only 83 (17%) of the detected bands were polymorphic. This value is in agreement with the data by Tavoletti & Iommarini (2007), and Tavoletti et al. (2011) who found 110 AFLP polymorphic bands using 13 primer combinations, in 20 populations of *L. sativus*. However, although the number of polymorphic fragments was low, they were able to distinguish all the accessions.

In another class of molecular markers, SSRs (Simple Sequence Repeats) or microsatellites, polymorphism originates from a different number of repetitive core motifs present at one *locus,* and is a valid tool for landrace fingerprinting. Recently, there has been a growing interest towards the development and characterization of this class of molecular markers as testified by the increasing number of papers produced in the last years. SSRs are regions in the genome of tandem repeated DNA segments with each repeat comprising up to six bases. SSRs are inherited in a co-dominant manner, and are multiallelic. They can be analyzed using a convenient PCR-based method, which makes them easy to screen a large number of individuals. Because of their high reproducibility, relatively high polymorphism, and abundance in genome, SSRs have become one of the most widely used molecular markers for genetic analysis (Park et al. 2009; Ohtsuki et al. 2011). Microsatellite markers published so far for grass pea have been obtained searching in the expressed sequence tag (EST) in the European Molecular Biology Laboratory (EMBL) or in the National Center for Biotechnology Information (NCBI) nucleotide sequence databases (Lioi et al. 2011; Ponnaiah et al. 2011; Sun et al. 2012). EST sequences available in the public database provide a cost effective and valuable genome resource for the development of molecular markers. EST-SSR markers are derived from transcribed regions of the DNA, for this reason they are expected to be conserved in the microsatellite flanking regions, where primers are designed. In addition, some EST-SSR markers were developed exploiting the cross-species transferability of markers from related species such as *Lathyrus japonicus* Willd., or from the model species *Medicago truncatula* L. (Lioi et al. 2011; Shiferaw et al. 2012). As underlined before, molecular breeding research is still lagging behind in *Lathyrus* species compared to other grain legumes, such as soybean and common bean. For this reason the development of new SSR *loci* is of great importance. The development of suitable SSR markers represents a priority for obtaining codominant molecular markers useful for linkage map constructions and comparative mapping with related species.

Conclusion

Generally, grain legumes contain substantial amounts of antinutritional compounds that are indigestible, unpalatable, or toxic. Attempts to eliminate them are probably old as mankind. Today several strategies are available to minimize these compounds, but before making a decision, it should be carefully evaluate the biological consequences and economic constrains of removing or reducing the antinutritional compounds. Grass pea is an emblematic example. The development of toxin-free varieties versus the complete ban of this crop becomes a viable option, also in relation to the presence in seeds of compounds having potential healthy effects. A systematic analysis of the minor natural compounds of seeds might produce an unpredictable bounty of health substances. However, as with all crops, there is a limit to the extent that yield can be sacrificed for improving seed quality, being farmers greatly interested in grain yield.

References

Aletor, V.A., El-Moneim, A.A., & Goodchild, A.V. (1994). Evaluation of the seeds of selected lines of three *Lathyrus sativus* spp. for β-N-oxalyl-amino-Lalanine (BOAA), tannins, trypsin inhibitor activity and certain in vitro characteristics. *Journal Science Food Agriculture, 65*, 143-151.

Asmussen, C.B., & Liston, A. (1998). Chloroplast DNA characters, phylogeny, and classification of *Lathyrus* (Fabaceae). *American Journal Botany, 85,* 387-401.

Badr, A., El Shazly, H., El Rabey, H., & Watson, L.E. (2002). Systematic relationships in *Lathyrus* sect. *Lathyrus* (Fabaceae) based on amplified fragment length polymorphism (AFLP) data. *Canadian Journal Botany, 80*, 962-969.

Belaïd, Y., Chtourou-Ghorbel, N., Marrakchi, M., & Trifi-Farah, N. (2006). Genetic diversity within and between populations of *Lathyrus* genus (Fabaceae) revealed by ISSR markers. *Genetic Resources Crop Evolution, 53*, 1413–1418.

Campbell, C.G., Mehra, R.B., Agrawal, S.K., Chen, Y.Z., Abd El-Moneim, A.M., Khawaja, H.I.T., Yadav, C.R., Tay, J.U., & Araya, W.A. (1994). Current status and future strategy in breeding grass pea (*Lathyrus sativus*). *Euphytica, 73*, 167-175.

Chtourou-Ghorbel, N., Lauga, B., Combes, D., & Marrakchi, M. (2001). Comparative genetic diversity studies in the genus *Lathyrus* using RFLP and RAPD markers. *Lathyrus Lathyrism Newsletter, 2*, 62-68.

Clemente, A., Marín-Manzano, M., Arques, M., & Domoney, C. (2013). Bowman-Birk inhibitors from legumes: utilisation in disease prevention and therapy. In: Hernández-Ledesma B. editor, *Bioactive Food Peptides in Health and Disease*, ISBN: 978-953-51-0964-8, Available from: http://www.intechopen.com/books/bioactive-food-peptides-in-health-and-disease/bowman-birk-inhibitors-from-legumes-utilisation-in-disease-prevention-and-therapy

Croft, A.M., Pang, E.C.K., & Taylor, P.W.J. (1999). Molecular analysis of *Lathyrus sativus* L. (grasspea) and related *Lathyrus* species. *Euphytica, 107*, 167-176.

Dadi, L.H., Teklewold, H., Aw-Hassan, A., Abd El-Moneim, A.M., & Bejiga, G. (2003). The socio-economic factors affecting grass pea consumption and the incidence of lathyrism in Ethiopia. Integrated Natural Resources Management ICARDA, *Technical Research Report Series 4*, Aleppo, Syria.

De Bruyn, A., Becu, C., Lambein, F., Kebede, N., Abegaz, N., & Nunn, P. (1994). The mechanism of the rearrangement of the neurotoxin β-ODAP to α-ODAP. *Phytochemistry, 36*, 85-89.

De Falco, E., Piergiovanni, A.R., Zaccardelli, M., Lupo, F., Carbonaro, M., Sparvoli, F., & Giordano, I. (2007). Yield and qualitative characterization of a collection of *Lathyrus sativus* L. CIGR Section VI, 3rd International Symposium, 24-26 September 2007, Naples, Italy, 245.

De la Rosa, L. & Martin, I. (2001). Morphological characterisation of Spanish genetic resources of *Lathyrus sativus* L. *Lathyrus Lathyrism Newsletter, 2*, 31-34.

Dixit, S., Khanna, S.K., & Das, M. (2008). Non-uniform implementation of ban on *Lathyrus* cultivation in Indian states leading to unwarranted exposure of customers. *Current Science, 94*, 570-572.

Enneking, D. (2011). The nutritive value of grass pea (*Lathyrus sativus*) and allied species, their toxicity to animals and the role of malnutrition in neurolathyrism. *Food Chemical Toxicology, 49*, 694-709.

Fikre, A., Yami, A., Kuo, Y.H., Ahmed, S., Gheysen, G., & Lambein, F. (2010). Effect of methionine supplement on physical responses and neurological symptoms in broiler chicks fed grass pea (*Lathyrus sativus*) based starter ration. *Food Chemical Toxicology, 48*, 11-17.

Girma, A., Tefera, B., & Dadi, L. (2011). Grass pea and neurolathyrism: farmers' perception on its consumption and protective measure in north Shewa, Ethiopia. *Food Chemical Toxicology, 49*, 668-672.

Granati, E, Bisignano, V., Chiaretti, D., Crinò, P., & Polignano, G.B. (2003). Characterization of Italian and exotic *Lathyrus* germplasm for quality traits. *Genetic Resources Crop Evolution, 50*, 273–280.

Grela, E.R., Studzinski, T., & Matras, J. (2001). Antinutritional factors in seeds of *Lathyrus sativus* cultivated in Poland. *Lathyrus Lathyrism Newsletter, 2*, 101-104.

Hanbury, C.D., White, C.L., Mullan, B.P., & Siddique, K.H.M. (2000). A review of the potential of *Lathyrus sativus* and *L. cicera* grain for use as animal feed. *Animal Feed Science Technology, 87*, 1-27.

Jiao, C.J., Xu, Q.L., Wang, C.Y., Li, F.M., Li, Z.X., & Wang, Y.F. (2006). Accumulation pattern of toxin β-ODAP during lifespan and effect of nutrient elements on β-ODAP content in *Lathyrus sativus* seedlings. *Journal Agricultural Sciences, 144*, 369-375.

Kenicer, GJ., Kajita, T., Pennington, R.T., & Murata, J. (2005). Systematics and biogeography of *Lathyrus* (Leguminosae) based on internal transcribed spacer and cpDNA sequence data. *American Journal Botany, 92*, 1199-1209.

Kislev, M.E. (1989). Origin of the cultivation of *Lathyrus sativus* L. and *L. cicera* (Fabaceae). *Economic Botany, 43*, 262-270.

Kumar, S., Bejiga, G., Ahmed, S., Nakkoul, H., & Sarker, A. (2011). Genetic improvement of grass pea for low neurotoxin (β-ODAP) content. *Food Chemical Toxicology, 49*, 589-600.

Lambein, F., Khan, J.K., & Kuo, Y.H. (1992). Free amino acids and toxins in *Lathyrus sativus* seedlings. *Planta Medica, 58*, 380-381.

Lambein, F., Ongena, G., & Kuo, Y.H. (1990). B-isoxazolinone-alanine is involved in the biosynthesis of the neurotoxin, β-N-oxalyl-L-α,β-diaminopropionic acid. *Phytochemistry, 29*, 3793-3796.

Lioi, L., Sparvoli, F., Sonnante, G., Laghetti, G., Lupo, F., & Zaccardelli, M. (2011). Characterization of Italian grasspea (*Lathyrus sativus* L.) germplasm using agronomic traits, biochemical and molecular markers. *Genetic Resources Crop Evolution, 58*, 425-437.

Mera, M., Montenegro, A., Espinoza, A., & Guete, N. (2000). Research backs grass pea exports by small Chilean farmers. *Lathyrus Lathyrism Newsletter, 1*, 31.

Mehta, T., Parker, A.J., Cusick, P.K., Zarghami, N.S., & Haskell, B.E. (1983). The *Lathyrus sativus* neurotoxin: resistance of the squirrel monkey to prolonged oral high doses. *Toxicology Applied Pharmacology,* 69, 480-484.

Milczak, M., Pędziński, M., Mnichowska, H., Szwed-Urbaś, K., & Rybiński, W. (2001). Creative breeding of grass pea (*Lathyrus sativus* L.) in Poland. *Lathyrus Lathyrism Newsletter, 2,* 18-23.

Misra, R., Martin, A., & Gowda, L. (2009). Detection of 3-N-oxalyl-L-2,3-diaminopropanoic acid in thermally processed foods by reverse phase high performance liquid chromatography. *Journal Food Composition Analysis, 22*, 704-708.

Molsov, V.V., & Valueva, T.A. (2005). Proteinase inhibitors and their function in plants: a review. *Applied Biochemical Microbiology, 41*, 227-246.

Ohtsuki, T., Kaneko, Y., Mitsui, Y., & Setoguchi, H. (2011). Isolation and characterization of microsatellite loci in the beach pea, *Lathyrus japonicus* (Fabaceae), in Japan. *American Journal Botany, 98,* e375-e377.

Pastor-Cavada, E., Juan, R., Pastor, J.E., Alaiz, M., & Vioque, J. (2009). Antioxidant activity of seed polyphenols in fifteen wild *Lathyrus* species from South Spain. *LWT-Food Science Technology, 42*, 705-709.

Park, Y,-J., Lee, J.K., & Kim, N.-S. (2009). Simple sequence repeat polymorphism (SSRPs) for evaluation of molecular diversity and germplasm classification of minor crops. *Molecules, 14*, 4546-4569.

Peoples, M.B., Brockwell, J., Herridge, D.F., Alves, B.J.R., Urquiaga, S., Boddey, R.M., Dakora, F.D., Bhattarai, S., Maskey, S.L., Sampet, C., Rerkasem, B., Nielsen, H.H., & Jensen, E.S. (2008). Biological fixation by food legumes. In: Kharkwal MC, editor. *Food legumes for nutritional security and sustainable agriculture*. Vol. 1, Indian Society of Genetics and Plant Breeding, New Delhi, India, 435-453.

Piergiovanni, A.R., & Damascelli, A. (2011). L-homoarginine accumulation in grass pea (*Lathyrus sativus* L.) dry seeds. A preliminary survey. *Food and Nutrition Sciences, 2*, 207-213.

Piergiovanni, A.R., Lupo, F., & Zaccardelli, M. (2011). Environmental effect on yield, composition and technological seed traits of some Italian ecotypes of grass pea (*Lathyrus sativus* L.). *Journal Science Food Agriculture, 91*, 122-129.

Piergiovanni, A.R., & Pignone, D. (2003). Effect of year-to-year variation and genotype on trypsin inhibitor level in common bean (*Phaseolus vulgaris* L.) seeds. *Journal Science Food Agriculture, 83*, 473-476.

Polignano, G.B., Bisignano, V., Tomaselli, V., Uggenti, P., Alba, V., & Della Gatta, C. (2009). Genotypes x environment interaction in grass pea (*Lathyrus sativus* L.) lines. *International Journal Agronomy*, vol. 2009, Article ID 898396, 7 pages, doi: 10.1155/2009/898396.

Ponnaiah, M., Shiferaw, E., Pè, M.E., & Porceddu, E. (2011) Development and application of EST-SSRs for diversity analysis in Ethiopian grass pea. *Plant Genetic Resources, 9*, 276-280.

Rao, S.L.N. (1978). Entry of β-N-oxalyl-L-α,β-diaminopropionic acid, the *Lathyrus sativus* neurotoxin into the central nervous system of the adult rat, chick and the rhesus monkey. *Journal Neurochemistry, 30*, 1467-1470.

Rao, S.L.N. (2011). A look at the brighter facets of β-N-oxalyl-L-α,β-diaminopropionic acid, homoarginine and the grass pea. *Food Chemical Toxicology, 49*, 620-622.

Rao, S.L.N., Adiga, P.R., & Sharma, P.S. (1964). Isolation and characterisation of β-N-oxalyl-L-α,β-diaminopropionic acid, a neurotoxin from seeds of *Lathyrus sativus*. *Biochemistry, 3*, 432-436.

Robertson, L.D., Singh, K.B., Erskine, W., & Abd El-Moneim, A.M. (1996). Useful genetic diversity in germplasm collections of food and forage legumes from West Asia and North Africa. *Genetic Resources Crop Evolution, 43*, 447–460.

Shamin, M.Z., Hossain, M.S., Islam, K., Yusuf, H.K.M., & Lambein, L. (2002). Mechanism of ODAP toxicity in one-day-old chicks. *Dhaka University Journal Biology Science, 11*, 1-7.

Shiferaw, E., Pè, M.E., Porceddu, E., & Ponnaiah, M. (2012). Exploring the genetic diversity of Ethiopian grass pea (*Lathyrus sativus* L.) using EST-SSR markers. *Molecular Breeding, 30*, 789-797.

Skiba, B., Ford, R., & Pang, E.C.K. (2003). Amplification and detection of polymorphic sequence-tagged sites in *Lathyrus sativus*. *Plant Molecular Biology Report, 21*, 391-404.

Skiba, B., Ford, R., & Pang, E.C.K. (2004). Construction of a linkage map based on a *Lathyrus sativus* backcross population and preliminary investigation of QTLs associated with resistance to ascochyta blight. *Theoretical Applied Genetics, 109*, 1726-1735.

Srivastava, S., & Khokhar, S. (1996). Effects of processing on the reduction of β-ODAP (β-N-oxalyl-L-α,β-diaminopropionic acid) and anti-nutrients of khesari dhal, *Lathyrus sativus*. *Journal Science Food Agriculture, 71*, 50-58.

Stodolak, B., & Starzynska-Janiszewska, A. (2008). The influence of tempeh fermentation and conventional cooking on anti-nutrient level and protein bioavailability (*in vitro* test) of grass pea seeds. *Journal Science Food Agriculture, 88,* 2265-2270.

Sun, X.L., Yang, T., Guan, J.P., Ma, Y., Jiang, J.Y., Cao, R., Burlyaeva, M., Vishnyakova, M., Semerova, E., Bulyntsev, S., & Zong, X.-X. (2012). Development of 161 novel EST_SSR markers from *Lathyrus sativus* (Fabaceae). *American Journal Botany*, e379-e390.

Tadesse, W., & Bekele, E. (2003). Variation and association of morphological and biochemical characters in grass pea (*Lathyrus sativus* L.). *Euphytica, 130*, 315-324.

Tavoletti, S., & Iommarini, L. (2007). Molecular marker analysis of genetic variation characterizing a grasspea (*Lathyrus sativus*) collection from Central Italy. *Plant Breeding, 126*, 607-611.

Tavoletti S., Iommarini, L., Crinò, P. & Granati, E. (2005). Collection and evaluation of grass pea (*Lathyrus sativus* L.) germplasm of central Italy. *Plant Breeding, 124*, 388–391.

Tavoletti, S., Iommarini, L., & Mogliani, L. (2011). Agronomic qualitative (β-ODAP) and molecular variability in grasspea populations of the Marche region (central Italy). *Food Chemical Toxicology, 49*, 601-606.

Tekle Haimanot, R., Kidane, Y., Wuhib, E., Kassina, A., Endeshaw, Y., Alemu, T., & Spencer, P.S. (1993). The epidemiology of lathyrism in north and central Ethiopia. *Ethiopian Medical Journal, 31,* 15-24.

Urga, K., Fufa, H., Biratu, E., & Husain, A. (2005). Evaluation of *Lathyrus sativus* cultivated in Ethiopia for proximate composition, mineral and β-ODAP and antinutritional components. *African Journal Food Agriculture Nutritional Development, 5*, 1-15.

Yan, Z.Y., Spencer, P.S., Li, Z.X., Liang, Y.M., Wang, Y.F., Wang, C.Y., & Li, F.M. (2006). *Lathyrus sativus* (grass pea) and its neurotoxin ODAP. *Phytochemistry, 67*, 107-121.

In: Legumes
Editors: Hiroto Satou and Ren Nakamura
ISBN: 978-1-62808-280-7

Chapter 7

LENTILS (*LENS CULINARIS* L.): LINKING WHOLE FOODS FOR BETTER HUMAN HEALTH

Dil Thavarajah[1*], Pushparajah Thavarajah[2] and Shiv Kumar[3]

[1] School of Food Systems, North Dakota State University, Fargo, North Dakota
[2] School of Food Systems, North Dakota State University, Fargo, North Dakota, USA
[3] International Centre for Agricultural Research in the Dry Areas, Rabat-Institutes, Rabat, Morocco

ABSTRACT

The failure to link current agricultural production with human nutritional needs has led to the development of unhealthy food systems, which cause both malnutrition and chronic diseases. Malnutrition is responsible for the deaths of 30% of children under 5 years of age and accounts for 11% of the global disease burden. Poor diets and nutrition resulting from unhealthy food systems also create huge economic burdens for healthcare systems with negative consequences on child development and long-term direct impacts on sustainable development. Therefore, sustainable agricultural systems are essential for providing food energy and sufficient daily nutrients for humans, and in so doing support health and general well-being. Cool season food legumes, especially lentils (*Lens culinaris* L.), have the potential to provide adequate amounts of protein, prebiotic carbohydrates, and essential micronutrients; a 50 g serving can provide 3.7-4.5 mg iron (Fe), 2.2-2.7 mg zinc (Zn), 22-34 μg of selenium (Se), 50-250 μg of beta-carotene, and 50-300 μg of folates. Unlike other grains, lentils are very low in phytic acid (2.5-4.4 mg g^{-1}), which binds Fe and Zn and thus renders these nutrients poorly bioavailable. The present chapter provides an overview of current global lentil production, global malnutrition issues, and the promise of pulse crops, particularly lentil, as a whole food solution to combat global malnutrition issues. In addition, the superior mineral bioavailability profiles and genetic potential of lentil for further enrichment of bioavailable mineral micronutrients will be briefly discussed with respect to developing sustainable food systems.

Keywords: Lentils, biofortification, food systems, micronutrient malnutrition, minerals.

---

[*] Corresponding author: dilrukshi.thavarajah@ndsu.edu.

Global Hidden Hunger

The failure to successfully link agricultural production with human nutrition and health has led to the development of unhealthy food systems. These food systems can result in basic malnutrition and high-calorie diets, both of which are linked to common non-communicable diseases. Malnutrition and diet-related diseases together are the leading cause of deaths worldwide, accounting for approximately 20 million deaths per year (Kennedy et al. 2003). In particular, micronutrient malnutrition causes five million childhood deaths per year (Kennedy et al. 2003).

"Malnutrition" refers to a lack of essential nutrients necessary for human health and general wellbeing. The two types of malnutrition caused by unhealthy food systems are protein-energy and micronutrient malnutrition. Micronutrient malnutrition, also known as "hidden hunger", affects more than 2 billion people worldwide. "Hidden hunger" can impact physical and mental growth, organ function, immune system development, and can cause neurological and cognitive impairments. Women and preschool children have been found to be the most vulnerable (WHO, 2013). The present chapter primarily focuses on micronutrient malnutrition.

Stunting, underweight, and wasting are the major global health concerns linked to micronutrient malnutrition. Approximately 165 million children (26%) around the world are stunted (i.e. height-for-age below-2SD), 101 million (16%) are underweight (i.e. weight for age below-2SD), and 52 million (8%) are wasted (i.e. weight for height below-2SD) (de Onis et al. 2012). Some nutrient deficiencies are more common than others and the effects may vary. It is estimated that over 60% of the world's 7 billion people are Fe deficient, over 30% Zn deficient, 30% I deficient, and more than 15% are selenium (Se) deficient. Approximately 3 million children around the world develop xerophthalmia (damage to the cornea of the eye) and every year more than half a million children lose eyesight; both diseases are caused by vitamin A (beta-carotene) deficiency (WHO, 2013).

What really caused the present-day micronutrient malnutrition? In the mid-20th century a set of initiatives were undertaken resulting in a global phenomenon called the "green revolution." One of the green revolution's purposes was to produce calorie-rich foods to prevent mass starvation in populations around the globe. The green revolution mostly achieved its goals by developing and expanding cereal crop production to meet much of the world's calorie-based energy demands. However, while these initiatives did prevent mass starvation, they had other unintended consequences. The expansion of cereal crop production pushed the production of food legume crops to less-endowed marginal areas, thus negatively impacting global legume production levels (Bouis and Welch 2010; Graham and Welch 2000). Since 1970 South Asian cereal production has increased four times (Welch and Graham 1999). As a consequence, pulse production has decreased significantly in developing countries during the last 50 years.

In the present day, these calorie-rich, micronutrient-deficient diets are not only affecting people's health but are also placing a burden onto the economy. Food and nutrition insecurity leads to energy and nutritional deficiencies, infectious diseases, and chronic diseases, all of which negatively affect the agricultural productivity of smallholder farmers in Asia and Africa. These diseases also affect the health care industry, where malnutrition is straining health care resources. These issues can be resolved by developing sustainable food production with healthy and diverse crops such as traditional pulse crops. Various measures can enable these traditional pulse crops to serve as a foundation for economic development and healthful food systems. These measures include biofortification (breeding for micronutrient-enhanced crops), crop selection, critical agricultural inputs, and the demonstration of these crops' economic and nutritional benefits. Among all the possible target crops, the incorporation of nitrogen-fixing pulses provides the greatest promise, as they can potentially accomplish all of these goals.

Linking Food Systems to Human Health

Every human requires at least 50 nutrients including water, carbohydrates, essential fatty acids, amino acids, vitamins, trace elements, and minerals to maintain a healthy life style (Table 1; Welch and Graham 2005). However, most food systems are not well-constructed around these nutrients, leading to micronutrient malnutrition, obesity-linked diseases, and the economic burdens discussed earlier. These issues could be reduced by the integration of traditional pulse crops with cereals, oils seeds, fruits, and vegetables. An ideal conceptual food system is shown in Figure 1. A healthy food system approach links the processes of production, acquisition, and utilization in regards to nutritious foods, leading to multiple benefits including healthier human activities and improved sustainable crop production (Welch and Graham 2005). Agriculture is the primary source of all micronutrients for human consumption; for that reason, the integration of crops and farming systems for greater micronutrient delivery through agriculture might be a sustainable solution to reduce micronutrient deficiencies.

By linking agriculture to nutrition research, food systems can be established which boost food security globally. This can be achieved by: (1) determining current ecological and health-related opportunities and challenges faced in different food systems around the world; (2) determining the impact of improving healthful food systems on diet and health; (3) developing sustainable agricultural and food security approaches to produce healthy foods by introducing traditional legume crop systems with animal husbandry; (4) developing appropriate health food marketing approaches for value-added functional foods; and (5) increasing public awareness of the relationship between sustainable ecology, food systems, diet, and health (Welch and Graham 2005).

Past attempts to address micronutrient malnutrition include food fortification, dietary supplementation, dietary diversification, and agronomic-fortification of staple crops, with each having only limited success (Broadley et al. 2006; Combs, Jr. 2001; Welch 2002). Sustainable solutions to micronutrient malnutrition call for approaches linking food systems with human dietary needs. Biofortification, the process of enriching micronutrients into staple food crops through conventional plant breeding and modern biotechnology, provides a sustainable, long-term solution to global micronutrient malnutrition. For example, the biofortification of food crops with highly bioavailable iron (Fe), zinc (Zn), selenium (Se), iodine (I), beta-carotene, and folic acid is urgently needed to combat micronutrient malnutrition in both the US and the developing world. HarvestPlus has made progress on the biofortification of rice (*Oryza sativa* L.), wheat (*Triticum aestivum* L.), cassava (*Manihot esculenta* Crantz), maize (*Zea mays* L.), pearl millet (*Pennisetum americanum* Leeke), and common bean (*Phaseolus vulgaris* L.) (Saltzman et al., 2013). Limited work has been done on the biofortification and bioavailability of micronutrients in pulse crops.

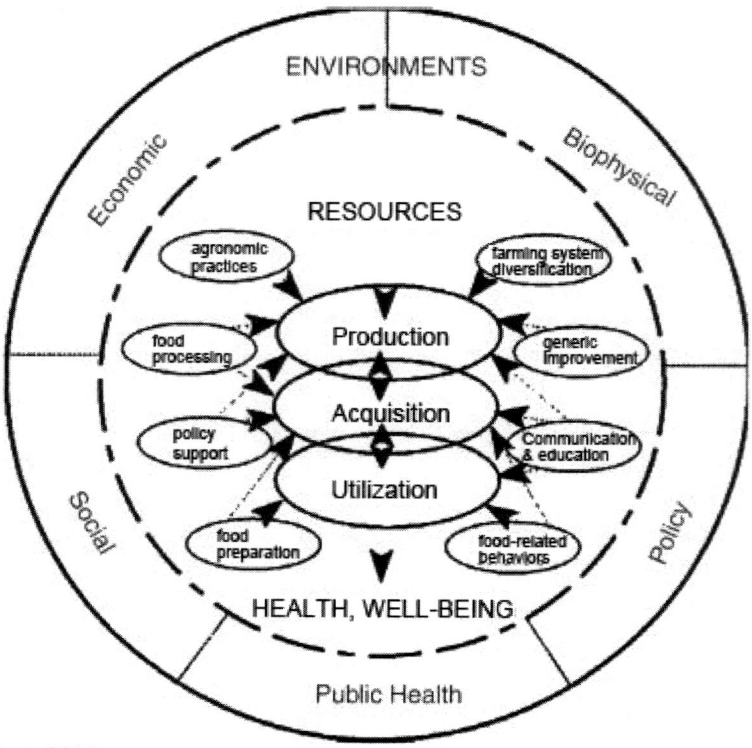

Welch and Graham 2005.

Figure 1. Conceptual model of food system. Adapted with permission.

Several agricultural tools are now available to increase the level of micronutrients in staple food crops so as to meet human nutrition requirements. These include: (1) agronomic biofortification including fertilization, crop diversification, rotation, and location sourcing, and (2) genetic biofortification through conventional plant breeding and GMO techniques. Agronomic biofortification through micronutrient fertilization is a widely used method for Zn, I, Se, copper (Cu), and nickel (Ni). For example, an effective Se fertilization has been reported from Finland, Australia, and New Zealand (Lyons et al. 2009; Lyons et al. 2003; Lyons et al. 2005). In Australia, soil application of Se increased wheat grain Se levels by 20- to 133-fold, and foliar application increased levels by 6- to 20-fold in wheat (Lyons et al. 2004).

However, agronomic biofortification may not always be feasible, especially for Fe, because of the tight homoeostatic mechanisms that regulate Fe absorption, translocation, and redistribution within the plants (Grusak et al. 1999; Welch 1995). The regulation of the homoeostatic control mechanism in plants is not fully understood (Grusak et al. 1999). Furthermore, soil or foliar application of soluble ferrous form is not an effective method for increasing Fe levels in plants due to the rapid oxidation in soils and low mobility in phloem (Graham and Welch 2000). Chromium, boron, and vanadium are also ineffective on agronomic biofortification because of their low phloem mobility.

Table 1. The essential major- and micro-nutrients for humans

Category	Component
1. Energy	Carbohydrate
2. Water	Water
3. Proteins (amino acids)	Histidine, isoleucine, leucine, lysine, methionine, phenylalanine, threonine, tryptophan, valine
4. Lipids (fatty acids)	Linoleic acid, linolenic acid
5. Macronutrients	Na, K, Ca, Mg, S, P, Cl
6. Microelements	Fe, Zn, Cu, Mn, I F, B, Se, Mo, Ni, Cr, Si, As, Li, Sn, V, Co (B_{12})
7. Vitamins	A, D, E, K, C, B_1 (thiamin), B_2 (riboflavin), B_3 (pantothenic acid), B_6, folic acid, biotin, niacin, B_{12} (cobalamin)

Welch and Graham, 2005.

Global Lentil Production

Lentil (*Lens culinaris* ssp. *culinaris*) is an important food legume crop with various uses as food and feed because of its protein and micronutrient-rich grains and straw. Globally, it is cultivated as a rainfed crop on 4.08 million ha area with 4.36 million tons of production during 2009-2011 (Table 2; FAOSTAT, 2013). The major geographical regions of lentil production are North and South America (45%), South Asia and China (31%), Central and West Asia and North Africa (14%), Sub-Saharan Africa (2%), and Australia (2%).

South Asia grows lentil on 1.85 million ha with 1.36 million tons of production exclusively as a post-rainy season crop on residual moisture. West Asia and North Africa's (WANA) main producers are Turkey, Syria, Iran, and Morocco; these countries grow winter and spring-planted lentils on 0.58 million ha with 0.63 million ton production. In Sub-Saharan Africa, Ethiopia and Eritrea are the major producers with 0.11 million ton lentil from a 0.10 million ha area. In recent years, area under lentil has expanded manifolds in the Northern Great Plains of North America (Canada and USA), which produces 1.97 million ton lentil and has emerged as the foremost production base. Australia grows lentil on 0.16 million ha area with 0.22 million tons production. The average productivity of lentil in South Asia and WANA region is less than the world average mainly due to the prevalence of different diseases with varying intensity.

The top 11 countries, namely Canada, India, Turkey, USA, Australia, Nepal, China, Ethiopia, Syria, Iran, and Bangladesh account for 96% of the global production. World lentil production has risen steadily by nearly five times (475 %) in the past 50 years, from an average of 0.92 million tons in 1961-63 to 4.36 million tons in 2009-11. This growth is primarily from an expanding harvested area from 1.64 million ha in 1961-63 to 4.08 million ha in 2009-11. Additionally, this increase also reflects an improvement in productivity from an average yield of 560 kg/ha in 1961-63 to 1070 kg/ha in 2009-11 (FAOSTAT, 2013).

Table 2. Global trends in lentil production

Period	World			Developing countries			Developed countries		
	Area (m ha)	Production (m tonnes)	Yield (kg/ha)	Area (m ha)	Production (m tonnes)	Yield (kg/ha)	Area (m ha)	Production (m tonnes)	Yield (kg/ha)
1961-63	1.64	0.92	559	1.38	0.71	517	0.26	0.20	787
1971-73	1.81	1.08	596	1.53	0.82	540	0.28	0.25	905
1981-83	2.51	1.68	667	1.79	0.99	554	0.72	0.69	950
1991-93	3.33	2.67	802	2.18	1.54	703	1.14	1.13	990
2001-03	3.73	3.04	814	2.49	1.73	696	1.25	1.31	1049
2006-08	3.62	3.16	873	2.37	1.65	696	1.25	1.51	1207
2009-11	4.08	4.36	1070	2.34	1.73	740	1.74	2.63	1514

FAOSTAT, 2013.

Lentil consumption is highest in Asia, where seven of the top 11 lentil-producing countries lie. Most production (56%) is consumed locally with only 44% of the global production exported (FAOSTAT, 2013). The global import and export of lentil are 1.71 and 1.79 million tons, worth 1.55 and 1.48 million USD, respectively. Lentil remains in short supply in South Asia and WANA regions due to a gap between present production (1.19 and 0.48 million tons) and consumption (1.83 and 0.88 million tons). The demands for lentil in these two regions are expected to rise further due to population growth and rising income. By 2030, the world lentil consumption is estimated at 5.5 million tons, an increase of almost 1 million tons from the present level (Clancey 2009).

BIOFORTIFICATION OF LENTILS

Lentil belongs to the genus *Lens*. All members of the genus *Lens* are self-pollinating diploids (2n=2x=14) and the haploid genome size of the cultivated species, *Lens culinaris* ssp. *culinaris* is 4,063 Mbp (Baum et al. 2008). There are two types of cultivated lentils, small-seeded microsperma and large-seeded macrosperma. Lentil is a slender, semi-erect annual plant with compound leaves, white to pale blue/violet flowers, and small pods with one or two seeds. Seeds are generally lens shaped with a range of cotyledon colors including red, yellow, and green. The International Center for Agricultural Research in the Dry Areas (ICARDA) has a global mandate for research on lentil improvement for dry areas around the world (Baum et al. 2008). The following sections will discuss recently published lentil biofortification research efforts.

Minerals

Iron and Zinc

Lentils are rich in proteins (20-30%), low-digestible carbohydrates, dietary fiber, and a range of micronutrients (Bhatty 1988). Lentil provides protein nutritional benefits by virtue of its amino acid profiles, which complement those of cereal proteins to produce a dietary protein mixture of high biologic value. It is considered a "clean crop" due to its high nutrition quality, low levels of antinutrients, and flatulence-causing carbohydrates, and it produces a substantially low glycemic index when compared to cereals and other pulses (Bhatty 1988).

Mineral biofortification research on lentil was initiated in ICARDA in early 2000. Approximately 1000 lentil accessions were evaluated for Fe and Zn concentrations. The results indicated wide genetic variation for Fe (41-109 mg/kg) and Zn (22-78 mg/kg) concentrations in lentil seeds (Baum et al. 2008). Furthermore, multi-location testing for biofortified lentils began in 2009 in Bangladesh, Ethiopia, India, Nepal, and Syria. Since then, significant progress has been made; the high Fe and Zn lentil cultivar ILL 7723 is proposed to release soon (Saltzman et al. 2013). Currently available biofortified Fe and Zn varieties from ICARDA are listed in Table 3.

Although biofortified lentil cultivars have been released in Southeast Asia, its production has been almost stagnant during the last 50 years. This stagnation is particularly visible in developing countries, and can be attributed to the relegation of its cultivation to marginal soils resulting in poor yields. For example, in Bangladesh farmers are shifting from lentils to high-yielding cultivars of rice and wheat. The main drivers causing the shift from lentils to other crops are poor soil conditions, frequent outbreak stemphylium blight disease, and non-availability of extra-short duration lentil cultivars which can fit into cereal-based systems. Therefore, biofortification efforts should focus on agronomic traits such as yield, phenology, disease resistance, and stress tolerance.

Table 3. List of high Fe and Zn lentil varieties

Variety	Target mineral	Concentration (mg/kg)	Released country
Idlid-2(ILL 5883)	Fe	73	Syria
Idlib-3(ILL 6994)	Fe	72	Syria
Beleza (ILL 7711)	Fe	74 for Fe; 56 for Zn	Portugal
Alemaya	Fe and Zn	82 for Fe; 66 for Zn	Ethiopia
Sisir	Fe and Zn	98 for Fe; 64 for Zn	Nepal
Khajurah-2	Fe and Zn	94 for Fe; 54 for Zn	Nepal
Barimasur-5	Fe and Zn	86 for Fe; 59 for Zn	Bangladesh
Barimasur-6	Fe and Zn	86 for Fe; 63 for Zn	Bangladesh

Baum et al. 2008; Saltzman et al. 2013.

Recent research in Canada has highlighted the micronutrient values of Saskatchewan-grown pulse crops (Thavarajah et al. 2011a). Canadian lentils are naturally high in total Fe, Zn, and Se, especially when compared with legumes grown in other world production areas (Baum et al. 2008; Thavarajah et al. 2008-; Thavarajah et al. 2009a). For example, 100 g of Canadian-grown lentils provides a minimum of 50-100% of the daily recommended allowances (RDA) for these nutrients. Moreover, data from Canadian-grown pulses show lentil has the genetic potential for increased amounts of total Fe, Zn, Ca, and Mg. Broad-sense heritability estimates for these elements are high, demonstrating the possibility to breed lentil cultivars that accumulate these elements in seed despite environmental influences (Thavarajah et al. 2008-; Thavarajah et al. 2009a). While the Canadian data show the promise of lentils, similar trends were observed in US-grown pulses including lentils, field pea (*Pisum sativum* L.), and chickpea (*Cicer arietinum* L.) (Amarakoon et al. 2012; Johnson et al. 2013b; Thavarajah and Thavarajah 2012). In addition, several studies were reported for Fe and Zn values for lentils grown in Spain, France, and the United Kingdom (Table 4). Overall, lentil is naturally rich in Fe and Zn; however, the micronutrient concentrations can be further enhanced by location sourcing and the selection of specific genetic materials which have been shown to enrich micronutrients.

Table 4. Iron and Zn composition (mg/kg) of lentils with different origins

Origin	Fe (mg/kg)	Zn (mg/kg)	Reference
ICARDA	41-109	22-78	Baum et al. 2008
Canada	73-90	44-54	Thavarajah et al. 2009
USA	56-70	28-38	Johnson et al. 2013b
Spain	82	37	Campos-Vega et al. 2010
France	80	-	Campos-Vega et al. 2010
UK	111-122	39-48	Campos-Vega et al. 2010

Selenium

Selenium (Se) is another essential micronutrient for biofortification research as 30 to 100 million people are Se-deficient. This deficiency is mostly due to low concentrations of Se in commonly eaten staple foods (Combs, Jr. 2001; Ellis and Salt 2003). Selenium is known to be an essential component of 25 selenoproteins. The consumption of selenium has been shown to help reduce the risk of cancer and viral infections, and has also proven useful in HIV treatment and protection from heavy metals including arsenic, cadmium, and mercury (Combs, Jr. et al. 2002; Combs, Jr. 2001; Thavarajah et al. 2007).

Selenium application might be necessary for developing countries to maximize lentil quality. In the past, Se-enriched fertilizers have been used in Finland to increase Se concentration in their major food crops (Wang et al., 1995). The addition of Se to Finnish fertilizer (16 μg/g as sodium selenate) resulted in an increase in Finnish people's daily Se intake from 39 to 110 μg of Se per day (Wang et al., 1995).

The Se levels of lentils produced in different countries vary with soil Se content. The highest Se concentrations were found in lentils from Nepal and Australia (180 and 148 $\mu g\ kg^{-1}$, respectively) and the lowest Se concentrations were found in lentils from Syria, Morocco, and Turkey (22, 28, and 47 $\mu g\ kg^{-1}$, respectively; (Thavarajah et al. 2011b). The Se concentrations found in lentils from North America were considerably higher. The total Se concentrations of 19 lentil genotypes grown at eight locations for two years in Saskatchewan, Canada ranged from 425 to 673 $\mu g\ kg^{-1}$ (Thavarajah et al. 2008). In addition, the most recent data from North Dakota shows that mean Se concentrations from lentil ranged from 300-1445 $\mu g\ kg^{-1}$, respectively. Therefore, changes in soil Se concentration might play a greater role in determining the lentil seed Se concentration. Se uptake by lentil plant may be governed by several genes, and genetic selection for Se uptake might require minimum environmental influence. The broad sense heritability in Canadian-grown lentils was 40%, which could indicate significant environmental influence on Se uptake. However, there was 4-5-fold difference in Se uptake by lentil genotypes, indicating a significant genetic potential for enriching Se in lentils.

MINERAL BIOAVAILABILITY ISSUE

Biofortification efforts are meaningful only when the micronutrients are bioavailable. Generally, plant-based diets are rich in antinutrients that interfere with the absorption or utilization of minerals in human gut. Bouis and Welch (2010) reported that approximately 5% of the total Fe and 25% of the total Zn present in the seeds of staple food crops are bioavailable. Therefore, increasing the bioavailable form of Fe from 5 to 20% would be equivalent to increasing the total Fe by fourfold. In addition, mineral bioavailability is mainly dependent on the food matrix factors i.e. bioavailability promoters and inhibitors. Lentils have high concentrations of mineral bioavailability promoters (e.g., ascorbic acid, beta-carotene, and prebiotics) and low concentrations of inhibitors (e.g., phytic acid, and polyphenols).

Prebiotics

One benefit of the US-grown lentil in particular is its high quantity of prebiotics. In 1997, prebiotics were first defined as a "non-digestible food ingredient that beneficially affects the host by selectively simulating the growth and/or activity of one or a limited number of bacteria in the colon, and thus improves host health" (Gibson and Roberfroid 1995). Almost all food oligosaccharide and polysaccharide including dietary fiber have been identified as prebiotics; however, it is important to highlight that not all dietary carbohydrates are considered prebiotics. Therefore, clear guidelines have been established for classifying a food ingredient as prebiotic based on the following: (1) the food component or ingredient resists host digestion, absorption, and adsorption processes, (2) the food ingredient is fermented by the microflora colonizing the gastrointestinal system, and (3) the food ingredient selectively stimulates the growth and/or the activity of one or a limited number of bacteria within the gastrointestinal system (Gibson 2004).

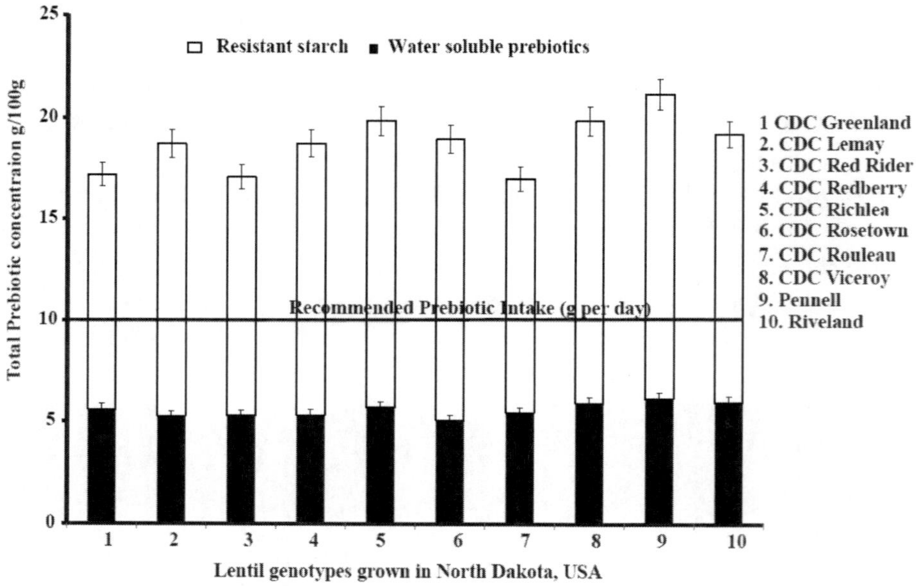

Original data from Jhonson et al., 2013a .

Figure 2. Mean resistant starch and water-soluble prebiotic concentrations of 10 lentil genotypes grown in North Dakota, USA.

Recent discoveries show that the consumption of prebiotics can lead to multiple benefits. For example, recent studies suggest the intestinal microbiome and a prebiotic-rich, low-caloric diet could play important roles in combating obesity and related diseases (Dumas et al. 2006; Wu et al. 2011). The intestinal microbiota ferments food ingredients that cannot be digested by the host in the small gut. These ingredients include resistant starch, non-digestible carbohydrates, oligosaccharides, proteins, lignin, and mucins (Gibson 2004). Two groups of bacteria are present in the human gut: the Bacteroidetes and the Firmicutes. The relative proportion of Bacteroidetes is decreased in obese individuals compared to lean individuals; however, the relative proportion of Bacteroidetes can be improved when weight loss occurs with a prebiotic-rich, low-caloric diet. Furthermore, the consumption of prebiotics like non-digestible, fermentable carbohydrates may stimulate the growth and activity of hind gut bacteria by producing short-chain fatty acids with a variety of benefits. These include providing an energy source for colonocytes, strengthening the gut mucosal barrier, and suppressing the colonization of pathogens (Gibson and Roberfroid 1995; Gibson 2004).

The major dietary sources of prebiotics are cereals, pulses, fruits, and vegetables. Prebiotics found in cereal grain products range from 1.12 g in wheat to 0 g in rice, 0.6 g in dark rye bread, 0.11 g in oats, and 0.05 g in potato chips per 100 g per portion size. Interestingly, US-grown lentils may provide over 13 g of prebiotics per 100 g serving (Johnson et al. 2013a; Figure 2). Therefore lentils offer new opportunities to increase prebiotic intake and as a potential food to combat major non-communicable diseases.

ANTINUTRIENTS

The nutritional value of pulses depends on the nutrient and antinutrient concentrations in a given portion size. Pulses, especially lentil, are a good source of a range of nutrients, including carbohydrate, protein, fats, and micronutrients like minerals and vitamins. However, the type and level of lentil antinutrients may affect micronutrient bioavailability. This section will address the type of antinutrients that may be present in lentil and other pulses and their impact on mineral and micronutrient bioavailability.

Antinutrients could be broadly defined as natural or synthetic substances that interfere with the biological absorption of nutrients in human gut. Nutrients must be present in bioavailable form to absorb under human digestive environments. These natural and synthetic compounds interact with nutrients leading to chemical changes, forming strong complexes or destroying bioavailable forms under human digestive environments. Among antinutrients, protease/lipase/amylase inhibitors and phytic acid are the major natural antinutrients found in commonly eaten foods.

Protease inhibitors are substances that inhibit trypsin, pepsin, and other proteases in the human digestive environments. Although these substances have been recognized to inhibit protein digestions, the latest results suggest a positive role of these substances in human health. In particular, Bowman-Birk serine protease inhibitors have been shown to possess anticarcinogenic properties. The Bowman-Birk serine protease inhibitors have been isolated from field pea, lentil, and chickpea (Alfonso, Gabriella and Clare, 2011). This suggests that these foods may have additional pro-nutritional and health-promoting benefits from this type of protease inhibitor. No studies have been performed on lipase or amylase inhibitors in lentils.

Phytic acid (PA; 1,2,3,4,5,6-hexakis myo-inositol), an inositol phosphate isomer, is considered an anti-nutrient in food, agriculture, and nutritional sciences. Phytic acid is able to complex with divalent cations including iron (Fe), zinc (Zn), magnesium (Mg), and calcium (Ca), thus reducing their bioavailability. Phytic acid accounts for 1-5% of the weight of cereals, pulses, tubers, fruits, and vegetables. PA accumulates during seed development and is utilized during germination to maintain a relatively constant level of inorganic phosphorous (P). The primary functions of PA in the seed are to (1) provide storage for cations and P; (2) act as a precursor for cell wall development; (3) act as a storehouse for potential energy; (4) enable cell signaling, and (5) provide antioxidant protection against free radicals and iron.

Thavarajah et al. 2009b has reported moderate levels of phytic acid in lentil, chickpea, and field pea ranging from 0.25 to 2%.

In order to reduce the negative effects of phytic acid on mineral bioavailability, phytic acid reduction efforts have taken place. However, while phytic acid is typically viewed as a food anti-nutrient, recent inositol phosphate research shows phytic acid and other inositol phosphates may provide a range of human health benefits. Moderate intakes of phytic acid result in positive health effects for humans (Konietzny & Greiner, 2003) and several animal and epidemiological studies demonstrate beneficial effects of dietary PA, including decreased risk of heart disease, renal stone formation, and colon cancer (Fox & Eberl, 2002).

Future Research

Scientific evidence suggests lentil as a promising, nutritious whole food with the ability to supply a range of micronutrients. However, lentil's potential to deliver nutrients is faced with challenges: (1) lower grain yields in developing countries, and (2) a lack or no quality nutritional data from most lentil-producing countries. Research into increasing yields is urgently needed to increase farm productivities (hence incentives to grow lentils) and as a mean to supply anticipated future lentil demands. Further research on the nutritional quality of lentil in all lentil-producing countries will enable lentil breeding, micronutrient biofortification, consumer education, and more meaningful efforts to deliver food-based nutrition solutions to micronutrient deficiencies. Along with other legumes, lentil could prove to be a valuable crop in the development of healthy and sustainable food systems.

Acknowledgments

Authors would like to thank Drs. Jerry Combs Jr. and Ross Welch for their discussion and guidance.

References

Alfonso, C., Gabrielle, S., and Claire, D. (2011). Bowman-Birk inhibitors from legumes and human gastrointestinal health: current status and perspectives. *Current Protein and Peptide Science,* 12 (5), 358-373.

Amarakoon, D., Thavarajah, D., McPhee, K. and Thavarajah, P. (2012). Iron-, zinc-, and magnesium-rich field peas (Pisum sativum L.) with naturally low phytic acid: A potential food-based solution to global micronutrient malnutrition. *Journal of Food Composition and Analysis* 27, 8-13.

Baum, M., Hamwieh, A., Furman Bonnie, Sarker, A. and Erskine William. (2008). Biodiversity, genetic enhancement and molecular approaches in lentils. In: Kumar Sharma Arun and Archana Sharma (Eds), *Plant Genome;biodiversity and evolution*, Enfield N. H, pp. 135-152.

Bhatty, R.S. (1988). Composition and Quality of Lentil (Lens-Culinaris Medik) - A Review. *Canadian Institute of Food Science and Technology* 21, 144-160.

Bouis, H. and Welch, R. (2010). Biofortification-a sustainable agricultural stratergy for reducing mincronutrient malnutrition in the global south. *Crop Science* 50, S20-S23.

Broadley, M.R., White, P.J., Bryson, R.J., Meacham, M.C., Bowen, H.C., Johnson, S.E., Hawkesford, M.J., McGrath, S.P., Zhao, F.J., Breward, N., Harriman, M. and Tucker, M. (2006). Biofortification of UK food crops with selenium. *Proceedings of Nutrition Socety* 65, 169-181.

Campos-Vega, R., Loarca-Pina, G. and Oomah, B.D. (2010). Minor components of pulses and their potential impact on human health. *Food Research International* 43, 461-482.

Clancey B. 2009. World pulse outlook: Report to the Saskatchewan *Pulse growers*. P 17.

Combs, G.F., Jr., Clark, L.C. and Turnbull, B.W. (2002). Cancer prevention by selenium. *Metal Ions in Biology and Medicine* 7, 600-603.

Combs, G.F., Jr. (2001). Selenium in global food systems. *British* Journal of *Nutrition* 85, 517-547.

de Onis, M., Brown, D., Blossner, M. and Borghi, E. (2012). Levels and Trends in Child Malnutrition. In: United Nations Children's Fund, World Health Organization, The world Bank, UNICEF-WHO-Wprld Bank Joint Child Malnutriton Estimates.

Dumas, M.E., Barton, R.H., Toye, A., Cloarec, O., Blancher, C., Rothwell, A., Fearnside, J., Tatoud, R., Blanc, V.+., Lindon, J.C., Mitchell, S.C., Holmes, E., McCarthy, M.I., Scott, J., Gauguier, D. and Nicholson, J.K. (2006). Metabolic profiling reveals a contribution of gut microbiota to fatty liver phenotype in insulin-resistant mice. *Proceedings of the National Academy of Sciences* 103, 12511-12516.

Ellis, D.R. and Salt, D.E. (2003). Plants, selenium and human health. *Current Opinion* in *Plant Biology* 6, 273-279.

FAOSTAT 2013. Food and Agriculture Organization of the United Nations. http://faostat.fao.org

Fox, C. H. and Eberl, M. (2002). Phytic acid (IP6), novel broad spectrum antineoplastic agent: a systematic review. *Complementary Therapies in Medicine* 10, 229–234.

Gibson, G.R. and Roberfroid, M.B. (1995). Dietary Modulation of the Human Colonic Microbiota - Introducing the Concept of Prebiotics. *Journal of Nutrition* 125, 1401-1412.

Gibson, G.R. (2004). Fibre and effects on probiotics (the prebiotic concept). In: pp. 25-31.

Graham, R.D. and Welch, R.M. (2000). Plant food micronutrient composition and human nutrition. *Communications* in *Soil Science* and *Plant Analysis* 31, 1627-1640.

Grusak, M.A., Pearson, J.N. and Marentes, E. (1999). The physiology of micronutrient homeostasis in field pea. *Field Crops Research* 60, 41-56.

Johnson, C., Thavarajah, D., Combs, G.F., Jr. and Thavarajah, P. (2013a). Lentil (Lens culinaris L.): a prebiotic rich whole food legume. *Food Research International* 52, 107-113.

Johnson, C., Thavarajah, D. and Thavarajah, P. (2013b). Lentil (*Lens culinaris* L.) phenolics and phytic acid food matric factors for iron bioavailability. *Journal of Food Composition and Analysis* (Accepted).

Kennedy, G., Nantel, G. and Shetty, P. (2003). The scourge of "hidden hunger": global dimensions of micronnutrient deficiencies. *Food Nutrition and Agriculture* 32, 8-16.

Konietzny, U. and Greiner, R. (2003). Phytic Acid/Nutritional Impact. *Phytochemistry* 4555–4563.

Lyons, G.H., Genc, Y., Soole, K., Stangoulis, J.C.R., Liu, F. and Graham, R.D. (2009). Selenium increases seed production in Brassica. *Plant Soil* 318, 73-80.

Lyons, G.H., Lewis, J., Lorimer, M.F., Holloway, R.B., Brace, D.M., Stangoulis, J.C.R. and Graham, R.D. (2004). High-selenium wheat: Agronomic biofortification strategies to improve human nutrition . *Food, Agriculture & Environment* 2, 171-178.

Lyons, G., Stangoulis, J. and Graham, R. (2003). High-selenium wheat: biofortification for better health. *Nutrition Research Reviews* 16, 45-60.

Lyons, G.H., Judson, G.J., Ortiz-Monasterio, I., Genc, Y., Stangoulis, J.C.R. and Graham, R.D. (2005). Selenium in Australia: selenium status and biofortification of wheat for better health. *Journal* of *Trace Elements* in Medicine and *Biology* 19, 75-82.

Saltzman, A., Birol, E., Bouis, H., Boy, E., De Moura, F.F., Islam, Y. and Pfeiffer, W.H. (2013). Biofortification: progress towards a more nourishing future. Global Food Security http://dx.doi.org/10.1016/j.gfs.2012.12.003.

Thavarajah, D., Ruszkowski, J. and Vandenberg, A. (2008) High Potential for Selenium Biofortification of Lentils (Lens culinaris L.). *Journal of Agriculture and Food Chememistry* 56, 10747-10753.

Thavarajah, D., Thavarajah, P., Wejesuriya, A., Rutzke, M., Glahn, R.P., Combs, G.F. and Vandenberg, A. (2011a). The potential of lentil (Lens culinaris L.) as a whole food for increased selenium, iron, and zinc intake: preliminary results from a 3 year study. *Euphytica* 180, 123-128.

Thavarajah, D., Vandenberg, A., George, G.N. and Pickering, I.J. (2007) Chemical Form of Selenium in Naturally Selenium-Rich Lentils (Lens culinaris L.) from Saskatchewan. *Journal of Agriculture and Food Chememistry* 55, 7337-7341.

Thavarajah, D. and Thavarajah, P. (2012) Evaluation of chickpea (Cicer arietinum L.) micronutrient composition: Biofortification opportunities to combat global micronutrient malnutrition. *Food Research International* 49, 99-104.

Thavarajah, D., Thavarajah, P., Sarker, A., Materne, M., Vandemark, G., Shrestha, R., Idrissi, O., Hacikamiloglu, O., Bucak, B. and Vandenberg, A. (2011b). A global survey of effects of genotype and environment on selenium concentration in lentils (*Lens culinaris* L.): Implications for nutritional fortification strategies. *Food Chemistry* 125, 72-76.

Thavarajah, D., Thavarajah, P., Sarker, A. and Vandenberg, A. (2009a). Lentils (*Lens culinaris* Medikus Subspecies culinaris): A Whole Food for Increased Iron and Zinc Intake. *J. Journal of Agriculture and Food Chememistry* 57, 5413-5419.

Thavarajah, P., Thavarajah D., and Vandenberg, A. (2009b). Low phytic acid lentils(*Lens culinaris* L.): a potential solution for increased micronutrient bioavailability. *Journal of Agriculture and Food Chememistry*, 57, 9044-9049.

Wang, W.C., Naentoe, V., Maekelae, A.L., Maekelae, P.(1995). Effect of nationwide selenium supplementation in Finland on selenium status in children with juvenile rheumatoid arthritis. a ten-year follow-up study. 120 ed.; pp 955-958.Welch, R. (1995). Micronutrient nutrition of plants. *Critical Reviews in Plant Sciences* 14, 49-82.

Welch, R. and Graham, R.D. (2005). Agriculture: the real nexus for enhancing bioavailable micronutrients in food crops. *Journal* of *Trace Elements* in Medicine and *Biology* 18, 299-307.

Welch, R.M. (2002). Breeding Strategies for Biofortified Staple Plant Foods to Reduce Micronutrient Malnutrition Globally. *Journal of Nutrition* 132, 495S-499S.

Welch, R.M. and Graham, R.D. (1999). A new paradigm for world agriculture: meeting human needs: Productive, sustainable, nutritious. *Field Crops Research* 60, 1-10.

Welch, R.M. and Graham, R.D. (2004). Breeding for micronutrients in staple food crops from a human nutrition perspective. *Journal of Experimanetal Botany* 55, 353-364.

WHO (2013). http://www.who.int/nutrition/topics/ida/en/. Accessed on April 3, 2013.

Wu, G.D., Chen, J., Hoffmann, C., Bittinger, K., Chen, Y.Y., Keilbaugh, S.A., Bewtra, M., Knights, D., Walters, W.A., Knight, R., Sinha, R., Gilroy, E., Gupta, K., Baldassano, R., Nessel, L., Li, H., Bushman, F.D. and Lewis, J.D. (2011). Linking Long-Term Dietary Patterns with Gut Microbial Enterotypes. *Science* (Washington, DC, U. S.) 334, 105-108.

In: Legumes
Editors: Hiroto Satou and Ren Nakamura

ISBN: 978-1-62808-280-7

Chapter 8

MEDICAGO TRUNCATULA: A MODEL ORGANISM TO STUDY THE BIOLOGY OF AGRICULTURALLY IMPORTANT LEGUME CROPS

Mauricio Alberto Reynoso, Virginia Savy, Marcos Lancia, Karen Hobecker, Flavio Blanco and María Eugenia Zanetti*

Instituto de Biotecnología y Biología Molecular - Facultad de Ciencias Exactas-Universidad Nacional de La Plata, La Plata, Argentina

ABSTRACT

Medicago truncatula (barrel medic) is a species native to the Mediterranean that has been adopted as a model organism for temperate legumes due to its relatively small size, short generation time, small diploid genome and the capacity to reproduce by self- and out-crossing. It is a close relative of alfalfa (*M. sativa*), which is used as forage, but it is also a model organism for other legume crops used for human consumption such as pea (*Pisum sativum*), lentil (*Lens culinaris*) and faba bean (*Vicia faba*). As many other legumes, *M. truncatula* establishes a nitrogen fixing symbiosis with soil bacteria called *Rhizobia*, which accounts for a significant proportion of the nitrogen incorporated into the ecosystems. In addition, it has the capacity to interact with fungi to form mycorrhiza, an ancient root endosymbiosis important for plant mineral nutrition. In the past decade, a significant amount of genetic and molecular tools has been developed to study the biology of this model species: the sequencing of nearly its complete genome, the generation of genome-wide expression datasets, the development of techniques for stable and hairy root transformation, and the establishment of mutant collections for genetic studies. This compendium of tools allowed the dissection of the genetic networks that

* Corresponding author: María Eugenia Zanetti. Instituto de Biotecnología y Biología Molecular - Facultad de Ciencias Exactas-Universidad Nacional de La Plata - CCT La Plata CONICET. Calle 115 y 49. C.P.1900 - La Plata Argentina. E-mail: ezanetti@biol.unlp.edu.ar.

govern symbiotic association with bacteria and fungi. The knowledge generated in this model legume is crucial to improve crop productivity in agriculture.

INTRODUCTION

Legumes constitute one of the largest families of flowering plants. Dietary legumes of worldwide importance include bean (*Phaseolus vulgaris*), soybean (*Glycine max*), pea (*Pisum sativum*), chickpea (*Cicer arietinum*), faba bean (*Vicia faba*), pigeon pea (*Cajanus cajan*) and lentil (*Lens culinaris*), whereas alfalfa (*Medicago sativa*), several clovers and some species of the *Lotus* genus are used as pasture.

In addition, some legumes are used for oil production in the industry. However, most of the crop and forage legumes are poor experimental systems for plant biology research. Cultivated legumes have large seeds and seedlings, preventing high density culture required for experimental research. Some of them have long regeneration times or are self-incompatible, difficult to transform or recalcitrant to regeneration. They have large genomes and some of them are tetraploids (e.g. alfafa and peanut) or have duplicated genomes (soybean).

For these reasons, *Medicago truncatula* and *Lotus japonicus* have been adopted by the international community as model organisms for legume research. *M. truncatula* is native to the Mediterranean and has been selected as the legume of reference for temperate legumes such as alfalfa and pea. It has a diploid genome of small size (550 Mbp), which has been nearly completely sequenced (Young et al. 2011), small seeds and seedlings, short regeneration time and the capacity to reproduce by self- and out-crossing. Since its adoption as model legume, a significant amount of genomic and genetic resources have been developed for *M. truncatula* (see box 1).

In this chapter, we will discuss how the tools and resources developed in this legume have highlighted our knowledge of fundamental biological processes established between legumes and benefic microorganisms (bacteria and fungi).

ROOT NODULE SYMBIOSIS

Most legume plants establish a nitrogen fixing symbiosis with soil bacteria, collectively known as rhizobia. The symbiotic interaction between legumes and rhizobia is responsible for most of the nitrogen incorporated biologically into ecosystems.

Nitrogen is an essential macronutrient required for the biosynthesis of protein and DNA/RNA precursors (amino acids and nucleotides, respectively). Although nitrogen is highly abundant in the atmosphere, only a small number of microorganisms can convert atmospheric nitrogen into ammonia through a high energy demanding reaction performed by the nitrogenase multienzymatic complex. Leguminous plants can take advantage of this ability associating with these nitrogen fixing bacteria to establish a symbiotic relationship. Most legumes allocate bacteria within new organs developed in their root systems, the nodules, where bacteria fix atmospheric dinitrogen into reduced forms that are readily available for the plant. In exchange, bacteria obtain a source of carbon.

Box 1. *M. truncatula* genetic and genomic resources

Genome sequences
HAPMAN PROJECT
http://www.medicagohapmap.org/?genome
This site contains version Mt3.0 and Mt3.5 of the *M. truncatula* genome (Young et al. 2011).
Expressed sequenced tags (ESTs)
DFCI gene index
http://compbio.dfci.harvard.edu/tgi/cgi-bin/tgi/gimain.pl?gudb=medicago. (Quackenbush et al. 2000)
The Gene Index Project contains Expressed Sequence Transcripts (ESTs) and gene sequences, along with the reference genomes wherever available, which provides an inventory of likely genes and their variants and information regarding the functional roles played by these genes and their products.
Plant Genome database (PlantGBD)
http://www.plantgdb.org/download/download.php?dir=/Sequence/ESTcontig/Medicago_truncatula. (Dong et al. 2004, Duvick et al. 2008)
PlantGDB provides sequence data for more than 70,000 plant species, EST assemblies (PUT) for over 150 species, web tools and plant genome browsers.
Expression data
Medicago truncatula Gene Expression Atlas (MtGEA)
http://mtgea.noble.org/v3/
This is a compendium of gene expression profiles for the majority of *M. truncatula* genes covering all its major organs (roots, nodules, stems, petioles, leaves, vegetative buds, flowers, seeds and seed pods) with detailed developmental time-series using the Affymetrix Medicago Gene Chip ® (Benedito et al. 2008, He et al. 2009).
Tilling mutants
REvGenUK
http://revgenuk.jic.ac.uk/TILLING.htm
This site offers single base mutants of *M. truncatula* in the gene of interest to allow reverse genetic approaches. A population of plants with random point mutations was generated by treatment of seeds with ehylmethanesulphonate (EMS). TILLING (Targeting Induced Local Lesions IN Genomes) technology relies on the ability of a special enzyme to detect mismatches in normal and mutant DNA strands when they are annealed.
Tnt1 mutants
http://www.noble.org/staff/mysore-kiran/
M. truncatula Tnt1-insertion mutant population available at The Samuel Robert Foundation is a collection of DNA insertional mutants established from a starter parental line Tnk88-7-7, which contain several copies of Tnt1 insertion transferred into wild type ecotype R108 by *Agrobacterium tumefaciens* transformation (d'Erfurth et al. 2003, Pislariu et al. 2012, Tadege et al. 2008).
Methods for transformation
Agrobacterium rhizogenes mediated transformation
http://www.noble.org/Global/medicagohandbook/pdf/AgrobacteriumRhizogenes.pdf (Boisson-Dernier et al. 2001).
This handbook provides two methods for generating composite plants comprising a transgenic hairy root system attached to non-transformed shoots and leaves. Composite plants generated by *A. rhizogenes* transformation are well-adapted for studies of root-specific interactions since they can be nodulated by *Sinorhizobium meliloti*, colonized by endomycorrhizal fungi and infected/infested by pathogenic/parasitic organisms such as root nematodes.
Agrobacterium tumefaciens mediated transformation
http://www.noble.org/Global/medicagohandbook/pdf/AgrobacteriumTumefaciens.pdf (Chabaud et al. 2003, Trieu et al. 2000)
This handbook describes four methods to generate stable transgenic plants. They are based on *A. tumefaciens* transformation using different genotypes and embryogenesis or organogenesis as regeneration process. Homozygous transgenic lines can be obtained containing well characterized insertion profiles consisting of a small number of T-DNA copies.

Root nodule symbiosis requires the activation of two highly coordinated morphogenetic programs in the plant: the organogenesis of the nodule and the bacterial infection (Oldroyd et al. 2011). Nodule organogenesis begins with the re-activation of cell divisions in the root cortex to form a nodule primordium, whereas bacterial infection occurs predominantly by a sophisticated intracellular mechanism that requires the attachment of the bacteria to actively growing root hairs, and the formation of a tubular structure called the infection thread (IT) that guides the bacteria from the epidermis to the nodule primordia (Figure 1). Once the IT reaches the developing nodule, bacteria are released into organelle-like structures called symbiosomes, where they differentiate to nitrogen fixing bacteroids. The legume-rhizobia interaction is initiated in the rhizosphere and depends on the exchange of signals between the macro (the plant) and the micro-symbiont (the bacterium). Roots of plants growing under low nitrogen availability exude phenolic compounds (usually flavonoids or isoflavonoids), which are perceived by the bacteria, leading to the biosynthesis and secretion of a signaling molecule that plays an essential role in nodulation, the Nod Factor (NF).

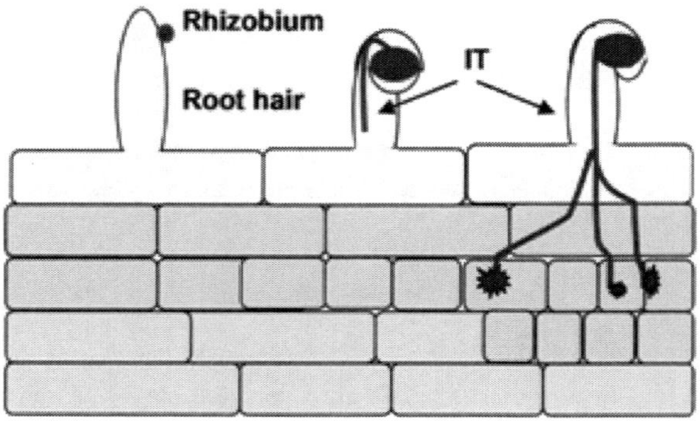

a

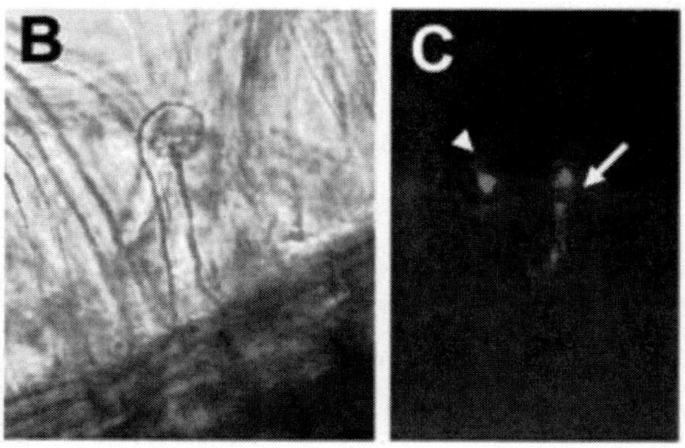

b c

Figure 1. Rhizobial infection during root nodule symbiosis. A) Schematic representation of the coordinated developmental events of infection thread formation (IT) in the epidermal cell layer (white) with the activation of cell division in the cortex (gray). The attachment of rhizobia results in the curling of the root hair that entraps the bacteria. The IT is initiated from the infection foci and grows toward the dividing cortical cells. B) A curled root hair of *M. truncatula* visualized by optical microscopy at 5 days post inoculation with *S. meliloti.* C) An infection foci (arrowhead) and an IT (arrow) developed in *M. truncatula* roots visualized by epifluorescence microscopy at 5 days post inoculation with a strain of *S. meliloti* expressing a fluorescent protein.

The NF is composed by a backbone of four or five 1,4-linked N-acetyl-glucosamine residues with several substitutions or "decorations" that determine which bacteria will be able to nodulate a particular legume species (Denarie et al. 1996). This signal molecule is perceived by receptor like kinases (RLK) with extracellular LysM motifs (LysM-RLKs) present in the plasma membrane of root cells (Limpens et al. 2003). Host-legume specificity is governed by the recognition and interaction between the bacterial NF and the cognate legume receptors. *M. truncatula* establishes symbiosis with the rhizobial species *Sinorhizobium meliloti* and *S. medicae*, being the first association the best characterized at molecular level. The screening of *M. truncatula* mutant collections led to the identification of two LysM-RLKs, named NFP (NF Perception) and LYK3 (LysM kinase 3), which are required for nodulation. *nfp* mutants do not show any of the early responses to NF treatment, including root hair deformation, calcium spiking and the induction of early nodulation (*ENOD*) genes (Arrighi et al. 2006).

Studies using RNA of interference (RNAi), revealed that knockdown of *LYK3* expression blocks IT formation, leading to the proposal that LYK3 might act as an entry receptor (Limpens et al. 2003). Later on, Smit et al. (2007) reported that mutants in *LYK3*, also known as *hcl* (hair curling) mutants, are unable to form curled root hairs in response to *S. meliloti*, a step that precedes IT formation.

Perception of NF by the LysM-RLKs activates a signaling cascade (see Figure 2) that involves another RLK with an extracellular domain containing Leucine Rich Repeats (LRR) named DMI2 (Does not Make Infection 2) (Endre et al. 2002). Loss of function mutants in *dmi2* show exacerbated root hair deformation in response to rhizobia, but are impaired in IT formation and progression.

In addition, *DMI1* is also required for rhizobial infection and encodes a putative ion channel with strong homologies to a calcium-gated K^+ channel (Ane et al. 2004). However, initial responses to NF such as root hair swelling are not altered in *dmi1* mutants. These four components of the signaling cascade, along with several nuclear pore proteins (NUPs), are required for the generation of nucleus-associated calcium spikes in response to NF (Groth et al. 2010, Kanamori et al. 2006, Saito et al. 2007). Several lines of evidence suggest that calcium spiking is decoded by a nucleus-localized calcium-calmoduline dependent kinase (CCaMK) identified as DMI3. *dmi3* mutants fail to form ITs but display normal spikes of calcium in the root hairs upon NF application (Lévy et al. 2004, Mitra et al. 2004). This CCaMK has been also involved in nodule organogenesis since expression of constitutive active forms of CCaMK in *M. truncatula* plants results in the formation of spontaneous nodules, i.e. development of nodules in the absence of rhizobia (Hayashi et al. 2010). A yeast two hybrid screening identified a protein with a coil-coiled motif that physically interacts with DMI3 (IPD3) in yeast and in the nuclei of *Nicothiana benthamiana* cells (Messinese et al. 2007). Mutants in *IPD3* gene are affected in IT growth and the invasion of the nodules by *S. meliloti* (Horváth et al. 2011).

Downstream of the DMI3-IPD3 nuclear complex there is a number of transcription factors which are required for nodule formation and/or infection (see Figure 2). Among them, two members of the GRAS family of transcription factors, Nodulation Signaling Pathway 1 and 2 (NSP1 and NSP2), are key components of the pathway. Mutants in these genes are compromised in the majority of molecular and physiological responses, but retain NF-induced calcium spiking (Kalo et al. 2005, Smit et al. 2005). It has been shown that NSP1 and NSP2 can form heterodimers that bind to the promoter of the *ENOD11* gene (Hirsch et al. 2009).

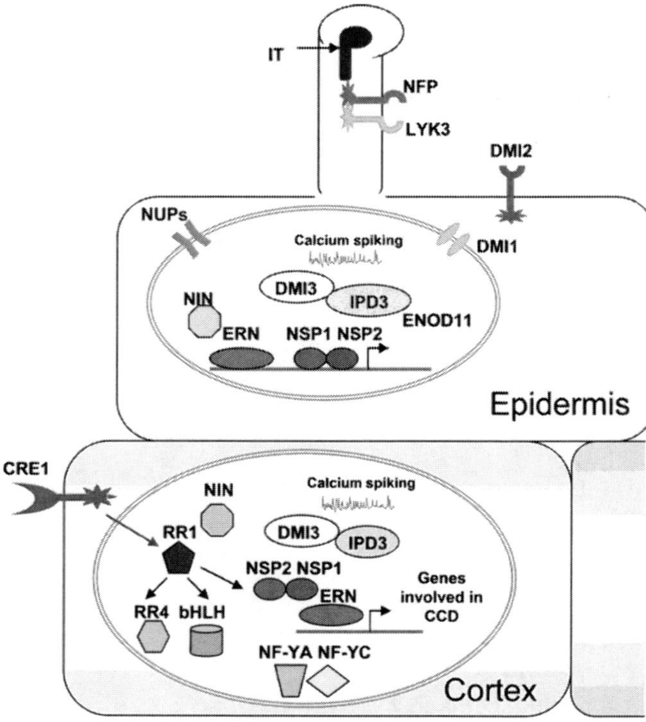

Figure 2. Root Nodule symbiotic pathway. Nod factor produced by rhizobia is perceived by LysM-RLK (NFP and LYK3) present in the root hair plasma membrane activating the LRR-RLK DMI2, the cation chanel DMI1, Nuclear pore proteins (NUPs) and triggering nuclear calcium oscillations presumably decoded by the CCaMK DMI3. DMI3 forms a complex with the nuclear protein IPD3. In the epidermis, transcription factors NSP1 and NSP2 together with ERN promote transcriptional activation of the Early nodulation gene ENOD11. NIN is another transcriptional factor that plays a role in nodulation in both epidermis and cortex. NF-YA and NF-YC play a role in the direct or indirect activation of genes that promote cortical cell divisions (CCD). Cytokinin signaling activated in the root cortex during root nodule symbiosis involves the CRE1 receptor, the Response Regulator RR1 and its primary targets RR4, NSP2 and a transcription factor of the bHLH family. IT: infection thread.

Nodule inception (NIN) is another putative transcription factor required for proper nodule development, but seems to affect later stages of the interaction since *nin* mutants undergo excessive root hair curling, but are impaired in infection and fail to form nodules after inoculation with *S. meliloti* (Marsh et al. 2007). Also, a Ethylene response factor Required for Nodulation (ERN) was identified by forward genetics. Mutation of this gene blocks the initiation and progression of ITs (Middleton et al. 2007). Interestingly, ERF binds to the *NF* box of the *ENOD11* promoter and activates the transcription of this gene (Andriankaja et al. 2007) The list of transcription factors involved in the root nodule symbiosis also includes two subunits of the nuclear factor Y family. These genes were characterized by reverse genetics revealing essential functions in IT initiation and progression, as well as at later stages of nodule development (Combier et al. 2008, Combier et al. 2006, Zanetti et al. 2010).

Nodule organogenesis is also governed by the action of different phytohormones such as cytokines and auxins. It was shown that silencing of the cytokinin receptor CRE1 blocked nodule formation, but allowed rhizobial infection in *M. truncatula*, indicating that cytokinin perception was required for proper nodule formation (Gonzalez-Rizzo et al. 2006). Moreover, a gain of function mutant of the *L. japonicus* ortholog of CRE1 resulted in spontaneous nodule formation (Tirichine et al. 2007). This result indicated that cytokinin signaling is necessary, but also sufficient to trigger nodule organogenesis.

The primary targets of the cytokinin signaling pathway during nodule formation are a group of genes known as Response Regulators (RR), which include the nodule-enhanced MtRR1 (Gonzalez Rizzo et al. 2006). More recently, the transcription factor NSP2, the response Regulator MtRR4 and a transcription factor of the basic helix-loop-helix family were identified as direct targets of MtRR1 (Ariel et al. 2012). Another effect of the CRE1 mediated signaling pathway during root nodule symbiosis is the change in polar auxin transport, which is suppressed by infection with rhizobia and by treatment with purified NF (Plet et al. 2011, Ruzicka et al. 2009). The role of auxin was initially investigated by exogenous application of auxin transport inhibitors, which were sufficient to induce nodule organogenesis and the expression of early nodulation genes in alfalfa (Mathesius et al. 1998). The suppression of auxin transport during nodule formation indicates that initiation of nodules is associated with low levels of auxin.

As summarized in this section, the analysis of mutants in single genes allowed the identification of key components of the nodulation signaling pathway (Figure 2). However, it is expected that many other components of the pathway remain to be discovered, since genetic redundancy limits the power of forward genetic strategies. The development of DNA microarray and, more recently, the transcriptomic based on deep sequencing technologies allowed the generation of large datasets of gene expression at different stages of legume-rhizobia symbiosis (Lohar et al. 2006, Maunoury et al. 2010, Mitra and Long 2004, Moreau et al. 2011). These approaches identified hundred or thousand of genes whose transcripts increase their abundance either at early infection steps or during the differentiation of bacteria to bacteroids within nodules. However, an increase of transcript levels does not necessarily guarantee the synthesis of the encoded protein.

Cytosolic transcripts can be either recruited to the translation machinery or stored in other cytosolic messenger ribonucleoprotein complexes (mRNPs). Therefore, the analysis of mRNA associated with translating ribosomes might provide a more accurate picture of changes in gene expression.

We recently reported the application of the Translating Ribosome Affinity Purification (TRAP) technology to the analysis of a set of genes required either for bacterial infection or nodule formation in *M. truncatula* (Reynoso et al. 2013). A schematic description of TRAP technology is presented in Figure 3 and for a detailed protocol see Mustroph et al. 2009.

Results obtained using this technique revealed that genes encoding NFP, DMI2 and CRE1 receptors and GRAS and NF-Y transcription factors are subjected to a positive regulation at translational, but not at transcriptional level, during IT formation and cortical cell divisions in *M. truncatula* roots. It is expected that the large scale analysis of the population of transcripts associated with translating ribosomes would help to identify additional genes regulated at this step and necessary for proper development of nitrogen fixing nodules.

ARBUSCULAR MYCORRHIZATION

The arbuscular mycorrhizal (AM) symbiosis is a widespread association established between the majority of the flowering plants and a large variety of fungi from the phylum *Glomeromycota*.

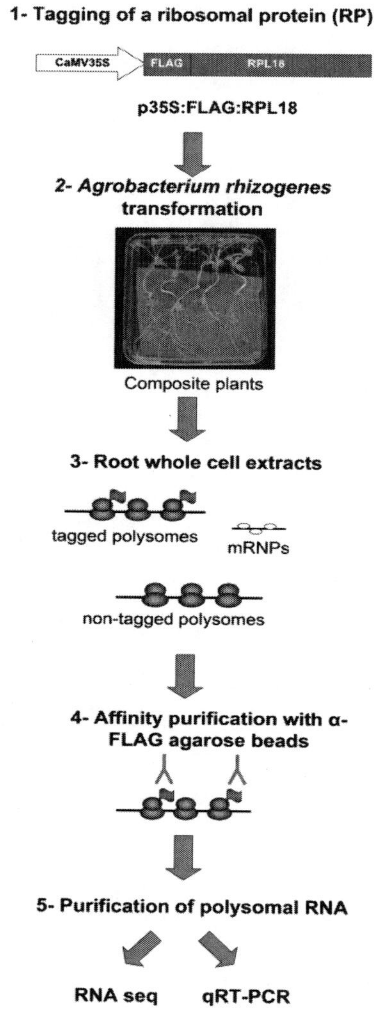

Figure 3. Translating Ribosome Affinity Purification (TRAP) in *M. truncatula*. Composite plants (containing a wild type shoot and transgenic roots) of *M. truncatula* were transformed with the ribosomal protein of the large subunit RPL18 fused to a FLAG epitope under the control of the Cauliflower mosaic virus 35S (CaMV35S) promoter. The FLAG tagged ribosomal protein is incorporated into ribosomes and polysomes (mRNAs with 2 or more ribosomes, considered as engaged in active translation). The epitope provides a mean to separate tagged polysomes from non-tagged polysomes and other messenger ribonucleoprotein complexes by affinity purification using anti-FLAG antibodies coupled to agarose beads. The RNA contained within these polysomes can be isolated and used to characterize the population of mRNAs actively translated in the cell either by quantitative reverse transcription followed by quantitative polymerase chain reaction (qRT-PCR) or by deep sequencing of cDNA libraries derived from fragmented mRNA (RNAseq).

This association is considered non-specific since most plants form AM associations with a broad range of fungi under laboratory conditions; however, in natural ecosystems, some plant species exhibit preferences for individual AM fungi (Smith et al. 2009). The interaction begins upon contact of the fungus with the root epidermis and the formation of a specialized penetration structure called the hyphopodium (see Figure 4). Perception of the fungus results in the formation of a novel plant derived structure comprising microtubules, microfilaments and endoplasmic reticulum that is called the pre-penetration apparatus (PPA). This structure is assembled in the epidermal cells and guides the infecting hyphae to the cortical cells. Once inside the root cortex, the fungus forms highly branched structures, called arbuscules, in the inner cortical cells of the root. These arbuscules enable nutrient exchange between both partners. The hyphae that remain in the soil increase the capacity of the plant to access nutrients (especially phosphates), whereas the fungus is supplied with a source of carbohydrates derived from plant photosynthesis (Smith et al. 2011).

Despite its ecological and agronomical importance, the molecular mechanisms governing AM symbiosis have been only partially characterized. The signaling molecule produced by the mycorrhizal fungus *G. intradices* was recently described by Maillet et al. (2011). This molecule, known as the Myc factor, is a lipochito-oligosaccharide molecule that could be in a sulphated and a non- sulphated form. Most of the information about the signaling pathway activated in the plant at early stages of the plant-fungus interaction was obtained by comparing root nodule and AM symbiosis in model legume species such as *M. truncatula* (Geurts and Vleeshouwers 2012).

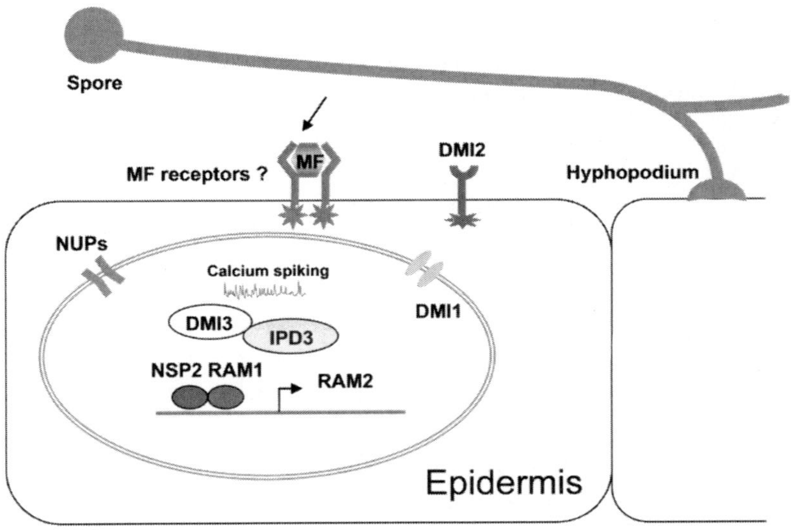

Figure 4. Arbuscular mycorrhiza symbiotic pathway. A fungal hypha in the surface of the host root epidermis forms a specialized structure called hyphopodium that serves for attachment and for the absorption of nutrients. At early stages of the interaction, the fungus secretes the Myc factor (MF) triggering symbiotic responses in the plant root. MFs are likely perceived by yet unknown plant receptors activating a signaling cascade that shares several components with the nodule symbiotic pathway (e.g. DMI2, DMI1, NUPs, calcium spiking, DMI3, IPD3, NSP2), but also contains AM specific components such as the GRAS transcription factor RAM1 and its downstream regulated gene RAM2 encoding an enzyme involved in the biosynthesis of cutin monomers.

Several studies led to the identification of genes that are required for the establishment of both root nodule and AM symbiosis. These genes are part of the so-called common symbiotic pathway and includes the RLK DMI2, the cation channel DMI1, nuclear pore proteins (NUPs), the CCaMK DMI3 and its partner IPD3, as well as the transcription factor of the GRAS family NSP2 (for review see Parniske 2008). More recently, Gobbato et al. (2012) identified another member of the GRAS family of transcription factors named RAM1 (Required for Arbuscular Mycorrhization 1) that plays a specific function during AM symbiosis. Plants mutated in *RAM1* form normal nodules, but are affected in the mycorrhizal fungal colonization. Experiments in the heteologous system *Nicotiana benthaminana* have shown that RAM1 protein is able to physically interact with NSP2, but not with NSP1, in the nuclei of the epidermal cells (Gobbato et al. 2012). Moreover, in this report the authors presented evidences that NSP1 and RAM1 may compete for the interaction with NSP2. In addition, a second mutant affected in AM but not in root nodule symbiosis, known as *ram2* was recently described (Wang et al. 2012). *RAM2* gene encodes a glycerol 3-phosphate acyl-transferase (GPAT), an enzyme required for the biosynthesis of cutin monomers. In this report the authors showed that mycorrhiza trigger the production of cutin monomers and that these monomers are sufficient to promote the formation of the hyphopodia. Interestingly, a chromatin immunoprecipitation experiment has shown that *RAM2* is a direct target of the transcriptional regulator RAM1 in *M. truncatula* (Gobbato et al. 2012) It is clear that a significant progress has been made in the last few years in the discovery of signals implicated in AM symbiosis; and the increasing number of mutagenized lines available would certainly help to identify additional components of the signaling pathway activated at early and late stages of AM symbiosis. Unraveling the genetic mechanisms governing AM symbiosis is crucial for the development of new strategies for optimization of nutrient acquisition in flowering plants, including agronomical important crop species.

CONCLUSION

The interest in *M. truncatula* as a model legume began in the early 1990`s with research focused on the symbiotic relationship with *S. meliloti*. Later on, *M. truncatula* was extensively used in studies of arbuscular mycorrhizal symbiosis, root arquitecture, abiotic and biotic stress. Since its adoption as a model legume, an increasing number of genetic tools has been developed allowing nowadays to perform genomic, transcriptomic, proteomic and functional genomics analysis in different aspect of legume biology. The biological functions of gene products can be discovered by forward genetics, i.e. the search of a particular phenotype in a mutagenized population of plants and the identification of the mutated gene by positional cloning.

However, TILLING technology and Tnt-1 insertion-mutant population have made possible reverse genetic screenings that link the gene of interest with a mutant phenotype. In addition, post-transcriptional gene silencing is also an attractive alternative for reverse genetics that have highlighted the function of genes in several biological processes (Crespi et al. 1994, Combier et al. 2006, Baier et al. 2007, Wang et al. 2008). The combination of "omics" technologies with forward and reverse genetic approaches will highlight our understanding of the fundamental biological process in *M. truncatula*.

Translation of this knowledge to agricultural important crops will undoubtedly help to increase the sustainability of agricultural systems dedicated to the production of human food, pastures and oils.

REFERENCES

Andriankaja, A., Boisson-Dernier, A., Frances, L., Sauviac, L., Jauneau, A., Barker, D. G., and de Carvalho-Niebel, F. (2007) AP2-ERF transcription factors mediate Nod factor dependent Mt ENOD11 activation in root hairs via a novel cis-regulatory motif. *Plant Cell*, 19, 2866-2885.

Ane, J. M., Kiss, G. B., Riely, B. K., Penmetsa, R. V., Oldroyd, G. E., Ayax, C., Levy, J., Debelle, F., Baek, J. M., Kalo, P., Rosenberg, C., Roe, B. A., Long, S. R., Denarie, J., and Cook, D. R. (2004) Medicago truncatula DMI1 required for bacterial and fungal symbioses in legumes. *Science*, 303, 1364-1367.

Ariel, F., Brault-Hernandez, M., Laffont, C., Huault, E., Brault, M., Plet, J., Moison, M., Blanchet, S., Ichante, J. L., Chabaud, M., Carrere, S., Crespi, M., Chan, R. L., and Frugier, F. (2012) Two direct targets of cytokinin signaling regulate symbiotic nodulation in Medicago truncatula. *Plant Cell*, 24, 3838-3852.

Arrighi, J. F., Barre, A., Ben Amor, B., Bersoult, A., Soriano, L. C., Mirabella, R., de Carvalho-Niebel, F., Journet, E. P., Gherardi, M., Huguet, T., Geurts, R., Denarie, J., Rouge, P., and Gough, C. (2006) The Medicago truncatula lysin [corrected] motif-receptor-like kinase gene family includes NFP and new nodule-expressed genes. *Plant Physiol.*, 142, 265-279.

Benedito, V. A., Torres-Jerez, I., Murray, J. D., Andriankaja, A., Allen, S., Kakar, K., Wandrey, M., Verdier, J., Zuber, H., Ott, T., Moreau, S., Niebel, A., Frickey, T., Weiller, G., He, J., Dai, X., Zhao, P. X., Tang, Y., and Udvardi, M. K. (2008) A gene expression atlas of the model legume Medicago truncatula. *Plant J.*, 55, 504-513.

Boisson-Dernier, A., Chabaud, M., Garcia, F., Becard, G., Rosenberg, C., and Barker, D. G. (2001) Agrobacterium rhizogenes-transformed roots of Medicago truncatula for the study of nitrogen-fixing and endomycorrhizal symbiotic associations. *Mol. Plant Microbe. Interact.*, 14, 695-700.

Combier, J. P., de Billy, F., Gamas, P., Niebel, A., and Rivas, S. (2008) Trans-regulation of the expression of the transcription factor MtHAP2-1 by a uORF controls root nodule development. *Genes Dev.*, 22, 1549-1559.

Combier, J. P., Frugier, F., de Billy, F., Boualem, A., El-Yahyaoui, F., Moreau, S., Vernie, T., Ott, T., Gamas, P., Crespi, M., and Niebel, A. (2006) MtHAP2-1 is a key transcriptional regulator of symbiotic nodule development regulated by microRNA169 in Medicago truncatula. *Genes Dev.*, 20, 3084-3088.

Crespi, M. D., Jurkevitch, E., Poiret, M., d'Aubenton-Carafa, Y., Petrovics, G., Kondorosi, E., and Kondorosi, A. (1994) enod40, a gene expressed during nodule organogenesis, codes for a non-translatable RNA involved in plant growth. *EMBO J.*, 13, 5099-5112.

Chabaud, M., de Carvalho-Niebel, F. and Barker, D. G. (2003) Efficient transformation of Medicago truncatula cv. Jemalong using the hypervirulent Agrobacterium tumefaciens strain AGL1. *Plant Cell Rep.*, 22, 46-51.

D'Erfurth, I., Cosson, V., Eschstruth, A., Lucas, H., Kondorosi, A., and Ratet, P. (2003) Efficient transposition of the Tnt1 tobacco retrotransposon in the model legume Medicago truncatula. *Plant J.*, 34, 95-106.

Denarie, J., Debelle, F. and Prome, J. C. (1996) Rhizobium lipo-chitooligosaccharide nodulation factors: signaling molecules mediating recognition and morphogenesis. *Annu. Rev. Biochem.*, 65, 503-535.

Dong, Q., Schlueter, S. D. and Brendel, V. (2004) PlantGDB, plant genome database and analysis tools. *Nucleic Acids Res.*, 32, 354-359.

Duvick, J., Fu, A., Muppirala, U., Sabharwal, M., Wilkerson, M. D., Lawrence, C. J., Lushbough, C., and Brendel, V. (2008) PlantGDB: a resource for comparative plant genomics. *Nucleic Acids Res.*, 36, D959-965.

Endre, G., Kereszt, A., Kevei, Z., Mihacea, S., Kalo, P., and Kiss, G. B. (2002) A receptor kinase gene regulating symbiotic nodule development. *Nature*, 417, 962-966.

Geurts, R. and Vleeshouwers, V. G. (2012) Mycorrhizal symbiosis: ancient signalling mechanisms co-opted. *Curr. Biol.*, 22, R997-999.

Gobbato, E., Marsh, J. F., Vernie, T., Wang, E., Maillet, F., Kim, J., Miller, J. B., Sun, J., Bano, S. A., Ratet, P., Mysore, K. S., Denarie, J., Schultze, M., and Oldroyd, G. E. (2012) A GRAS-type transcription factor with a specific function in mycorrhizal signaling. *Curr. Biol.*, 22, 2236-2241.

Gonzalez-Rizzo, S., Crespi, M. and Frugier, F. (2006) The Medicago truncatula CRE1 cytokinin receptor regulates lateral root development and early symbiotic interaction with Sinorhizobium meliloti. *Plant Cell.*, 18, 2680-2693.

Groth, M., Takeda, N., Perry, J., Uchida, H., Draxl, S., Brachmann, A., Sato, S., Tabata, S., Kawaguchi, M., Wang, T. L., and Parniske, M. (2010) NENA, a Lotus japonicus homolog of Sec13, is required for rhizodermal infection by arbuscular mycorrhiza fungi and rhizobia but dispensable for cortical endosymbiotic development. *Plant Cell*, 22, 2509-2526.

Hayashi, T., Banba, M., Shimoda, Y., Kouchi, H., Hayashi, M., and Imaizumi-Anraku, H. (2010) A dominant function of CCaMK in intracellular accommodation of bacterial and fungal endosymbionts. *Plant J.*, 63, 141-154.

He, J., Benedito, V. A., Wang, M., Murray, J. D., Zhao, P. X., Tang, Y., and Udvardi, M. K. (2009) The Medicago truncatula gene expression atlas web server. *BMC Bioinformatics*, 10, 441-450.

Hirsch, S., Kim, J., Muñoz, A., Heckmann, A. B., Downie, J. A., and Oldroyd, G. E. D. (2009) GRAS Proteins Form a DNA Binding Complex to Induce Gene Expression during Nodulation Signaling in Medicago truncatula. *Plant Cell*, 21, 545-557.

Horváth, B., Yeun, L. H., Domonkos, Á., Halász, G., Gobbato, E., Ayaydin, F., Miró, K., Hirsch, S., Sun, J., Tadege, M., Ratet, P., Mysore, K. S., Ané, J.-M., Oldroyd, G. E. D., and Kaló, P. (2011) Medicago truncatula IPD3 Is a Member of the Common Symbiotic Signaling Pathway Required for Rhizobial and Mycorrhizal Symbioses. *Mol. Plant Microbe Interact.*, 24, 1345-1358.

Kalo, P., Gleason, C., Edwards, A., Marsh, J., Mitra, R. M., Hirsch, S., Jakab, J., Sims, S., Long, S. R., Rogers, J., Kiss, G. B., Downie, J. A., and Oldroyd, G. E. (2005) Nodulation signaling in legumes requires NSP2, a member of the GRAS family of transcriptional regulators. *Science*, 308, 1786-1789.

Kanamori, N., Madsen, L. H., Radutoiu, S., Frantescu, M., Quistgaard, E. M., Miwa, H., Downie, J. A., James, E. K., Felle, H. H., Haaning, L. L., Jensen, T. H., Sato, S., Nakamura, Y., Tabata, S., Sandal, N., and Stougaard, J. (2006) A nucleoporin is required for induction of Ca2+ spiking in legume nodule development and essential for rhizobial and fungal symbiosis. *Proc. Natl. Acad. Sci. US*, 103, 359-364.

Lévy, J., Bres, C., Geurts, R., Chalhoub, B., Kulikova, O., Duc, G., Journet, E.-P., Ané, J.-M., Lauber, E., Bisseling, T., Dénarié, J., Rosenberg, C., and Debellé, F. (2004) A Putative Ca2+ and Calmodulin-Dependent Protein Kinase Required for Bacterial and Fungal Symbioses. *Science*, 303, 1361-1364.

Limpens, E., Franken, C., Smit, P., Willemse, J., Bisseling, T., and Geurts, R. (2003) LysM domain receptor kinases regulating rhizobial Nod factor-induced infection. *Science*, 302, 630-633.

Lohar, D. P., Sharopova, N., Endre, G., Penuela, S., Samac, D., Town, C., Silverstein, K. A., and VandenBosch, K. A. (2006) Transcript analysis of early nodulation events in Medicago truncatula. *Plant Physiol.*, 140, 221-234.

Maillet, F., Poinsot, V., Andre, O., Puech-Pages, V., Haouy, A., Gueunier, M., Cromer, L., Giraudet, D., Formey, D., Niebel, A., Martinez, E. A., Driguez, H., Becard, G., and Denarie, J. (2011) Fungal lipochitooligosaccharide symbiotic signals in arbuscular mycorrhiza. *Nature*, 469, 58-63.

Marsh, J. F., Rakocevic, A., Mitra, R. M., Brocard, L., Sun, J., Eschstruth, A., Long, S. R., Schultze, M., Ratet, P., and Oldroyd, G. E. (2007) Medicago truncatula NIN is essential for rhizobial-independent nodule organogenesis induced by autoactive calcium/calmodulin-dependent protein kinase. *Plant Physiol.*, 144, 324-335.

Mathesius, U., Schlaman, H. R., Spaink, H. P., Of Sautter, C., Rolfe, B. G., and Djordjevic, M. A. (1998) Auxin transport inhibition precedes root nodule formation in white clover roots and is regulated by flavonoids and derivatives of chitin oligosaccharides. *Plant J.*, 14, 23-34.

Maunoury, N., Redondo-Nieto, M., Bourcy, M., Van de Velde, W., Alunni, B., Laporte, P., Durand, P., Agier, N., Marisa, L., Vaubert, D., Delacroix, H., Duc, G., Ratet, P., Aggerbeck, L., Kondorosi, E., and Mergaert, P. (2010) Differentiation of symbiotic cells and endosymbionts in Medicago truncatula nodulation are coupled to two transcriptome-switches. *PLoS One*, 5, e9519.

Messinese, E., Mun, J.-H., Yeun, L. H., Jayaraman, D., Rougé, P., Barre, A., Lougnon, G., Schornack, S., Bono, J.-J., Cook, D. R., and Ané, J.-M. (2007) A Novel Nuclear Protein Interacts With the Symbiotic DMI3 Calcium- and Calmodulin-Dependent Protein Kinase of Medicago truncatula. *Mol. Plant Microbe Interact.*, 20, 912-921.

Middleton, P. H., Jakab, J., Penmetsa, R. V., Starker, C. G., Doll, J., Kalo, P., Prabhu, R., Marsh, J. F., Mitra, R. M., Kereszt, A., Dudas, B., VandenBosch, K., Long, S. R., Cook, D. R., Kiss, G. B., and Oldroyd, G. E. (2007) An ERF transcription factor in Medicago truncatula that is essential for Nod factor signal transduction. *Plant Cell*, 19, 1221-1234.

Mitra, R. M., Gleason, C. A., Edwards, A., Hadfield, J., Downie, J. A., Oldroyd, G. E., and Long, S. R. (2004) A Ca2+/calmodulin-dependent protein kinase required for symbiotic nodule development: Gene identification by transcript-based cloning. *Proc. Natl. Acad. Sci. US*, 101, 4701-4705.

Mitra, R. M. and Long, S. R. (2004) Plant and bacterial symbiotic mutants define three transcriptionally distinct stages in the development of the Medicago truncatula/ Sinorhizobium meliloti symbiosis. *Plant Physiol.*, 134, 595-604.

Moreau, S., Verdenaud, M., Ott, T., Letort, S., de Billy, F., Niebel, A., Gouzy, J., de Carvalho-Niebel, F., and Gamas, P. (2011) Transcription Reprogramming during Root Nodule Development in Medicago truncatula. *PLoS One*, 6, e16463.

Mustroph, A., Juntawong, P. and Bailey-Serres, J. (2009) Isolation of plant polysomal mRNA by differential centrifugation and ribosome immunopurification methods. *Methods Mol. Biol.*, 553, 109-126.

Oldroyd, G. E., Murray, J. D., Poole, P. S., and Downie, J. A. (2011) The rules of engagement in the legume-rhizobial symbiosis. *Annu. Rev. Genet.*, 45, 119-144.

Parniske, M. (2008) Arbuscular mycorrhiza: the mother of plant root endosymbioses. *Nat. Rev. Microbiol.*, 6, 763-775.

Pislariu, C. I., Murray, J., Wen, J., Cosson, V., Duvvuru Muni, R. R., Wang, M., Benedito, V., Andriankaja, A., Cheng, X., Torres Jerez, I., Mondy, S., Zhang, S., Taylor, M., Tadege, M., Ratet, P., Mysore, K. S., Chen, R., and Udvardi, M. K. (2012) A Medicago truncatula tobacco-retrotransposon (Tnt1)-insertion mutant collection with defects in nodule development and symbiotic nitrogen fixation. *Plant Physiol.*, 159, 1686-1699.

Plet, J., Wasson, A., Ariel, F., Le Signor, C., Baker, D., Mathesius, U., Crespi, M., and Frugier, F. (2011) MtCRE1-dependent cytokinin signaling integrates bacterial and plant cues to coordinate symbiotic nodule organogenesis in Medicago truncatula. *Plant J.*, 65, 622-633.

Quackenbush, J., Liang, F., Holt, I., Pertea, G., and Upton, J. (2000) The TIGR gene indices: reconstruction and representation of expressed gene sequences. *Nucleic Acids Res.*, 28, 141-145.

Reynoso, M. A., Blanco, F. A., Bailey-Serres, J., Crespi, M., and Zanetti, M. E. (2013) Selective recruitment of mRNAs and miRNAs to polyribosomes in response to rhizobia infection in Medicago truncatula. *Plant J.*, 73, 289-301.

Ruzicka, K., Simaskova, M., Duclercq, J., Petrasek, J., Zazimalova, E., Simon, S., Friml, J., Van Montagu, M. C., and Benkova, E. (2009) Cytokinin regulates root meristem activity via modulation of the polar auxin transport. *Proc. Natl. Acad. Sci. US*, 106, 4284-4289.

Saito, K., Yoshikawa, M., Yano, K., Miwa, H., Uchida, H., Asamizu, E., Sato, S., Tabata, S., Imaizumi-Anraku, H., Umehara, Y., Kouchi, H., Murooka, Y., Szczyglowski, K., Downie, J. A., Parniske, M., Hayashi, M., and Kawaguchi, M. (2007) NUCLEOPORIN85 is required for calcium spiking, fungal and bacterial symbioses, and seed production in Lotus japonicus. *Plant Cell*, 19, 610-624.

Smit, P., Limpens, E., Geurts, R., Fedorova, E., Dolgikh, E., Gough, C., and Bisseling, T. (2007) Medicago LYK3, an entry receptor in rhizobial nodulation factor signaling. *Plant Physiol.*, 145, 183-191.

Smit, P., Raedts, J., Portyanko, V., Debelle, F., Gough, C., Bisseling, T., and Geurts, R. (2005) NSP1 of the GRAS protein family is essential for rhizobial Nod factor-induced transcription. *Science*, 308, 1789-1791.

Smith, F. A., Grace, E. J. and Smith, S. E. (2009) More than a carbon economy: nutrient trade and ecological sustainability in facultative arbuscular mycorrhizal symbioses. *New Phytol.*, 182, 347-358.

Smith, S. E., Jakobsen, I., Gronlund, M., and Smith, F. A. (2011) Roles of arbuscular mycorrhizas in plant phosphorus nutrition: interactions between pathways of phosphorus uptake in arbuscular mycorrhizal roots have important implications for understanding and manipulating plant phosphorus acquisition. *Plant Physiol.*, 156, 1050-1057.

Tadege, M., Wen, J., He, J., Tu, H., Kwak, Y., Eschstruth, A., Cayrel, A., Endre, G., Zhao, P. X., Chabaud, M., Ratet, P., and Mysore, K. S. (2008) Large-scale insertional mutagenesis using the Tnt1 retrotransposon in the model legume Medicago truncatula. *Plant J.*, 54, 335-347.

Tirichine, L., Sandal, N., Madsen, L. H., Radutoiu, S., Albrektsen, A. S., Sato, S., Asamizu, E., Tabata, S., and Stougaard, J. (2007) A gain-of-function mutation in a cytokinin receptor triggers spontaneous root nodule organogenesis. *Science*, 315, 104-107.

Trieu, A. T., Burleigh, S. H., Kardailsky, I. V., Maldonado-Mendoza, I. E., Versaw, W. K., Blaylock, L. A., Shin, H., Chiou, T. J., Katagi, H., Dewbre, G. R., Weigel, D., and Harrison, M. J. (2000) Transformation of Medicago truncatula via infiltration of seedlings or flowering plants with Agrobacterium. *Plant J.*, 22, 531-541.

Wang, E., Schornack, S., Marsh, J. F., Gobbato, E., Schwessinger, B., Eastmond, P., Schultze, M., Kamoun, S., and Oldroyd, G. E. (2012) A common signaling process that promotes mycorrhizal and oomycete colonization of plants. *Curr. Biol.*, 22, 2242-2246.

Wang, H., Chen, J., Wen, J., Tadege, M., Li, G., Liu, Y., Mysore, K. S., Ratet, P., and Chen, R. (2008) Control of compound leaf development by FLORICAULA/LEAFY ortholog SINGLE LEAFLET1 in Medicago truncatula. *Plant Physiol.*, 146, 1759-1772.

Young, N. D., Debelle, F., Oldroyd, G. E. D., Geurts, R., Cannon, S. B., Udvardi, M. K., Benedito, V. A., Mayer, K. F. X., Gouzy, J., Schoof, H., Van de Peer, Y., Proost, S., Cook, D. R., Meyers, B. C., Spannagl, M., Cheung, F., De Mita, S., Krishnakumar, V., Gundlach, H., Zhou, S., Mudge, J., Bharti, A. K., Murray, J. D., Naoumkina, M. A., Rosen, B., Silverstein, K. A. T., Tang, H., Rombauts, S., Zhao, P. X., Zhou, P., Barbe, V., Bardou, P., Bechner, M., Bellec, A., Berger, A., Berges, H., Bidwell, S., Bisseling, T., Choisne, N., Couloux, A., Denny, R., Deshpande, S., Dai, X., Doyle, J. J., Dudez, A.-M., Farmer, A. D., Fouteau, S., Franken, C., Gibelin, C., Gish, J., Goldstein, S., Gonzalez, A. J., Green, P. J., Hallab, A., Hartog, M., Hua, A., Humphray, S. J., Jeong, D.-H., Jing, Y., Jocker, A., Kenton, S. M., Kim, D.-J., Klee, K., Lai, H., Lang, C., Lin, S., Macmil, S. L., Magdelenat, G., Matthews, L., McCorrison, J., Monaghan, E. L., Mun, J.-H., Najar, F. Z., Nicholson, C., Noirot, C., O/'Bleness, M., Paule, C. R., Poulain, J., Prion, F., Qin, B., Qu, C., Retzel, E. F., Riddle, C., Sallet, E., Samain, S., Samson, N., Sanders, I., Saurat, O., Scarpelli, C., Schiex, T., Segurens, B., Severin, A. J., Sherrier, D. J., Shi, R., Sims, S., Singer, S. R., Sinharoy, S., Sterck, L., Viollet, A., Wang, B.-B., Wang, K., Wang, M., Wang, X., Warfsmann, J., Weissenbach, J., White, D. D., White, J. D., Wiley, G. B., Wincker, P., Xing, Y., Yang, L., Yao, Z., Ying, F., Zhai, J., Zhou, L., Zuber, A., Denarie, J., Dixon, R. A., May, G. D., Schwartz, D. C., Rogers, J., Quetier, F., Town, C. D., and Roe, B. A. (2011) The Medicago genome provides insight into the evolution of rhizobial symbioses. *Nature*, 480, 520-524.

Zanetti, M. E., Blanco, F. A., Beker, M. P., Battaglia, M., and Aguilar, O. M. (2010) A C Subunit of the Plant Nuclear Factor NF-Y Required for Rhizobial Infection and Nodule Development Affects Partner Selection in the Common Bean-Rhizobium etli Symbiosis. *Plant Cell*, 22, 4142-4157.

In: Legumes
Editors: Hiroto Satou and Ren Nakamura

ISBN: 978-1-62808-280-7

Chapter 9

ANTIHYPERTENSIVE POTENTIAL OF PROTEIN HYDROLYSATES FROM THE VELVET BEAN (*MUCUNA PRURIENS*)

Maira Segura-Campos, Saulo Galicia-Martínez, Luis Chel-Guerrero and David Betancur-Ancona*
Facultad de Ingeniería Química, Universidad Autónoma de Yucatán, México

ABSTRACT

Lifestyle-related diseases are becoming increasingly more prevalent worldwide, and arterial hypertension is among the most common. High blood pressure is a risk factor for cardiovascular disease, but can be controlled in many ways from inhibiting enzymes that promote it to mitigating oxidation processes. Angiotensin I-converting enzyme (ACE) controls blood pressure by converting angiotensin I into angiotensin II, a pressor hormone that causes peripheral vasoconstriction, sodium reabsorption in the kidneys and increased body fluid retention. Produced by decomposition of angiotensin II, angiotensin III controls aldosterone production and raises blood pressure. Inhibiting ACE is one of the most effective methods for suppressing increased blood pressure. Recent studies on the biological activities and potential health benefits of peptides from food proteins have shown that these peptides may provide specific health benefits, such as lowering blood pressure. Alkaline extraction is a method for protein extraction from legumes. The present chapter describes the inhibitory activity of ACE in enzymatic hydrolysates from *Mucuna pruriens* protein concentrates. The concentrates were prepared by wet fractionation, and enzymatically hydrolyzed using different enzymes (Alcalase®, Flavourzyme®, pepsin, pancreatin and the sequential system Alcalase®-Flavourzyme®) for different lengths of time (5–120 min). Their biological potential was evaluated based on the IC_{50} value (i.e. the hydrolysate concentration required to produce 50% ACE inhibition). Hydrolysis of *M. pruriens* protein concentrates with commercial and gastric enzymes generates extensive hydrolysates. These can be used to produce low molecular

* E-mail:bancona@uady.mx. Periférico Nte. Km. 33.5, Tablaje Catastral 13615, Col. Chuburná de Hidalgo Inn, 97203 Mérida, Yucatán, México. Tels. +52 (999) 9460956, 946-09-81 and 946-09-89; Fax. +52 (999) 946-09-94.

weight peptides, which may possess hypotensive and antihypertensive properties. Protein hydrolysates with these properties have commercial potential as 'health-enhancing ingredients' in the manufacture of functional foods aimed at preventing and/or treating cardiovascular diseases or oxidative stress, while also providing added value to this legume.

INTRODUCTION

The genus *Mucuna*, belonging to the Fabaceae family, consists of approximately 100–150 species of annual and perennial legumes. The study taxon (*Mucuna* spp.) is known by various names in different tropical regions of the world: "Picapica mansa" (Veracruz, México) and "Nescafé" (Tabasco, México) since it is occasionally used as a coffee substitute; "frijol de abono" (manure bean in Honduras); "frijol de mula" (mule's bean in Guatemala); "haba de terciopelo" (Velvet broad bean in Puerto Rico); "poroto aterciopelado" (Velvetbean in Argentina); "ojo de venado" (deer's eye in Spain); "haricot velouté" (France); "makhmali sem" (India); "Stizolobia" (Italy); Velvet bean" (USA); and banana stock pea (Australia). This legume is native to Malaysia, South China, China and India, but nowadays is widely distributed in many tropical regions. Cultivated and wild varieties from America and Africa were originally introduced and propagated by humans along various commercial routes (Ortiz Ceballos et al., 2012).

The main differences among cultivated species are in the character of the pubescence on the pod, seed color, and number of days to harvest of the pod. To date, improved cultivars have only been produced in Australia (White, Mauritius, Black Mauritius, Somerset, Marbilee, Smith and Jubilack), the USA (Georgia, Alabama, Osceola, Yokohama and Florida), and Zimbabwe (Bengal, White Stigless, SES 30, SES 45, SES 68 SES 74 and SES 108). Throughout its distribution range, the species has also been given various scientific names, such as *M. deeringianum* Bort, *M. utilis* (Wall) Baker ex Burck, *M. pruriens* (L.) DC, *M. cochichinensis* (Lour) Burk, *M. nivea* (Roxb.) Kuntze, *M. capitata, M. aterrima* Piper & Tracy, *M. hassjo, M. diabolica, M. cinerum, M. haltonii,* and *M. sloanei* Fawcett & Rendle, some of which are simply synonyms while others may represent valid names referring to different taxa (Ortiz Ceballos et al., 2012).

The dry weight composition of *Mucuna* seeds is as follows: 10.0% moisture, 19.0–37.5% protein, 4.7–9.0% fat, 51.5% nitrogen-free extract, 81.7% total digestible nutrients, 5.3–11.5% raw fiber and 2.9–5.7% ash. Studies on mineral composition of *Mucuna* seeds indicate that they contain potassium (778–1846 mg/100 g), calcium (104–900 mg/100 g), iron (1.3–15 mg/100 g), manganese (0.6–9.3 mg/100 g), zinc (1.0–15.0 mg/100 g), copper (0.3–4.3 mg/100 g), magnesium (85–477 mg/100 g), phosphorous (98–498 mg/100 g) and sodium (12.7–150.0 mg/100 g). Among amino acids found in the seeds, aspartic and glutamic acids are predominant (8.9–19 and 8.6–14.4%, respectively), whereas the levels of other amino acids are low. The seeds of *Mucuna* also contain many antinutritional factors, such as total free phenolics, tannins (3.1–4.9%), L-dopa (4.2–6.8%), lectins (0.31–0.71%), protease inhibitors (trypsin and chymotrypsin), phytic acid, flatulence factors (e.g. oligosaccharides), saponins (1.15–1-31%), hydrogen cyanide (58 mg/kg), and alkaloids. Seeds are used for human consumption in some parts of Africa and India, which is reasonable given their high protein and mineral content despite low fiber content (Pugalenthi et al., 2005).

Alkaline extraction is a technique for protein isolation from *M. pruriens*. Recent studies have evaluated the biological activities and potential health benefits of food protein-derived peptides. Peptides extracted from partial enzymatic hydrolysates of food proteins may provide specific health benefits, such as antihypertensive or antioxidant activities. Many protein hydrolysate peptides exhibit inhibition of angiotensin I-converting enzyme (ACE; Dipeptidyl carboxypeptidase, EC. 3.4.15.1), which plays an important physiological role in blood pressure regulation.

Hypertension is a common health problem in all Western countries. It is a major risk factor for cardiovascular disease, including coronary heart disease, peripheral artery disease and stroke. Untreated hypertension can cause significant health problems, as well as significant costs to society. In 2000, the worldwide prevalence of hypertension was approximately 26% in adults (Kearney et al., 2005). It is estimated that this prevalence is rising among the aging population, along with increasing obesity and less physical activity. The Framingham Heart Study suggests that normotensive subjects aged 55–65 have a 90% risk of developing hypertension during their lifetime (Vasan et al., 2002). According to the recent National Health and Nutrition Examination Survey (NHANES IV), about 29% of the adult population in the USA are hypertensive (>140/90 mmHg) or use antihypertensive medication. There is a clear and independent relationship between blood pressure and cardiovascular disease (CVD) events. In recent studies, blood pressure variability has been shown to be a more important determinant of target organ damage than high blood pressure itself (Tatasciore et al., 2007). In the Framingham Heart Study, the risk ratio for coronary heart disease, stroke, peripheral artery disease and cardiac failure was approximately 2–3 times higher for hypertensive subjects compared to normotensive subjects (Kannel et al., 2004). Hypertension was also associated with nearly a 4-fold increase in risk of atherothrombotic brain infarction compared to normotension (Kannel et al., 1996). Clinical trials have shown that in hypertensive subjects, lowering of blood pressure reduces the risk of cardiovascular disease (Collins et al., 1990). Decreases in diastolic blood pressure of 5, 7.5 and 10 mmHg have been associated with at least 34%, 46% and 56% reductions in the incidence of stroke, respectively. Similarly, decreases in diastolic blood pressure of 5, 7.5 and 10 mmHg have been associated with at least 21%, 29% and 37% lower rates of coronary heart disease, respectively (MacMahon et al., 1990). On the other hand, a 20 mmHg increase in systolic blood pressure or a 10 mmHg increase in diastolic blood pressure doubles the risk of CVD in subjects with blood pressure ranging from 115/75 to 185/115 mmHg (Lewington et al., 2002).

According to the European Society of Hypertension (ESH), European Society of Cardiology (ESC), Joint National Committee (JNC 7) and the WHO-ISH 2003, the main goal in the treatment of hypertension is to reduce the overall risk of cardiovascular morbidity and mortality, as well as general morbidity and mortality (Mancia et al. 2007). Lifestyle modification should be included for all patients, especially before antihypertensive medication is prescribed, and also for patients who require pharmacological treatment. Lifestyle modification includes weight reduction in obese and overweight subjects, dietary treatment, physical activity and moderation in alcohol consumption. Pharmacological treatment is needed if the desired blood pressure values cannot be achieved with only lifestyle modification. Blood pressure lowering medication reduces the risk of complications from hypertension. ACE inhibitors are one of the main classes of antihypertensive medication, acting to inhibit ACE and reduce circulating levels of vasoconstrictor angiotensin II.

Angiotensin-I-converting enzyme (EC 3.4.15.1; ACE) is a metalloprotease, and zinc is necessary for its activity. ACE cleaves dipeptides from the carboxyl terminus of oligopeptides, playing an important physiological role in blood pressure regulation. Functional foods containing ACE inhibitory peptides may provide moderate control of blood pressure (Wu et al., 2013). For the above mentioned, the objective of the current chapter was to evaluate the antihypertensive potential of *M. pruriens* protein hydrolysates by measuring the hypotensive and antihypertensive effect in experimental animal models.

Materials and Methods

Materials

Mucuna pruriens seeds were obtained from the 2007 harvest in the state of Yucatan, Mexico. Reagents were analytical grade and purchased from J.T. Baker (Phillipsburg, NJ, USA), Sigma Chemical Co. (St. Louis, MO, USA), Merck (Darmstadt, Germany) and Bio-Rad Laboratories Inc. (Hercules, CA, USA). Alcalase® 2.4L FG and Flavourzyme® 500 MG enzymes were purchased from Novo Laboratories (Copenhagen, Denmark).

Protein Extraction

A single protein extraction was done from legume seeds. Impurities and damaged seeds were removed. Sound seeds were milled in a Mykros impact mill until passing through a 20-mesh screen (0.85 mm), and then milled further in a Cyclotec 1093 (Tecator, Sweden) mill until passing through a 60-mesh screen (0.24 mm). The resulting flour was processed using the wet fractionation method previously described by Betancur-Ancona et al. (2004). Briefly, whole flour was suspended in distilled water at a 1:6 (w/v) ratio, and pH was adjusted to 11 using 1M NaOH. The mixture was then dispersion stirred for 1 h at 400 rpm using a mechanical agitator (Caframo Rz-1, Heidolph Schwabach, Germany). This suspension was wet-milled using a Kitchen-Aid® mill, and the fiber solids were separated from the starch and protein mix by straining through 80- and 150-mesh sieves followed by 5 washes with distilled water. The protein-starch suspension was allowed to sediment for 30 min at room temperature to recover the starch and protein fractions. The pH of the separated solubilized protein was adjusted to its isoelectric point (pH 4.5) with 1M HCl. The suspension was then centrifuged at 1317 $\times$ g for 12 min (Mistral 3000i, Curtin Matheson Scientific Inc., Morris Plains, NJ, USA). The supernatants were discarded, and the precipitates were freeze-dried at -47°C and 13 $\times$ 10−3 mbar until further analysis.

Enzymatic Hydrolysis

Hydrolysis of protein extracts was done using a completely randomized design. Treatments were the enzymatic systems: Alcalase® 2.4L FG, Flavourzyme® 500 MG (Novo Laboratories, Copenhagen, Denmark), pepsin from porcine gastric mucosa (Sigma, P7000-

100G), pancreatin from porcine pancreas (Sigma, P3292-100G) and Alcalase®-Flavourzyme® sequential system. The response variable was degree of hydrolysis (DH).

M. pruriens protein concentrate (5 g/100 mL) was hydrolyzed for 5, 15, 30, 45, 60, 90 and 120 min with Alcalase®, Flavourzyme® and the Alcalase®-Flavourzyme® sequential system. Similarly, the protein concentrate was hydrolyzed for 5, 10, 30, 60, 90 and 120 min with pepsin and pancreatin. Each hydrolysis reaction was done under controlled conditions in a reaction vessel equipped with a stirrer, thermometer and pH electrode. Hydrolysis parameters for individual treatments with Alcalase® 2.4L FG and Flavourzyme® 500 MG were: enzyme/substrate ratio, 0.3 AU g-1 for Alcalase® and 50 LAPU g-1 for Flavourzyme®; pH 8 for Alcalase® and pH 7 for Flavourzyme®; and 50°C (Pedroche et al., 2002). Hydrolysis parameters for pepsin and pancreatin were: enzyme/substrate ratio, 1:10; pH 2 for pepsin and pH 7.5 for pancreatin; and 37°C (Megías et al., 2004). In all treatments, the reaction was stopped by heating to 80°C for 20 min, followed by centrifuging at 9880 × g for 20 min to remove the insoluble portion.

Degree of Hydrolysis

Degree of hydrolysis (DH) was calculated by determining free amino groups with o-phthaldialdehyde following the methodology described by Nielsen et al. (2001): DH= h/htot * 100, where htot is the total number of peptide bonds per protein equivalent and h is the number of hydrolyzed bonds. Dependent on the amino acid composition of the raw material, htot was determined by reverse-phase high performance liquid chromatography (RP-HPLC) (Alaiz et al., 1992). Samples (2–4 mg protein) were treated with 4 mL 6M HCl, placed in hydrolysis tubes and gassed with nitrogen at 110°C for 24 h. Samples were then dried in a rotavapor and suspended in 1M sodium borate buffer at pH 9.0. Amino acid derivatization was performed at 50°C using diethyl ethoxymethylenemalonate. Amino acids were separated using HPLC with a reversed-phase column (300 x 3.9 mm, Nova Pack C18, 4 mm; Waters), and a binary gradient system with 25mM sodium acetate containing 0.02 g/L sodium azide at pH 6.0 (A) and acetonitrile (B) as solvent. Flow-rate was 0.9 mL min-1. The elution gradient was: time 0.0–3.0 min, linear gradient A:B (91:9) to A-B (86:14); time 3.0–13.0 min, elution with A-B (86–14); time 13.0–30.0 min, linear gradient A-B (86:14) to A-B (69:31); time 30.0–35.0 min, elution with A-B (69:31).

ACE Inhibiting Activity

ACE inhibiting activity of protein hydrolysate was analyzed as previously described by Hayakari et al. (1978). Specifically, ACE hydrolyzes hippuryl-L-histidyl-L-leucine (HHL) to yield hippuric acid and histidyl-leucine. This method relies on the colorimetric reaction of hippuric acid with 2,4,6-trichloro-s-triazine (TT) in a 0.5 mL incubation mixture containing 40 μmol potassium phosphate buffer (pH 8.3), 300 μmol sodium chloride, 40 μmol 3% HHL in potassium phosphate buffer (pH 8.3) and 100 mU/mL ACE. The mixture was incubated at 37°C for 45 min. The reaction was terminated by addition of TT (3% v/v) in dioxane and 3 mL 0.2M potassium phosphate buffer (pH 8.3). After centrifugation at 10 000 × g for 10 min, enzymatic activity was measured in the supernatant based on absorbance at 382 nm. All runs were performed in triplicate. ACE inhibiting activity was quantified by a regression analysis of ACE inhibiting activity (%) versus peptide concentration. IC50 (i.e. the peptide concentration in µg protein/mL required to produce 50% ACE inhibition under the described conditions) was calculated as follows:

ACE inhibiting activity (%) = (A − B)/(A − C) × 100
where A represents absorbance in the presence of ACE and sample, B is absorbance of the control and C is absorbance of the reaction blank.

IC50 = (50 − b)/m
where b is the intersection and m is the slope.

Hypotensive Effect of Legume Hydrolysates in Experimental Animal Models

Thirty-two normotensive male Wistar-Kyoto rats (8–10 weeks old; 240–320 g initial weight) were housed in individual cages under controlled conditions (12:12 h light:dark period; 25 ± 1°C; 65 ± 5% relative humidity) (NOM-062-ZOO-1999). Free access was provided to feed and water. After a 1-week adaptation period, rats were randomly divided into 8 test groups (4 rats per group) and 1 control group. Three hydrolysate concentrations were evaluated: 5, 10 and 15 mg/kg body weight, diluted in 0.3 ml physiological saline solution. Captopril and physiological saline solution were used as a positive and negative control, respectively. Prior to the determination of blood pressure, animals were anesthetized with a single dose (30 mg/kg) of sodium barbital by intraperitoneal injection. Measurements of systolic (SBP) and diastolic (DBP) blood pressure were taken every 130 min using an electric sphygmomanometer on the tail.

Antihypertensive Effect of Legume Hydrolysates in Experimental Animal Models

Adult male Wistar rats (8–10 weeks old) weighing 240–320 g were used in the present study. The animals were housed at 25 ± 2°C on a 12:12 h light: dark cycle and had free access to standard rat chow and drinking water. Twenty eight rats were used for each treatment and these were divided into groups (4 rats per group). Hypertension was induced by oral administration of the NOS inhibitor, N□-nitro-L-arginine methyl ester (L-NAME; 25 mg•kg-1•day-1), dissolved in drinking water for 6 wk (Kuru et al., 2009). A group of 5 animals received normal tap water throughout the experiment (control group). The hydrolysates with higher hypotensive effect were defined as antihypertensive. Captopril was used as a positive control. Prior to the determination of blood pressure, animals were anesthetized with a single dose (30 mg/kg) of sodium barbital by intraperitoneal injection. Measurements of systolic (SBP) and diastolic (DBP) blood pressure were taken every 130 min using an electric sphygmomanometer on the tail.

Statistical analysis

One-way analysis of variance (ANOVA) was used to evaluate protein extract hydrolysis data and in vitro ACE inhibiting activity. A Duncan's multiple range test was applied to determine differences between treatments. All analyses were done as previously described by Montgomery (2004) and processed using the Statgraphics Plus software (version 5.1).

RESULTS AND DISCUSSION

Protein hydrolysates from M. pruriens were obtained by treatment with the endoprotease Alcalase®, the endo/exoprotease Flavourzyme®, the gastrointestinal enzymes pepsin and pancreatin, as well as sequential systems of Alcalase®-Flavourzyme® and Pepsin-Pancreatin. Theses proteases are used in the food industry to improve the functional and nutritional properties of protein preparations. According to Pedroche et al. (2003), the controlled liberation of biologically active peptides from proteins by enzymatic hydrolysis is one of the most promising trends concerning medical applications of the protein hydrolysates with degree of hydrolysis higher than 10%, while hydrolysates with a low degree of hydrolysis (lower than 10%) are used for the improvement of functional properties of flours or protein isolates. Therefore, the results suggest that M. pruriens proteins are an appropriate substrate for producing these bioactive peptides when hydrolyze enzymatically (Figure 1).

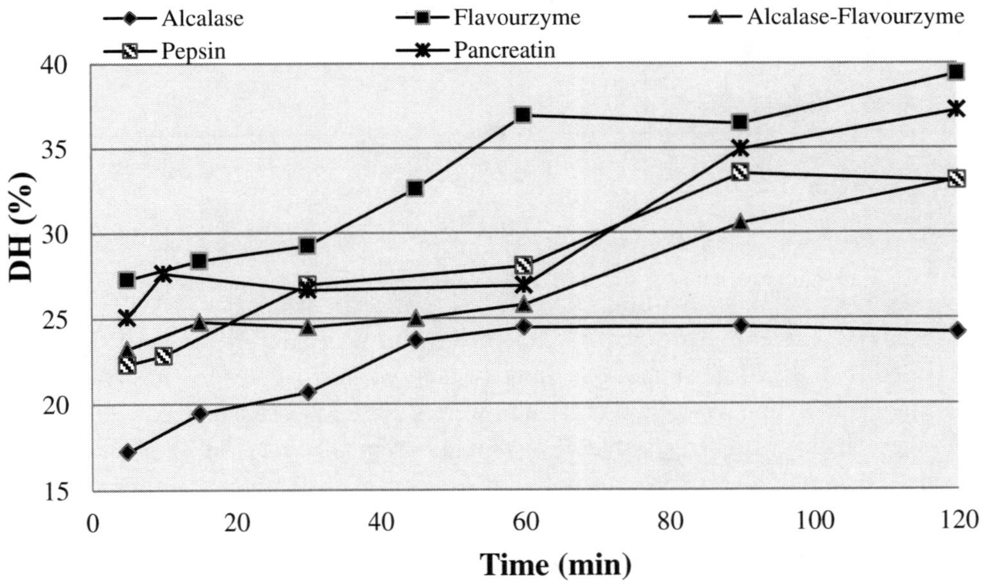

Figure 1. Degree of hydrolysis (DH) of Velvet bean (*M. pruriens*) protein hydrolysates.

The highest DH (39.39%) for *M. pruriens* protein concentrate was produced by Flavourzyme® after 120 min reaction time, while the lowest DH (17.23%) was produced with Alcalase® at 5 min reaction time (Figure 1). Variation in DH values was likely due to variation in protease specificity. The highest DH produced using Flavourzyme® to hydrolyze *M. pruriens* protein concentrate resulted because this is a protease complex produced by

Aspergillus oryzae, which contains endoproteinases and exoproteinases. This fungal protease complex has broader specificity that, when combined with its exopeptidase activity, produces DH values as high as 50% and dipeptides in the hydrolysate (Pommer, 1995). The highest DH obtained in *M. pruriens* protein concentrate by hydrolysis with Flavourzyme® was lower than previously reported by Pedroche et al. (2002) for chickpea hydrolysates produced sequentially with Alcalase® and Flavourzyme® (65% DH) at 150 min, and lower than reported by Vioque et al. (1999) for rapeseed protein hydrolysates (60% DH) produced with a mixture of Alcalase® and Flavourzyme® at 180 min. However, DHs recorded by all *M. pruriens* protein hydrolysates were higher than reported by Hong et al. (2005) for mung bean (*Vigna radiata*) protein hydrolysates produced with neutrase (12%) after 10 h, as well as values reported by Qi et al. (1997) for soy protein hydrolysates produced with pepsin for 60 min (11% DH) and 180 min (17% DH).

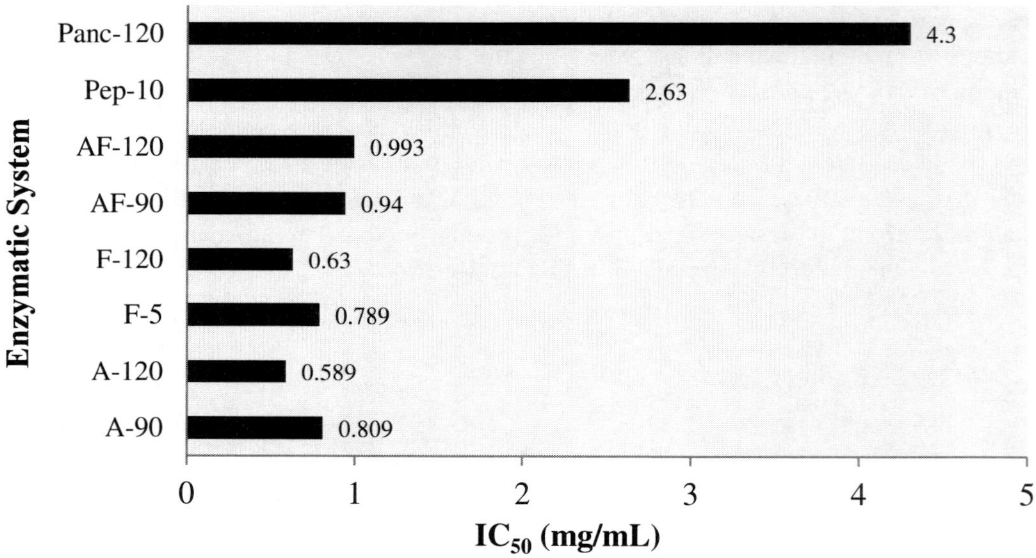

Figure 2. ACE-I inhibitory activity (IC50) of protein hydrolysates from M. pruriens.

M. pruriens protein hydrolysates obtained with Alcalase® (90 and 120 min), Flavourzyme® (5 and 120 min), the sequential system Alcalase®-Flavourzyme® (90 and 120 min), pepsin (10 min) and pancreatin (120 min) were selected to evaluate the ACE inhibitory activity. These hydrolysates registered IC_{50} ranging from 0.589 to 0.993 mg protein/mL, similar values than reported for enzymatic hydrolysates from different protein sources (IC_{50} = 0.200 to 2.467 mg protein/mL) with antihypertensive activity in spontaneously hypertensive rats (Hong et al., 2005). After extensive hydrolysis with Alcalase® (DH = 24.5 and 24.14% for 90 and 120 min, respectively), *M. pruriens* concentrate had a 0.809 and 0.589 mg/mL IC_{50}, while Flavourzyme® (DH = 27.31 and 39.39% for 5 and 120 min, respectively) had an IC_{50} of 0.789 and 0.630 mg/mL after 5 and 120 min, respectively. When *M. pruriens* concentrate was hydrolyzed for 90 and 120 min using the Alcalase®-Flavourzyme® sequential system (DH = 30.54 and 33.14%, respectively), the resulting IC_{50} was 0.940 and 0.993 mg/mL, respectively. However, pepsin and pancreatin had IC_{50} values of 2.63 and 4.3 mg protein/mL, respectively, which indicates that large quantities of these protein hydrolysates are required to inhibit 50% of ACE activity (Figure 2).

Hypotensive Effect of *M. pruriens* Hydrolysates in Experimental Animal Models

M. pruriens protein hydrolysates obtained with Alcalase® (90 and 120 min), Flavourzyme® (5 and 120 min) and the sequential system Alcalase®-Flavourzyme® (90 and 120 min) were selected to evaluate their hypotensive effect in animal models. Humans have used animals for centuries to understand their own biology. In cardiovascular research, animal models have enabled the study of early stage cardiovascular disease, as well as investigation of the mechanisms of pathogenesis of cardiovascular disease and the effects of drug intervention. According to Doggrell and Brown (1998), the aim of these studies is to provide clear concepts for selected investigations in humans. An ideal animal model for any cardiovascular disease in humans should have 5 characteristics: (I) mimic the human disease, (II) allow studies in chronic and stable disease, (III) produce symptoms which are predictable and controllable, (IV) satisfy economic, technical and animal welfare considerations, and (V) allow measurement of relevant cardiac, biochemical and haemodynamic parameters.

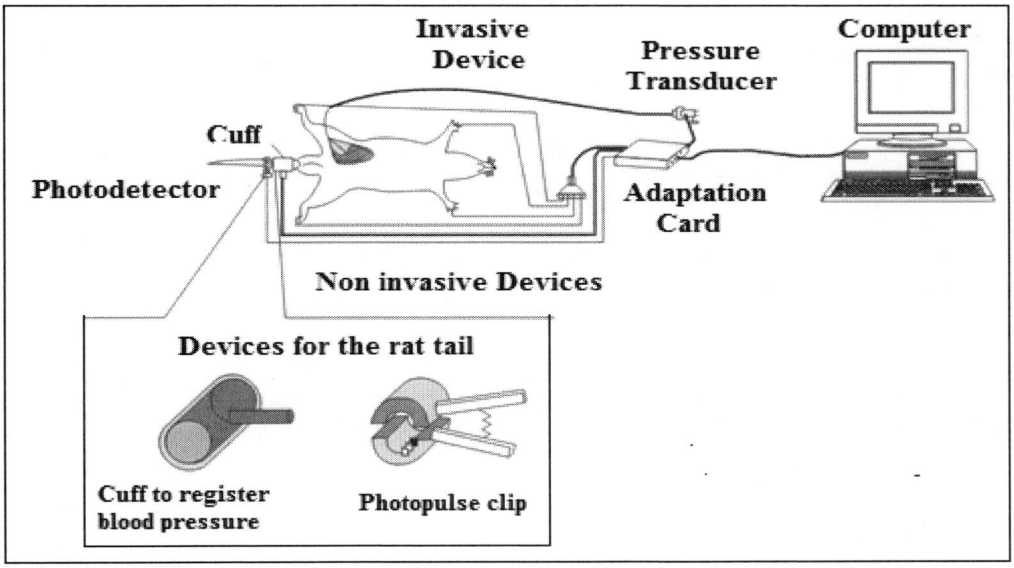

Figure 3. Tail-cuff method for measurement of systolic (SBP) and diastolic (DBP) blood pressure (Flores et al., 2007).

In the present study, the hypotensive effect of *M. pruriens* protein hydrolysates was evaluated using a trial where normotensive rats were administered a single dose of hydrolysates (5, 10 and 15 mg/kg) by intraperitoneal injection. The tail-cuff method was used to measure blood pressure (Figure 3).

The use of rats as animal models is rational from an economic viewpoint, and many techniques have been developed to measure relevant functional parameters. However, there is often insufficient consideration as to whether the other criteria, especially the mimicry of human disease, are satisfied with the choice of rat models for the study of cardiovascular disease in humans. According to Doggrell and Brown (1998), 3 examples illustrate the problems. First, cardiovascular diseases, such as hypertension and heart failure in humans, usually slowly develop with a wide range of neurohumoral adaptations compared to acute onset of symptoms in many surgical or drug-induced rat models. Second, cardiovascular disease is uncommon in young humans but markedly increases with age, yet most models of hypertension and heart failure only use young adult rats. Third, even in the presence of sustained high blood lipid levels, the development of atherosclerosis is very rare in most rat strains. Conversely, in humans, atherosclerosis is common and an important risk factor for hypertension and heart failure. These limitations of using rat models indicate that experiments in living animals are not appropriate for all investigations of the cardiovascular system. Cell dispersion from adult hearts, as well as culture of neonatal or adult cardiac fibroblasts and neonatal cardiomyocytes, have become increasingly more common since these allow studies of a single cell type in the absence of homeostatic mechanisms over a wide range of experimental conditions. However, while these studies have enabled characterization of cell properties, they cannot reproduce the working heart muscle (Doggrell and Brown, 1998).

In response to diametrically opposed opinions on the necessity of animals in research, the use of animal models may be limited by ethical concerns and legislation. Nevertheless, cardiovascular disease remains a major cause of death in Western countries. According to Doggrell and Brown (1998), the use of relevant models for human cardiovascular disease may provide useful information allowing a greater understanding of the cause and progression of the disease state, as well as potential therapeutic interventions. In this respect, the most common pathophysiological changes in the human cardiovascular system, such as hypertension, cardiac hypertrophy and heart failure, have been successfully reproduced in rat models

Peptides in the 6 analyzed *M. pruriens* protein hydrolysates exhibited hypotensive properties in experimental animal model, except Flavourzyme® hydrolysate at 5 min administered at a concentration of 5mg/kg (Figure 4). The hydrolysate generated with the Alcalase® system at 90 min led to reductions in SBP ranging from 12.36–31.11% after 130 min with a maximum reduction of 26.42% in DBP. The hydrolysate produced with the Alcalase® system at 120 min led to reductions ranging from 29.37–47.43% in SBP and from 23.04–43.47% in DBP. When the Flavourzyme® system at 5 min was injected, a maximum reduction of 24.06 and 25.15% in SBP and DBP, respectively, was observed. The hydrolysate produced with the Flavourzyme® system at 120 min led to reductions in SBP ranging from 1.09–12.81% and a maximum reduction of 12.85% in DBP. When the Alcalase®-Flavourzyme® sequential system at 90 and 120 min was injected, a maximum reduction of 27.95 and 30.62%, respectively, was observed in SBP, while a reduction of 32.30 and 30.66%, respectively, was seen in DBP. As expected, the negative control (i.e. physiological saline solution) registered a non-significant hypotensive effect (SBP = 0.6%, DBP = 1.19%), while the positive control (i.e. Captopril) dose-dependently affected SBP and DBP reductions ranging from 23.54–76.45% and of 22.68–61%, respectively.

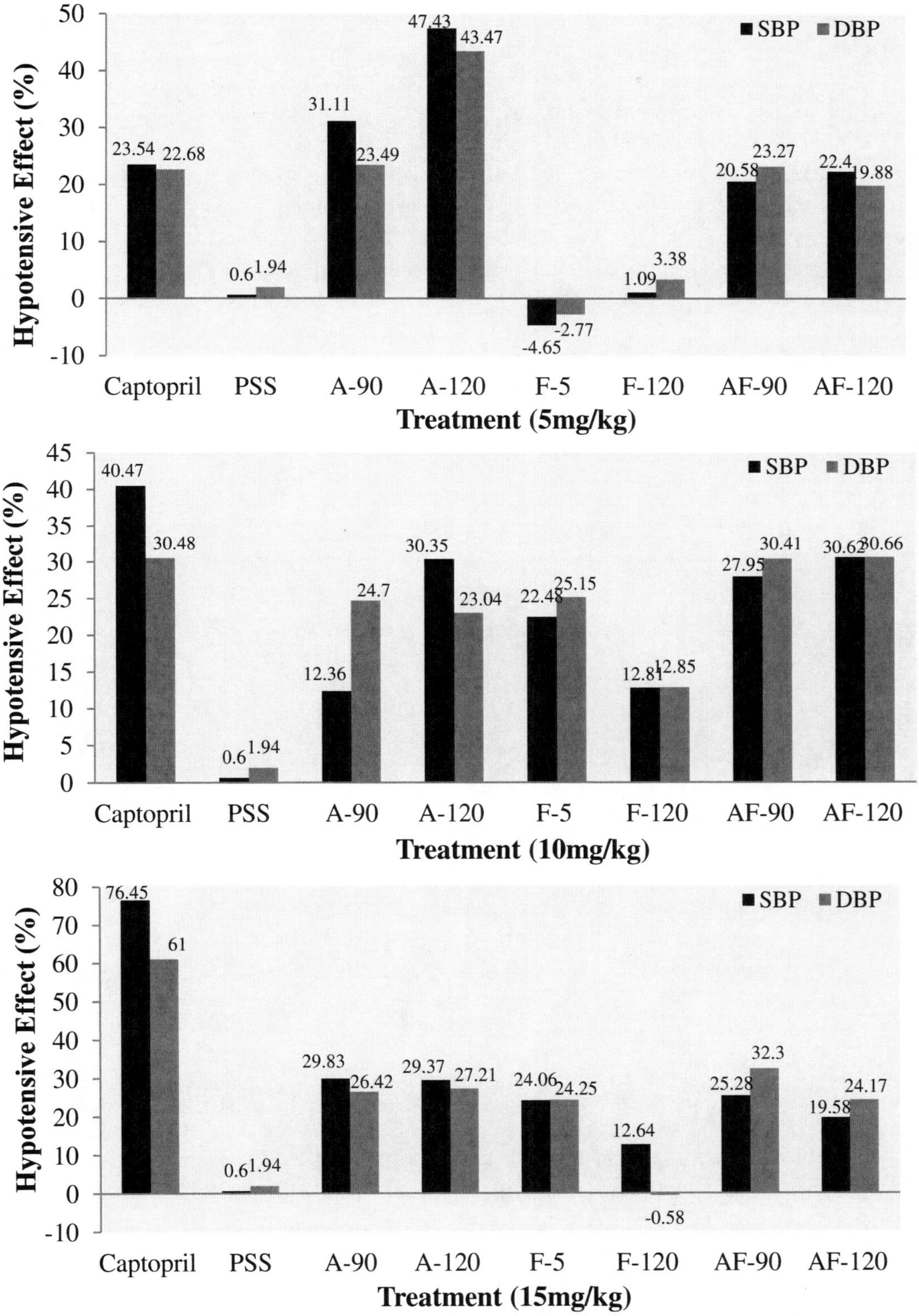

Figure 4. SBP and DBP reduction of captopril, physiological saline solution (PSS) and *M. pruriens* protein hydrolysates administered as a single dose (5, 10 and 15 mg/kg) by intraperitoneal injection into normotensive rats. A-90, A-120: Alcalase® at 90 and 120 min; F-5, F-120: Flavourzyme® at 5 and 120 min; AF-90, AF-120: Alcalase®-Flavourzyme® at 90 and 120 min.

Antihypertensive Effect of *M. pruriens* Hydrolysates in Experimental Animal Models

In the present study, hypertension was induced by oral administration of the nonselective NOS inhibitor L-NAME (25 mg/kg day) for 6 wk. According to Doggrell and Brown (1998), nitric oxide (NO, endothelium-derived relaxing factor), a paracrine vasodilator, has been implicated in regulation of vascular tone and myocardial contractility, as well as inhibition of platelet aggregation. Therefore, NO may be critical in the development of hypertension and atherosclerosis. At least 3 forms of NO synthase (NOS) exist. The cytokine-inducible NOS2 cosegregates with an increased blood pressure in the Dahl salt-sensitive rat, while endothelial cell NOS3 does not appear to be involved in human hypertension. Hypertensive effect by intraperitoneal administration of the NOS inhibitor N^{ω}-nitro-L-arginine methyl ester (L-NAME; 25 mg•kg^{-1}•day^{-1}) to hypertensive and normotensive rats showed that NOS was pharmacologically blocked to not registered significant increases in SBP and DBP in the hypertensive rats group in contrast with the normotensive rats group (Figure 5).

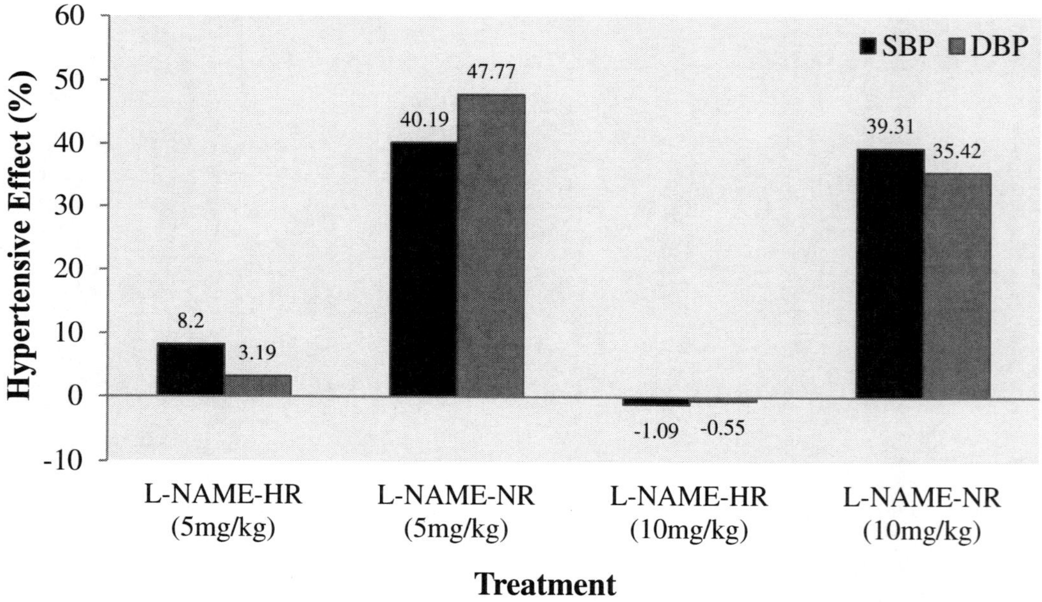

Figure 5. Hypertensive effect by intraperitoneal administration of the NOS inhibitor N□-nitro-L-arginine methyl ester (L-NAME; 25 mg•kg-1•day-1) to hypertensive (HR) and normotensive rats (NR).

The antihypertensive effect of *M. pruriens* protein hydrolysates that possessed the greatest hypotensive effect were evaluated in a trial, in which rats with induced arterial hypertension were administered a single dose of hydrolysates by intraperitoneal injection (Figure 6). The Alcalase® hydrolysate at 90 and 120 min (5mg/kg) produced maximum reductions of 25.53 and 27.29%, respectively, in SBP and 23.01 and 29.37%, respectively, in DBP. The Flavourzyme® hydrolysate at 5 min and the Alcalase®-Flavourzyme® system at 90 and 120 min showed reductions of 14.96, 22.22 and 8.84%, respectively, in SBP then 20.85, 25.58 and 16.10%, respectively, in DBP. Antihypertensive peptides from *M. pruriens* are not as potent as synthetic hypertension drugs (i.e. Captopril), but hold promise as safe and natural therapeutic agents free of adverse side effects. Results from the current study demonstrate the potential of using *M. pruriens* protein hydrolysates as alternative treatments to prevent/control hypertension and related illnesses.

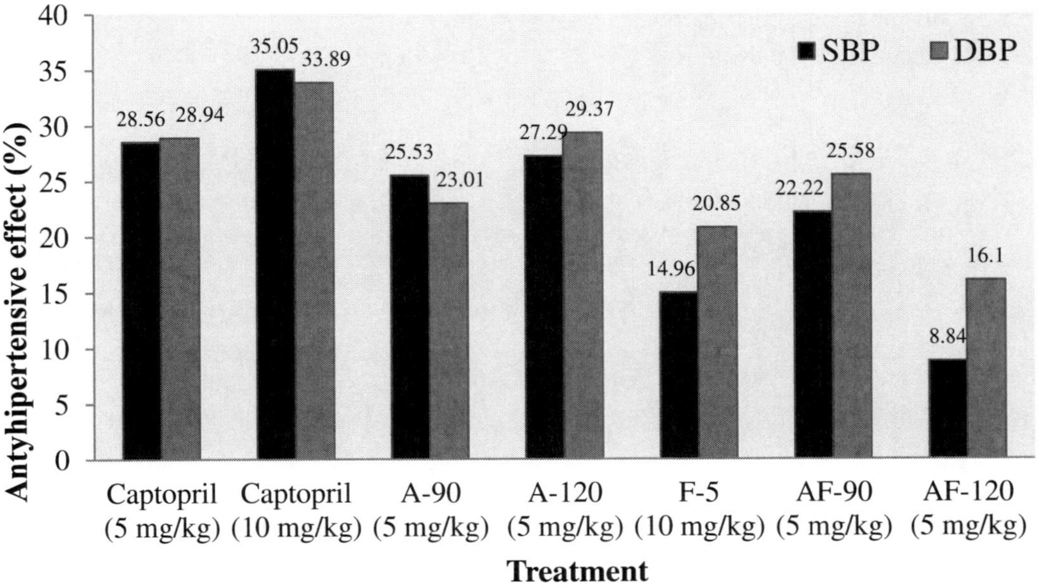

Figure 6. SBP and DBP reduction by captopril and M. pruriens protein hydrolysates administered as a single dose (5 or 10 mg/kg) by intraperitoneal injection into hypertensive rats. A-90, A-120: Alcalase® at 90 and 120 min; F-5: Flavourzyme® at 5min; AF-90, AF-120: Alcalase®-Flavourzyme® at 90 and 120 min.

CONCLUSION

Results suggest that *M. pruriens* protein hydrolysates have potential applications as flavor-active peptides in natural health products, nutraceuticals and functional foods or, possibly, as drugs. Antihypertensive peptides from *M. pruriens* are not as potent as synthetic hypertension drugs, but hold promise as safe and natural therapeutic agents free of adverse side effects. The current study demonstrated the potential of using *M. pruriens* protein hydrolysates as alternative treatments to prevent/control hypertension and related illnesses.

ACKNOWLEDGMENTS

This chapter was produced as part of CONACyT-Mexico project 154307 "Investigación científica dirigida al desarrollo de derivados proteínicos de *Mucuna pruriens* con potencial actividad biológica para la prevención y/o tratamiento de enfermedades crónicas asociadas al sobrepeso y la obesidad" and the Network "Bioactividad de Péptidos e Hidrolizados" supported by POMEP-SEP, México.

REFERENCES

Alaiz, M., Navarro, J.L., Giron, J., Vioque, E. (1992). Amino acid analysis by high-performance liquid chromatography after derivatization with diethylethoxy methylenemalonate. *J Cromatogr, 591*, 181–186.

Betancur-Ancona, D., Gallegos-Tintoré, S. and Chel-Guerrero, L. (2004). Wet fractionation of *Phaseolus lunatus* seeds: partial characterization of starch and protein. *J Sci Food Agric, 84*, 1193–1201.

Collins, R., Peto, R., MacMahon, S., Hebert, P., Fiebach, N.H., Eberlein, K.A. (1990). Blood pressure, stroke, and coronary heart disease. Part 2, Short-term reductions in blood pressure: overview of randomised drug trials in their epidemiological context. *Lancet, 335*, 827-838.

Doggrell, S.A., Brown, L. (1998). Rat models of hypertension, cardiac hypertrophy and failure. *Cardio Res, 39*, 89-105.

Flores, P., Santos, L., Martinéz, R., Cortés, S., Sanchez, G., y Infante, O. (2007). Confiabilidad de la presión arterial sistémica determinada por un método no invasivo en ratas normotensas. *Revista del Instituto Nacional de Enfermedades Respiratorias Ismael Cosio Villegas, 20*(4), 247-254.

Hayakari, M., Kondo, Y., Izumi, H. (1978). A rapid and simple espectrophotometric assay of angiotensin-converting enzyme. *Anal Biochem, 84*, 361–369.

Hong, L.G., Wei, L., Liu, H. & Hui, S.Y. (2005). *Mung-bean* protein hydrolysates obtained with alcalase exhibit angiotensin I-converting enzyme inhibitory activity. *Food Sci Technol Int, 11*, 281–287.

Kannel, W.B. (1996). Blood pressure as a cardiovascular risk factor: prevention and treatment. *JAMA, 275*, 1571-1576.

Kannel, W.B., Wilson, P.W., Nam, B.H., D'Agostino, R.B., Li, J. (2004). A likely explanation for the J-curve of blood pressure cardiovascular risk. *Am J Cardiol, 94*, 380-384.

Kearney, P.M., Whelton, M., Reynolds, K., Muntner, P., Whelton, P.K., He, J. (2005). Global burden of hypertension: analysis of worldwide data. *Lancet, 365*, 217-23.

Kuru, O., Kemal Sentürk, U., Koçer, G., Özdem, S., Baskurt, O.K., Çetin, A., Yesilkaya, A., Gündüz, F. (2009). Effect of exercise training on resistance arteries in rats with chronic NOS inhibition. *J Appl Physiol 107*, 896-902.

Lewington, S., Clarke, R., Qizilbash, N., Peto, R., Collins, R. (2002). Age-specific relevance of usual blood pressure to vascular mortality: a meta-analysis of individual data for one million adults in 61 prospective studies. *Lancet, 360*, 1903-1913.

MacMahon, S., Peto, R., Cutler, J., Collins, R., Sorlie, P., Neaton, J. (1990). Blood pressure, stroke, and coronary heart disease. Part 1, Prolonged differences in blood pressure: prospective observational studies corrected for the regression dilution bias. *Lancet, 335*, 765-774.

Mancia, G., De Backer, G., Dominiczak, A., Cifkova, R., Fagard, R., Ger,mano, G. (2007). Guidelines for the Management of Arterial Hypertension: The Task Force for the Management of Arterial Hypertension of the European Society of Hypertension (ESH) and of the European Society of Cardiology (ESC). *J Hypertens, 25*, 1105-1187.

Ortiz Ceballos, A.I., Aguirre Rivera, J.R., Osorio Arce, M.M., Peña Valdivia, C. Velvet bean (*Mucuna pruriens* var. *utilis*) a cover crop as bioherbicide to preserve the environmental services of soil. Herbicides-Environmental impact studies and management approaches. pp.167-184.

Pedroche, J., Yust, M.M., Girón-Calle, J., Alaiz, M., Millán, F. & Vioque, J. (2002). Utilization of chickpea protein isolates for the production of peptides with angiotensin I converting enzyme inhibitory activity. *J. Sci Food Agric, 2002, 82*, 960–965.

Pommer, K. (1995). New proteolytic enzymes for the production of savory ingredients. *Cereal Foods World, 40*, 745–748.

Qi, M., Hettiarachychy, N. & Kalapathy, U. (1997). Solubility and emulsifying properties of soy protein isolates modified by pancreatin. *J. Food Sci, 62,* 1110–1115.

Megías, C., Yust, M.M., Pedroche, J., Lquari, H., Girón-Calle, J., Alaiz, M., Girón-Calle, J., Millán, F. & Vioque, J. (2004). Purification of an ACE inhibitory peptide after hydrolysis of sunflower (*Helianthus annuus* L.) protein isolates. *J Agric Food Chem, 52*, 1928–1932.

Montgomery, D. (2004). Diseño y análisis de experimentos, Mexico City, Mexico. Limusa-Wiley.

Nielsen, P., Petersen, D., Dammann, C. (2001) Improved method for determining food protein degree of hydrolysis. *J Food Sci, 66(5)*, 642-646.

NORMA Oficial Mexicana NOM-062-ZOO-1999, Especificaciones técnicas para la producción, cuidado y uso de los animales de laboratorio.

Ortiz Ceballos, A.I., Aguirre Rivera, J.R., Osorio Arce, M.M., Peña Valdicia, C. (2012). Velvet vean (*Mucuna pruriens* var. *utilis*) cover crop as bioherbicide to preserve the environmental services of soil. In: Herbicides-Environmental Impact Studies and Management Approaches. Edited by Ruben Alvarez-Fernandez. InTech.

Pedroche, J., Yust, M.M., Girón-Calle, J., Alaiz, M., Millán, F. & Vioque, J. (2002). Utilization of chickpea protein isolates for the production of peptides with angiotensin I converting enzyme inhibitory activity. *J Sci Food Agric, 82*, 960–965.

Pedroche, J., Yust, M., Girón-Calle, J., Vioque, J., Alaiz, M. & Millán, F. (2003). Plant protein hydrolysates and tailor-made foods. *Electron J Environ Agric Food Chem, 2*, 233–235.

Pugalenthi, M., Vadivel, V. & Siddhuraju, P. (2005). Alternative food/feed perspectives of an underutilized legume *Mucuna pruriens* var. *utilis* - A review. *Plant Foods Hum Nutrition 60,* 201-218.

Tatasciore, A., Renda, G., Zimarino, M., Soccio, M., Bilo, G., Parati, G. (2007). Awake Systolic Blood Pressure Variability Correlates With Target-Organ Damage in Hypertensive Subjects. *Hypertension, 50*, 325-32.

Vasan, R.S., Beiser, A., Seshadri, S., Larson, M.G., Kannel, W.B., D'Agostino, R.B. (2002). Residual lifetime risk for developing hypertension in middle-aged women and men: The Framingham Heart Study. *JAMA, 287*, 1003-1010.

Vioque, J., Sánchez-Vioque, R., Clemente, A., Pedroche, J., Bautista, J., Millán, F. (1999). Production and characterization of an extensive rapeseed protein hydrolysate. *J Am Oil Chem Soc, 76,* 819–823.

Wu, J.C., Cheng, J., Shi, X.L. (2013). Preparation of ACE inhibitory peptides from Mytikus coruscus hydrolysate using uniform design. *BioMed Res Int*, 1-6.

In: Legumes
Editors: Hiroto Satou and Ren Nakamura

ISBN: 978-1-62808-280-7

Chapter 10

USE OF HERBICIDE (ROUNDUP®): POTENTIAL INFLUENCE ON SOIL METALS LABILITY AND ON RADISH METAL UPTAKE

Fabienne Baraud and Lydia Leleyter*
Université de Caen Basse-Normandie, Unité de Recherche Aliments Bioprocédés Toxicologie Environnements (UR ABTE) Bd Maréchal Juin, Caen, France

ABSTRACT

Important quantities of metals can be released from soils, accumulate in plants and then enter the food chain. Thus, the consumption of vegetables grown on contaminated soils has become one of the major routes of human exposure to potentially toxic metals. It is then crucial to better understand the various processes controlling metal mobilization / immobilization, uptake, and sequestration by plants, for better risk assessment. The aims of this work were to determine the potential influence of a commercial herbicide (Roundup®) on metals lability (mobility) and phytoavailability. Metals lability was estimated by single chemical extraction (HCl or EDTA) and phytoavailability by determination of the metals quantities actually taken by plants grown the on studied soils. Radish, an important, common vegetable in direct contact with soil, known as a useful indicator of metal phytoavailability, as a moderate accumulator (Davies and Hougton, 1984; Baraud and Leleyter 2012) was selected as a test plant for this purpose.

INTRODUCTION

There is a growing awareness and concern over food safety, due to accumulation of various chemicals, e.g. heavy metals, in agricultural soils (Peijnenburga et al., 2007; Salazar et al., 2012), resulting from urban, industrial and agricultural activities such as atmospheric deposition, waste disposal, waste incineration, urban effluent, vehicle exhaust, fertilizer

* Corresponding author: lydia.reinert@unicaen.fr, Tel: 33 2 31 56 74 91 - Fax: 33 2 31 56 73 03.

application and long term application of sewage sludge in agricultural land (Salazar et al., 2012; Douay et al., 2012). Soils contaminated to some extent with heavy metals are frequently used for vegetable growing (Moreno et al. 2003; Wang et al. 2004, Douay et al., 2012; Waterlot et al., 2013). When metals accumulate at toxic level, it can affect plant development and/or contaminate the food chain; food consumption is now well-recognized as a major pathway of metals to humans (Salazar et al., 2012). Excess dietary intake of potential toxic metals, such as Cd and Pb, might be deleterious to human health (Zhang et al., 2002). The soil-plant transfer of metals is a very complex system of chemical and biological processes and their interaction (Laurie and Manthey, 1994; Enst, 1996), subjected to antagonistic, additive and synergetic effects (Murakami and Ae, 2009; Salazar et al., 2012). In soil, only the metal present in the soil solution as free ions or soluble complexes, as well as possibly some forms adsorbed on the mineral surfaces, are readily available for the plant uptake. Numerous factors (pH, redox conditions, organic matter, interstitial water composition, microorganism activities…) control the mobility and (phyto) availability of metals. Plants have also evolved different strategies to mobilize the metals when necessary. For instance, they can modify the chemical composition of the rhizosphere, through acidification, electron transfer and release of root exudates (Marschner, 1986; Barona et al. 1996).

Metal chelation is recognized to be an important factor in metal uptake by plants, regulating to a large degree the metal solubility and availability (Laurie and Manthey, 1994; Salt et al., 1998; Shadi et al., 2012). Chelating agents present in the soil solution are generally assumed to increase metal solubility by their ability to form highly soluble metal chelates, preventing them from precipitation and sorption (Salt et al., 1995). The chelates can also affect the metal bioavailability. Different organic ligands can be present in soils, such as biochemical compounds, including microbial chelating agents, as well as root exudates (Baraud et al., 2005), but also anthropogenically derived organic chemical from industrial and urban activities or agriculture practices, such as herbicide application.

For instance, due to low production costs and high herbicidal efficiency, glyphosate (or 'GPS' : [N-(phosphonomethyl)glycine]) is a herbicide extensively used in agriculture (Morillo et al., 2007; Service, 2007; Bott et al., 2008). GPS is an organophosphonate, characterized by the presence of stable covalent carbon to phosphorus (C-P) bond (Bott et al., 2008), known as a chelating agent for metals (Subramaniam and Hoggard, 1988; Tsui et al., 2005). The chelation is mainly due to the 3 chemical groups (amine, carboxylate and phosphonate) that can form strong coordination bonds with metal ions to give bidentate and tridentate complexes (Morillo et al., 2007). GPS also presents a zwitterion structure depending on pH (Tsui et al., 2005; Morillo et al., 2007).

GPS is the active ingredient of the commonly used herbicide Roundup® and other brand names (Kools et al., 2005; Santos et al., 2011). This wide-spectrum herbicide acts as a non-selective total herbicide by interfering upon the metabolism of essential aromatic compounds in the shikimic acid pathway (Hernandez et al., 1999; Kools et al., 2005; Bott et al., 2008), causing impairment of general metabolic processes, such as protein synthesis and photosynthesis (Bott et al., 2008), exhibiting then specific growth reduction (Kools et al., 2005). It is usually applied at already developed plants (Santos et al., 2011).

Few studies deal with the potential joint toxicity of GPS-containing herbicide and other pollutants such as metals, although GPS is widely used in environments that can present significant metal contents (Tsui et al., 2005), such as agricultural soils.

Due to its ability as a chelator, GPS can possibly complex metals present in the interstitial solution or bound to the soil, then could desorb and mobilize soil-bound metals, possibly influencing the metal phytoavailability to some extent. Some authors (Bott et al., 2008; Wang et al., 2008) observed potential side effects of GPS on plant growth and micronutrient (Zn, Mn) status on GPS-resistant soybean variety. Hydroponic experiments demonstrated a reduction of Fe, Zn and Mn uptake and translocation in sunflowers, with low levels of GPS (Eker et al., 2006; Oztruk et al., 2008). Fe, Zn and Mn deficiencies induced by GPS application at field scale were also observed (Jolley et al., 2004; Huber, 2006). Tsui et al. (2005) demonstrated a reduction of the toxicity and bioavailability to *Ceriodaphnia dubia,* in water, of some metals (i.e. Ag), but an increase for others (i.e. Hg), when the animals were simultaneously exposed to metals and GPS or Roundup®. GPS was also used with success to enhance hyperaccumulation of copper and other metals in plants shoots for phytoremediation of contaminated soils (Ensley et al., 1999; Mathis and Kayser, 2001; Schimdt, 2003).

It is clear that potential synergism or antagonism effects of mixture of chemicals such as herbicide and metal occurs (Santos et al., 2011), depending on various parameters. Such possible interactions have to be further investigated, as the studies are still rare.

Our interest focuses on the potential influence of the commercial herbicide Roundup® on metals mobility and phytoavailiability. Radish was chosen as a test plant as this common dicotyledonous vegetable has its storage root-stem portion in direct contact with soil (Souza and Rauser, 2003). Being a useful indicator of metal availability it has been widely used in experiments on metals bioavailability or bioaccessability (Sauvé et al., 1996; Li et al., 2006; Intawongse and Dean, 2008; Gaw et al., 2008).

MATERIALS AND METHODS

1. Soils

For this preliminary test, three soils were selected in Normandy (France). Two soils (soil 1 and soil 2) were collected from private gardens, in rural areas. The third one (soil 3) was sampled from on ancient steel factory. The surface soil (15 cm) samples were air-dried then sieved (2 mm) before further use.

One week before seedling, soils were treated with herbicide. The soils were placed in plastic pots, holding about 2 kg each. The glyphosate formulation Roundup® was used in all the experiment, at the recommended dosage to better simulate real agricultural conditions and to include the potential effects of the other compounds (surfactant…) present in the commercial formulation.

One spray solution (40 mL of commercial solution for 1L) was prepared according to the product label. The prepared Roundup® solution was applied with a hand-held sprayer. One week after the herbicide treatment, soil subsamples were collected, subsequently air-dried, ground in an agate mortar and kept at 4°C before further analyses.

The soil composition was determined before and after the herbicide treatment, on dry soil, after grounding. About 0,2g of soil was digested with 10 mL of aqua regia (HCl/HNO_3, 3:1) in a Berghof speedwave MWS-2 microwave oven.

The acid digests obtained were analysed by atomic emission spectrometry (ICP-AES) in order to get the pseudo-total concentration of the chemical elements of interest, in the treated and untreated soils.

Single chemical extraction, a widely used method for evaluation of the bioavailability (Zhang et al., 2002; Baraud and Leleyter, 2012; Leleyter et al., 2012), was also performed on treated and untreated soils to evaluate the lability, then potential bioavailability of the elements studied. This involved mixing 10 mL of the chemical extractant (0.2 $mol.L^{-1}$ HCl or 0.05 $mol.L^{-1}$ EDTA) with 1g of dry soil, shaking at room temperature for 1h. The residue was then filtered on 0.45 μm pore size Millipore filter (polyvinylidene of fluoride (CHR-CFF-)n) and the leachate was stored in a polypropylene bottle at 4°C until chemical analyses.

2. Plant Growth Experiment

One week after the herbicide treatment, the treated soils (designed as 1T, 2T, 3T) and untreated soils (1nT, 2nT and 3nT: no herbicide treatment) were used for the growth experiments. 15 seeds were sown on each soil and the pots were placed under natural field conditions. Plants were watered every alternate day, or more frequently as required. No fertilizer was added. For this preliminary study, only one pot per soil was prepared.

After 3 weeks, the plants were harvested, washed with tap water, then rinsed thoroughly with deionised water and separated into shoots and roots. Such a washing process should have no significant influence on the root metal content (Wang et al., 2003).

Only the edible part of roots (referred to hereafter as the roots) was kept, as it was not possible to remove the adhered soil particles from the very thin fibrous roots that were then eliminated.

The tissues were oven-dried at 80°C for 48 h before weighing and analysis. The roots and shoots tissues were then ground in an agate mill and about 0.2 g of tissue was digested with 10mL of 70% HNO_3 in a micro-wave oven (Berghof speedwave MWS-2). The digests were analysed by ICP-AES to get the concentrations of the chemical elements of interest.

3. Chemical Analyses

The analyses of soils leachates, tissues and soils acid digests were performed by Inductively Coupled Plasma Atomic Emission Spectrometry (ICP-AES) (Varian Vista-MPX). The quality control of ICP-AES data was assessed by the analysis of blank reagents and calibration standards prepared with commercially available solutions (Varian solution standards). Accuracy of ICP-AES measurements was determined using various certified reference materials. These data and the limits of detection are presented elsewhere (Baraud and Leleyter, 2012).

RESULTS

1. Soils Analyses

1. 1. Pseudo-Total Concentrations

Table 1 reports the pseudo-total concentration of the elements of interest, in the soils before and after Roundup® application. Regarding the composition of the untreated soils (1nT-S, 2nT-S, 3nT-S) some differences can be noticed. Not surprisingly, regarding the respective sampling areas, the calcium content is far less important for soil 1 (granite area) than for the two other soils (calcareous area). Soil 1, that origins from a rural area poorly influenced by anthropogenic activities, presents the lowest contents in Cd, Co, Cr, Fe, Ni, Pb and Zn. On the contrary, soil 3, sampled in an urban area on an ancient industrial site, is quite contaminated with various metals : the highest concentrations of Cd, Cr, Pb, as well as Fe are observed for this soil. Soil 2, compared to the two other soils, presents intermediate concentration values for several elements (Cd, Cr, Pb) but is the most concentrated in Zn and Ni.

One week after the herbicide treatment, for each soil, no variation was observed, regarding the pseudo-total concentrations of the elements. These results suggest that no significant amount of metal were co-added with the herbicide solution. The main differences noticed are for Cr and Pb in soil 3, with a variation of + 39% of Cr, and – 63 % of Pb after Roundup® treatment. However, the Cr increase being only observed for this soil, it can not be attributed to the Roundup® addition. Moreover, the important decrease of the Pb content observed is difficult to explain. These two variations are probably the results of a heterogeneous distribution of Cr and Pb in the soil samples, rather than a consequence of the herbicide treatment.

1.2. Elements Availability

Table 2 reports, for each soil, the quantities of metals extracted by HCl 0,2 $mol.L^{-1}$ or by EDTA 0,05 $mol.L^{-1}$. The results are expressed as the ratio, in percentage, of element extracted to the pseudo-total content in soil (table 1), for each soil, before and after herbicide application. For most of the elements, the percentages extracted by HCl or EDTA are quite similar. This is particularly true for Cr, Fe, and Ni, with differences between HCl and EDTA extraction inferior to 5 % for Cr, Fe and Ni, and to 10 % for Mg and Mn, whatever the soil is. Ca is better extracted by HCl than by EDTA, on the soil 1, with or without herbicide applied. This suggests that Ca might be mainly present as calcium carbonate, which is very sensitive to pH and then, better solubilised by HCl 0,2 $mol.L^{-1}$ (pH <1) than by EDTA 0,05 $mol.L^{-1}$ (pH around 5). On the other hand, Pb, Cu and Cd are better extracted by EDTA than by HCl, for soil 2 and 3, with or without herbicide, whereas they are better extracted by HCl for soil 1. This suggests that the chemical or geochemical-speciation of these elements are similar in soils 2 and 3, but differs in soil 1. Whatever the soil and the extractant are, Ca is the most mobile element, whereas Cr, Fe and Mg are not extracted, as well as Ni except on the untreated soil 1. Significant percentages are observed for Pb, Zn, Cd and Cu, on soil 1 and soil 2, with or without herbicide.

Table 1. Pseudo-total concentrations of elements in the 3 soils before and after Roundup®treatment

		Al	Ca	Cd	Cr	Cu	Fe	Mg	Mn	Ni	Pb	Si	Zn	*Test name*
Soil 1	*no*	19637	1046	0,5	32	18	22824	4408	744	14	20	1114	84	*1nT-S*
	T	20502	819	0,5	32	17	22960	4462	743	13	19	1110	83	*1T-S*
Soil2	*no*	10967	45156	0,8	83	15	24666	2353	911	29	45	1078	168	*2nT-S*
	T	10575	53581	0,8	88	14	25194	2212	922	30	37	1069	151	*2T-S*
Soil3	*no*	9512	61686	1,7	120	19	48877	4505	2710	20	130	961	148	*3nT-S*
	T	9134	66215	1,7	166	16	50650	5577	3036	19	48	930	153	*3T-S*

Table 2. Percentage of metal extracted by HCl 0,2 $mol.L^{-1}$ or EDTA 0,05 $mol.L^{-1}$ from soils before and after herbicide application

		Ca		Cd		Cr		Cu		Fe		Mg		Mn		Ni		Pb		Zn		*Test name*
		EDTA	*HCl*	*EDTA*	*HCl*	*EDTA*	*HCl*	*EDTA*	*HCl*	*EDTA*	*HCl*	*EDTA*	*HCl*	*EDTA*	*HCl*	*EDTA*	*HCl*	*EDTA*	*HCl*	*EDTA*	*HCl*	
Soil 1	*no*	65	80	19	20	0	1	15	20	1	1	0	2	18	17	18	19	26	30	8	12	*1nT-S*
	T	46	68	17	21	0	1	13	19	1	1	0	1	15	10	4	6	23	24	9	12	*1T-S*
Soil2	*no*	26	35	27	14	0	0	26	0	1	0	4	12	19	12	5	1	36	0	33	8	*2nT-S*
	T	23	30	33	15	0	0	28	0	1	0	3	10	21	15	6	2	50	0	38	11	*2T-S*
Soil3	*no*	19	25	12	8	0	0	13	0	0	0	3	9	6	6	2	2	9	0	13	0	*3nT-S*
	T	19	23	13	2	0	0	17	0	0	0	2	7	6	5	2	1	24	0	13	0	*3T-S*

«no»= no herbicide application; «T» = Treatment = with herbicide application.

The herbicide treatment has no influence (difference between «nT» and «T» tests <8%) on the extractability of the studied elements, except for Ca and Ni in soil 1: they both appear less extracted to some extent, when Roundup® has been applied. For the soil 3, the percentage of Pb extracted by EDTA is also increased, from 9 to 24 %, after the herbicide treatment (but remains close to 0 with HCl). But this variation might be somehow the consequence of the possible sample heterogeneity suspected previously (table 1) for this soil. These results agree with the conclusions of Barrett and McBride (2006) on the application of Roundup® spray to soils. Indeed, they noticed that the elemental leaching of metals increased in highly metal-contaminated soils but not in soils with normal background concentrations of heavy metals, corresponding to the case of our studied soils.

2. Plants

2.1. Biomass Production

After 3 weeks, visually toxic symptoms were developed by plants grown on the untreated soil 3 (3nT) and on Roundup® treated soils (1T, 2T and 3T). This included reduction of the root growth and leaves expansion as well as some kind of chlorosis (figure 1).

First, biomass production is clearly influenced by the soil nature, when plants are grown on the untreated soils (figures 1 and 2). A drastic reduction of root and shoot biomass is noticed for the plants grown on soil 3, compared to soil 1 or 2, with a reduction about 80% of the shoot dry weight and about 95 % of the root dry weight per plant. Comparing plants grown on soils 1 and 2, the main difference for the root production, to a lesser extent, which is 20% reduced for soil 2, whereas shoot dry weight per plant is 7% higher than on soil 1.

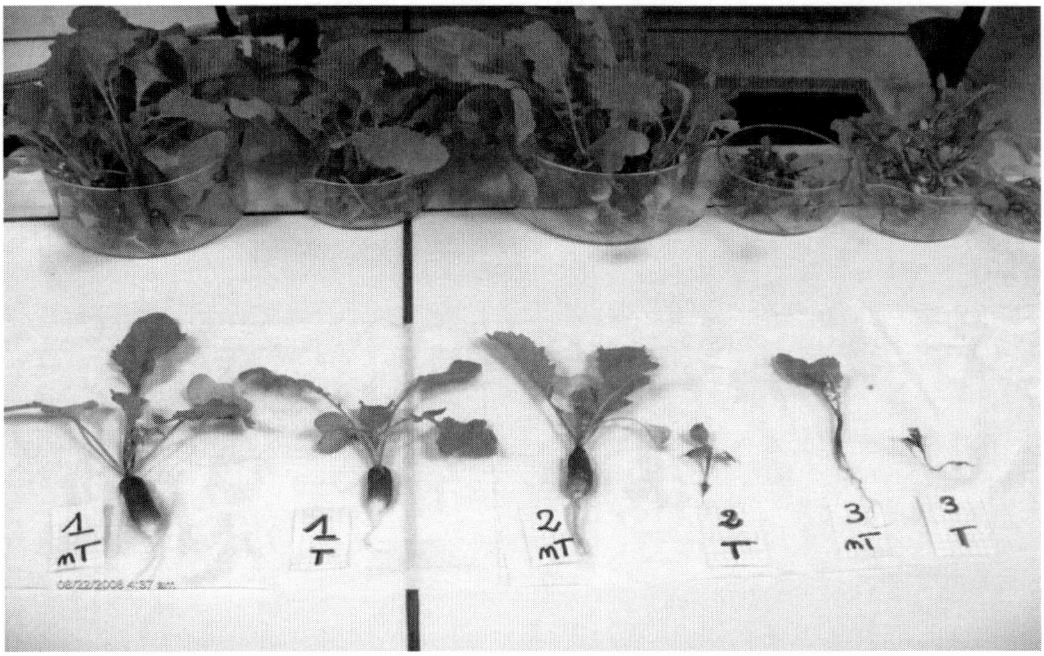

Figure 1. Picture of radishes grown on treated (T) and untreated (nT) soils.

This suggests that the soil 3 and, to a lesser extent, the soil 2, were less adapted to the radish growth, and/or contained elements available at toxic levels. For each soil, when comparing the biomass produced on the herbicide treated-soil and on the untreated soil, an important decrease of root and shoot dry weight per plant is also observed (Figures 1 and 2), which might be attributed to some residual herbicide effects. This reduction is less important on soil 1 (-30% and -75% of dry weight respectively for shoots and roots), than on soils 2 and 3 (dry weight reduction between 82 and 98% for shoots and roots).

2.2. Metals Contents in the Tissues

The collected roots (edible part) and shoots were finally digested with HNO_3 in a micro-wave oven and analysed for their metals contents (table 3). The metals contents vary from one soil to another. As for the biomass production, first, differences can be observed for the plants grown on the untreated soils, even if the concentration measured remain within the same range for each element of interest, as it can be observed in table 3.

For instance, all the metals of interest are more concentrated in the shoots and roots of plants grown on soils 3 than on soil 2. But when comparing with plants grown on soil 1, other observations can be made, i.e., Cr, Cu, Fe and Pb are more concentrated in the shoots and roots of plants grown on soil 3, whereas Cd, Mn, Ni and Zn are more concentrated in the plants grown on soil 1. However, all the concentrations observed are in the same range of the concentrations analysed in a bunch of radishes bought from a supermarket (table 4), except Cr which seems much more concentrated in the commercial radishes.

For a given soil, these first results suggests that the herbicide treatment could also induce variabilities in the metals uptake, as the metals content of the tissues differ with growing conditions (treated or untreated). These variabilities depend on the metal of interest as well as on the considered soil.

Figure 3 reports the ratios between the concentration in the plants grown on treated-soils to the concentration of the plants grown on the untreated soils, for each element and each soil. Contrary to reported results on reduced micronutrient uptake by plants (Bott et al., 2008), the general trend observed for the 3 tested soils is an increase of most of the metal content in the tissues, as a side effect of the Roundup® treatment, by factors from 2 to 6. However, in some cases, a decrease is noticed, for example, Zn (soil 2) and Pb (soil 1) in roots or Cd and Mg (soil 3) in shoots.

Heavy metals are known to be potentially toxic for plants, resulting in chlorosis, weak growth (as observed in this study) or disorders in plant metabolism (Guala et al., 2012). However, the phytoxicity observed, can also be attributed to GPS, that is known to be toxic to plants, causing impairment of general metabolic processes. Several studies report growth depression, chorosis, leaf necrosis, after GPS application, effects generally attributed to phytotoxic metabolite of GPS (i.e. AMPA: aminomethylphosphonic acid), or ingredients and surfactants (i.e. POEA: polyoxyethylene amine) present in the herbicide formulation (Bott et al., 2008).

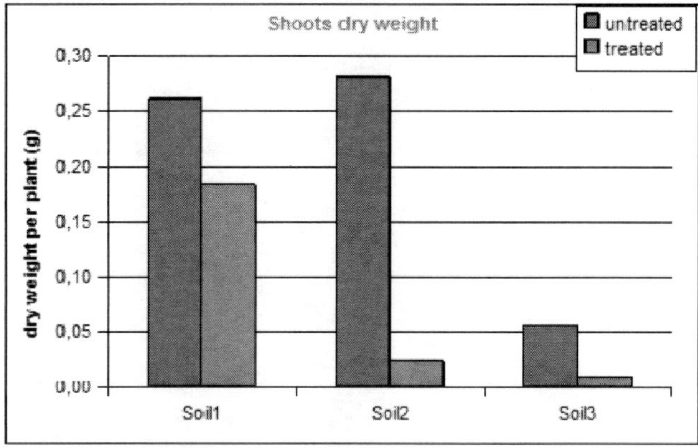

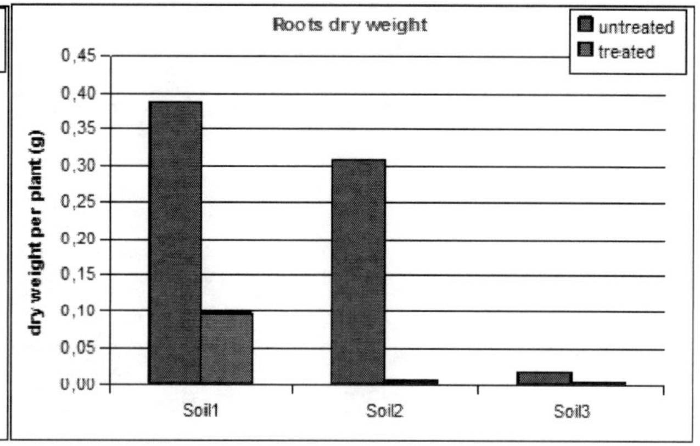

Figure 2. Roots and shoots dry weight (g).

Table 3. Metal content (in $\mu g.g^{-1}$ of tissue dry weight) in roots and shoots of the radish grown on untreated and Roundup® treated soils

		Ca		Cd		Cr		Cu		Fe		Mg		Mn		Ni		Pb		Zn		Test name
		Root	*Shoot*	*Root*	*Shoot*	*Root*	*Shoot*	*Root*	*Shoot*	*Root*	*Shoot*	*Root*	*Shoot*	*Root*	*Shoot*	*Root*	*Shoot*	*Root*	*Shoot*	*Root*	*Shoot*	
Soil 1	*no*	2643	25487	0,2	0,6	0,1	0,4	1	2	45	251	802	3563	51	543	2	2	0,5	1	51	143	*1nT-S*
	T	4917	21900	0,3	0,8	0,3	0,6	3	4	100	332	1359	3069	154	853	2	2	0,1	1	100	210	*1T-S*
Soil2	*no*	3585	24045	0,1	0,1	0,2	0,4	2	2	37	201	811	1452	5	39	0,1	0,2	0,5	1	30	40	*2nT-S*
	T	26282	23349	0,1	0,1	1	2	3	2	328	660	1575	1251	15	29	1	1	1	2	22	24	*2T-S*
Soil3	*no*	9204	21070	0,1	0,2	1	2	2	3	344	849	1562	2629	21	83	0	1	1	2	25	36	*3nT-S*
	T	24050	17764	0,1	0,1	4	7	7	5	1664	3886	3097	1989	77	138	5	3	5	7	32	36	*3T-S*

«no»= no herbicide application; «T» = Treatment = with herbicide application.

Table 4. metal content (in µg.g^{-1} of tissue dry weight) in roots and shoots of commercial radish

	Ca	Cd	Cr	Cu	Fe	Mg	Mn	Ni	Pb	Zn
Root	8518	0,1	52	3,8	71	3944	7,0	0,3	2	29
+/-	*274*	*0,1*	*	*0,7*	*6*	*92*	*0*	*0,1*	*0,1*	*1*
Shoot	28729	0,5	3,8	4,3	339	5908	59	0,9	0,82	56
+/-	*222*	*0,2*	*3,8*	*0,5*	*28*	*37*	*1*	*0,2*	*0,03*	*2*

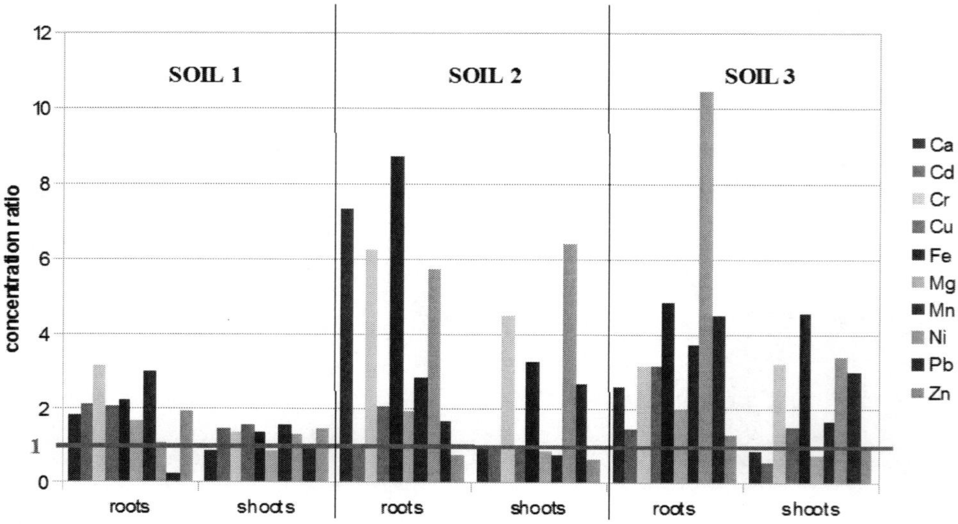

Figure 3. concentration ratio (T/nT) in the roots and shoots of the radish grown on untreated and Roundup® treated soils.

2.3. Bioconcentration Factors

As precised by Li et al. (2006), bioconcentrations factors (BCF) have been and are still widely used to estimate phyotavailability of metals on contaminated soils. They are expressed as the ratio of partial soil metal concentration separated by the plants (represented by total plant metal concentration) to total soil metal content (Kubova et al., 2008; Spiter et al., 2008) and are sometimes designed as soil–plant transfer coefficients (Kubova et al., 2008; Hao et al., 2009). BCF are generally considered as an indicator of the tendency for an element to accumulate in plants, which is highly dependent on metals and soils characteristics as well as on the plant genotype. The higher the value is, the more easily the plants absorb and accumulate the metal from the soil (Hao et al., 2009; Yang et al., 2009).

The pseudo-total concentration in soils, presented in table 1, and the element concentration estimated in the whole plant (root +shoot) were used is this study. BCF values are reported in table 5.

Table 5. BCF values for the plants grown on untreated or Roundup® treated soils

BCF values		Ca	Cd	Cr	Cu	Fe	Mg	Mn	Ni	Pb	Zn	*Test name*
Soil 1	*no*	11	0,6	0,01	0,1	0,01	0,4	0,3	326	222	0,04	*1nT-S*
	T	20	1,3	0,02	0,2	0,01	0,6	0,8	318	384	0,08	*1T-S*
Soil2	*no*	0,3	0,1	0,00	0,1	0,00	0,5	0,02	30	136	0,01	*2nT-S*
	T	0,4	0,1	0,02	0,2	0,02	0,6	0,03	58	77	0,1	*2T-S*
Soil3	*no*	0,3	0,1	0,02	0,2	0,01	0,5	0,03	267	31	0,1	*3nT-S*
	T	0,3	0,1	0,04	0,3	0,07	0,4	0,04	180	134	0,2	*3T-S*

«no»= no herbicide application; «T» = Treatment = with herbicide application.

Table 6. TF values for the plants grown on untreated or Roundup®-treated soils

TF values		Ca	Cd	Cr	Cu	Fe	Mg	Mn	Ni	Pb	Zn	*Test name*
Soil 1	*no*	9,6	3,6	4,6	1,9	5,6	4,4	10,5	1,0	1,6	2,8	*1nT-S*
	T	4,5	2,5	1,9	1,4	3,3	2,3	5,5	1,2	5,8	2,1	*1T-S*
Soil2	*no*	6,7	2,0	2,3	1,5	5,4	1,8	7,3	1,9	1,2	1,3	*2nT-S*
	T	0,9	1,8	1,7	0,7	2,0	0,8	1,9	2,1	1,9	1,1	*2T-S*
Soil3	*no*	2,3	2,3	1,6	1,4	2,5	1,7	4,1	1,7	1,9	1,5	*3nT-S*
	T	0,7	0,9	1,6	0,7	2,3	0,6	1,8	0,5	1,2	1,1	*3T-S*

«no»= no herbicide application; «T» = Treatment = with herbicide application.

For the plants grown on untreated soils, BCF values are similar for the radish grown on the three soils for Ca (except on soil 1), Cd, Cr, Cu, Fe, Mg, Zn and Mn, with values inferior to 1, in accordance to BCF values reported for radish by Baraud and Leleyter (2012). For Ni and Pb, much higher values are noticed, suggesting an important uptake, then phytoavailability of these elements. The decreasing order of the BCF values is similar from one soil to another, and is as follow: (Ni, Pb) >> (Ca, Cu, Cd, Mg, Zn) > (Mn Cr, Fe). It is interesting to note that these results can not be correlated to any of the single extraction performed, that does not predict such a higher Pb or Ni uptake compared to the other elements. On the contrary, the low Cr and Fe phytoavailability is in accordance to the very weak extractability by HCl or EDTA. For soils 1 and 2, the Rounpdup application induces an increase of the BCF values for all the elements, except for Ni on soil 1 (that remained unchanged) and for Pb on soil 2 (reduced value compared to the untreated soil). This suggests that the metal uptake and/or accumulation in the tissues are enhanced. The decreasing order of the BCF values remains unchanged compared to the untreated soils. However, as no change regarding the EDTA or HCl extractability is significant, it might be supposed that accumulation process in the tissues is influenced. For the plants grown on Roundup® treated soil 3, BCF values are unchanged or reduced for Ca, Cd, Mg and Ni, but increased for the other elements.

2.4. Translocation Factors

The translocation factors (TF) (Marchiol et al., 2004) are used as indicators for element translocation from roots to shoots (table 6). TF are calculated as the ratio between the shoots and roots concentrations for a given element. Decreasing values would indicate increasing element accumulation in the roots and restrained movement to the shoot. The TF values calculated in our experiments are in the range of values reported for radish grown on non-contaminated and contaminated soils (Marchiol et al., 2004).

Regarding the plants grown on untreated soils, the more « translocated » elements are Mn, Ca, and Fe in all cases. However, differences are noticed depending on the soil nature. Higher values are observed for all the elements, except Ni, for the plants grown on soil 1. This suggests that these elements (except Ni) are more accumulated in the roots of the plants grown on soil 2 or 3.

The Roundup® treatment induces a reduction of the translocation (reduction of the TF values), whatever the soil is, for all the studied elements, except for Ni and Pb on soil 1 and 2. This suggests that the Roundup® application had somehow an influence on the metal sequestration mechanisms, restraining the metal from movement to the shoots. On contrary, the Roundup® application induces a moderate increase of the Ni and Pb TF values, on soil 1 and 2, indicating in that case an enhanced translocation to the shoots.

CONCLUSION

Roundup® application, previous to growing experiments, had an influence on the biomass production and on the metals uptake and sequestration in radish tissues. For all the studied soils, a reduction of the biomass produced and an increase of the metals concentrations in the tissues were observed to various degrees. However, the lability/phyto-availability estimated by EDTA or HCl single chemical extraction remained unchanged, for all the studied metals, when comparing the Roundup®-treated and untreated soils, in all cases. This suggests, on one hand, that the single chemical extractions used might not be adapted to predict the metals phytoavailability to radish. These preliminary results also support the idea that the herbicide influences the plant metabolism rather than the soil chemistry and might modify or interfere on the plant metal uptake mechanisms rather than on the metals chemical speciation in soil.

As suggested by Bott et al. (2008), several hypotheses should be investigated. First, the origin of the observed effect on the plants has to be determined: is it the result from metals and/or herbicide/herbicides metabolites/ingredients accumulation and toxicity? Indeed, these effects could result from external or internal complexation with glyphosate or other ingredients present in the herbicide formulation, that would enhance metal availability; side effect could be also induced on the soil microorganisms, that are known to be involved in the plant uptake mechanism; uptake and accumulation of GPS or Roundup® or their metabolites in the tissues could also induce toxic effects and consequently disturb the plant metal uptake process.

Further investigations are then necessary to confirm these hypotheses. The studies dealing with the interaction of glyphosate and metals generally only consider few metals. Cu-GPS complex is one of the most studied, probably due to its high stability. However, the results obtained are highly dependent on the pH of the solutions generally used in adsorption experiments, since pH influence glyphosate speciation. Moreover, tests are often performed at very high concentration of glyphosate that are not relevant of real environmental conditions (Tsui et al., 2008).

Very contrasted results are reported in the literature, as very complex interactions between soils, plants and microorganisms occur, depending on numerous parameters, as well as on the specific conditions developed in experiments.

It is necessary to conduct more studies to better understand the processes and factors that control the metals speciation and bioavailability; to get better knowledge on the soil-plant transfer mechanisms and then to clarify the role of GPS and other ingredients associated to the herbicide formulations, on the chemical speciation of metals in soils and in plants.

REFERENCES

Baraud F. and Leleyter L. (2012). Prediction of phytoavailability of trace metals to plants: comparison between chemical extractions and soil-grown radish. *CRAS-Géosciences, 344*, 385-395

Baraud F., Fan T. W-M., Higashi R. M. Effect of cadmium and humic acids on metal accumulation in plants, In: Springer Eds, *Environmental Chemistry* , Chapitre 19, 2005, 205-214.

Barona, A. and Romero F. (1996). Study of metal accumulation in wild plants using principal component analysis and ionic impulsions. *Toxicology and Environmental Chemistry, 54*, p. 75-86.

Barrett K.A. and McBride M.B. (2006). Trace element mobilization in soils by glyphosate, *Soil Science Society of America Journal, 70*, 1882-1888.

Bott S., Tesfamariam T., Candan H., Cakùak I., Römheld V. and Neumann G. (2008) Glyphosate-induced impairment of plant growth and micronutrient status in glyphosate-resitant soybean (Glycine max L.), *Plant Soil, 312*, 185-194.

Davies B.E. and Houghton N.J. (1984). Distance declined patterns in heavy metal contamination of soils and plants in Birmingham, England. *Urban Ecology, 8*, 285-294.

Douay F., Pelfrêne A, Planque J., Fourrier H., Richard A., Roussel H. and Girondelot B. (2012). *Environmental Monitoring Assessment*, DOI 10,1007/s10661-012-2818-3.

Eker S., Levent O., Yazici A., Erenoglu B, Römheld V. and Cakmak I. (2006). Foliar-applied glyphosate susbstantially reduced uptake and transport of iron and manganese in sunflower (Helianthus annuus L.) plants, *Journal of Agricultural and Food Chemistry, 54*, 10019-10025.

Ensley B.D, Blaylock M.J., Dushenkov S., Kumar N.P.B.A and Kapulnik T. (1999). Inducing hyperaccumulation of metals in plant shoots, US patent 5, 917, 117.

Enst W.H.O (1996) Bioavailability of heavy metals and decontamination of soils by plants. *Applied geochemistry, 11*, 163-167.

Gaw S.K., Kim N.D., Northcott G.L., Wilkins A.L. and Robinson G. (2008). Uptake of SDDT, arsenic, cadmium, copper and lead by lettuce and radish grown in contaminated horticultural soils. *Journal of Agricultural Food Chemistry, 56*, 6584-6593.

Guala S.D., Vega F.A and Covelo E.F. (2012). The dynamics of heavy metals in plant-soil interactions. *Ecological Modelling, 221*, 1148-1152.

Hao, X.-Z., Zhou, D.-M., Huang, D.-Q., Cang, L., Zhang, H.-L. and Wang, H., 2009. Heavy metal transfer from soil to vegetable in southern Jiangsu Province, China. Pedoshere, 19(3): 305-311.

Hernandez A., Garcia-Plazaola J.I. and Becerril JM. (1999). Glyphosate effects on phenolic metabolism of nodulated soybean (*Glycine max* L. Merr.). J*ournal of Agricultural and Food Chemistry, 47*, 2920-2925.

Huber D.M., Strategies to ameliorate glyphosate immobilization of manganese and its impact on the rhizosphere and disease. In: Lorenz N., Dick R. (eds) *Proceedings of the glyphosate potassium symposium,* 2006. Ohio State University, AG Spectrum, DeWitt Iowa.

Intawongse M. and Dean J.R. (2008). Use of the physiologically-based extraction test to assess the oral accessability of metals in vegetable plants grown on contaminated soil, *Environmental Pollution*, *152*, 60-72.

Kools S.A.E, van Roovert M., Van Gestel N., Van Straalen M. (2005). Glyphosate degradation as a soil health indicator for heavy metal polluted soils. *Soil Biology and Biochemistry*, *37*, 1303-1307.

Kubova, J., Matus, P., Bujdos, M., Hagarova, I. and Medved, J. (2008). Utilization of optimized BCR three-step sequential and dilute HCl single extraction procedures fro soil-plant metal transfer predictions in contaminated lands. *Talanta, 75*, 1110-1122.

Laurie, S.H. and Manthey A., The chemistry and role of metal ion chelation in plant uptake processes, In: J.A. Mantley, D.E. Crowley, and D.G. Luster, Editors. *Biochemistry of metal micronutrients in the rhizosphere*, 1994, CRC Press Inc. 27-62.

Leleyter L., Rousseau C., Biree L. and Baraud F. (2012) Comparison of EDTA, HCl and sequential extraction procedures, for selected metals (Cu, Mn, Pb, Zn), in soils, riverine and marine sediments. *Journal of Geochemical Exploration*, *116–117*, 51–59.

Li S., Liu R., Wang M., Wang X., Shan H. and Wang H. (2006). Phytoavailability of cadmium to cherry-red radish in soils applied composted chicken or pig manure. *Geoderma, 136*, 260-271.

Marchiol L., Assolari S., Sacco P. and Zerbi G. (2004). Phytoextraction of heavy metals by canola (Brassica napus) and radish (Raphanus sativus) grown on multicontaminated soil. *Environmental Pollution, 132*, 21 -27.

Marschner H. *Mineral nutrition in higher plants*, A. Press edition, London, 1986.

Mathis P. and Kayser A. (2001). Plants uptake of heavy metals following glyphosate treatment, In: Proceeding of the 6th International Conference on The Geochemistry of trace Elements, Guelph, ON, Canada, 482.

Mijangos I., Becerril J.M., Albizu I., Epelde L. and Gabisu C. (2009). Effects of glyphosate on rhizosphere soil microbial communities under two different plant compositions by cultivation-dependent and -independent methodologies. *Soil Biologyand Biochemistry, 41*, 505-513.

Moreno D.A., Villora G., Ruiz J.M. and Romero L. (2003). Growth conditions, elemental accumulation and induced physical changes in chinese cabbage. *Chemosphere, 52*, 1031-1040.

Morillo E. , Undabeytia T. , Maqueda , Ramos A. (2007). The effect of dissolved glyphosate upon the sorption of copper by three selected soils. *Chemosphere, 47*, 747-752.

Murakami M. and Ae N. (2009). Potential for phytoextraction of copper, lead, and zinc by rice (*Oryza savit* L.), soybean (*Glycine max* [L.] merr.;), and maize (*Zea mays* L.). Journal of Hazardous Materials, 162, 1185-1192.

Ozturk L. Yaciki A., Eker S, Gokmen O., Römheld V and Cakmak I. (2008). Glyphosate inhibition of ferric reductase activity in iron-deficient sunflower roots, *New Phytol, 177*, 899-906.

Peijnenburga W.J.G.M., Zablotskajaa M. and Vijver M.G. (2007). Monitoring metals in terrestrial environments within a bioavailability framework and a focus on soil extraction. *Ecotoxicology and Environmental Safety, 67*, 163–179.

Sahid M., Pinelli E. and Dumat C. (2012). Review of Pb availability and toxicity to plants in relation with metal speciation; role of synthetic and natural ligands. *Journal of Hazardous Materials, 219,* 1-12.

Salazar M.J., Rodriguez J.H., Nieto G.L. and Pignata M.L. (2012). Effects of heavy metal concentrations (Cd, Zn and Pb) in agricultural soils near different emission sources on quality, accumulation and food safety in soybean [*Glycine max* (L.) Merrill]. *Journal of Hazardous Materials, 233-234*, 244-253.

Salt, D.E. (1995). Phytoremediation : a novel strategy for the removal of toxic metals from the environments using plants. *Biotechnology, 13*, 468-474.

Salt, D.E., Smith R.D. and Raskin I. (1998). Phytoremediation. *Annu. Rev. Plant Physiol. Plant Mol. Biol., 49*, 643-668.

Santos M.J.G., Soares A.M.V.M. and Loureiro S. (2011). Joint toxicity of three plant protection products to *Triticium aestivum* (L.) and *Brassica rapa* (L.). *Journal of Soils and Sediments, 11*, 990-999.

Sauvé S., Cook N., Hendershot W.H. and Mc Bride M. (1996). Linking plant tissue concentrations and soil copper pools in urban contaminated soils. *Environmental Pollution, 94*, 153-157.

Schmidt U. (2003) Enhancing phytoextraction: the effect of chemical soil manipulation on mobility, plant accumulation, and leaching of heavy metals, *Journal of Environmental Quality, 32*, 1939-1954.

Service R.F. (2007). A growing threat down on the farm, *Science, 316*, 114-117.

Souza J.F. and Rauser W.E. (2003). Maize and radish sequester excess cadmium and zinc in different ways. *Plant Science, 165*, 1009-1022.

Spiter, E., Rozsa, E., Gruiz, K., Tatrai, E. and Morvai, V. (2008). Site-specific risk assessment in contaminated vegetable gardens, *Chemosphere, 71*, 1301-1307.

Subramaniam V. and Hoggard P.E. (1988). Metal complexes of glyphosate. *Journal of Agricultural and Food Chemsitry, 36*, 1326-1329.

Tsui M.T.K., Wang W-X and Chu L.M. (2005). Influence of glyphosate and its formulation (Roundup®) on the toxicity and bioavailability of metals to *Ceriodphnia dubia. Environmental Pollution, 138*, 59-68.

Wang X., Shan X., Zhang S. and Wen B. (2004). A model for evaluation of the phytoavailability of trace elements to vegetables under the field conditions. *Chemosphere, 55*, 811-822.

Wang Y.J., Zhou D-M., Sun R-J., Jia D-A., Zhu H-W. and Wang S-Q. (2008). Zinc adsorption on goethite as affected by glyphosate, *Journal of Hazardous Materials, 151*, 179-184.

Waterlot C., Bidar G., Pelfrêne A., Roussel H., Fourrier H. and Douay F. (2013) Contamination, fractionation and availability of metals in urban soils in the vicinity of former lead and zinc smelters, France. *Pedosphere, 23(2)*, 143-159.

Yang Y., Zhang F.-S., Li H.-F. and Jiang R.-F. (2009). Accumulation of cadmium in the edible part of six vegetable species grown in Cd-contaminated soils. *Journal of Environmental Management, 90*, 1117-1122.

Zhang S., Lu A., Shan X-q., Wang Z. and Wa S. (2002). Microwave extraction of heavy metals from wet rhizosphere soils and its application to evaluation of bioavailability. *Analyltical Bioanalytical Chemistry, 374*, 942–947.

In: Legumes
Editors: Hiroto Satou and Ren Nakamura
ISBN: 978-1-62808-280-7

Chapter 11

CHEMICAL AND STRUCTURAL COMPOSITION OF ARABINOGALACTAN PROTEINS (AGP) OF THE MESQUITE (PROSOPIS SPP)

Yolanda L. López-Franco[1,*], Faviola Bainori-Vasquez[1], Ana K. Sánchez-Castañeda[1], Karla Martínez-Robinson[1], Tomás Madera-Santana[2], Alma Rosa Toledo-Guillén[1] and Jaime Lizardi-Mendoza[1]

[1]CTAOA. Biopolymers Group. Research Center for Food and Development, (CIAD, A.C.) Hermosillo, Sonora, Mexico

[2]CTAOV, Research Center for Food and Development, (CIAD, A.C.) Hermosillo, Sonora, Mexico

ABSTRACT

Arabinogalactan proteins (AGPs) are commonly found in higher plants. The AGPs are usually located on cell membranes and gum exudates. The AGPs characterized so far have a protein content of less than 10% with more than 90% carbohydrate content. As a consequence of their protein component, these macromolecules can be classified in two groups according to their amino acids composition: classical AGPs and non-classical AGPs. Furthermore, structures similar to glycosylphosphatidylinositol (GPI) lipid anchor have been reported in the classical AGP type molecules.

Mesquite gum (MG) is known to be comprised of macromolecular complexes of type II AG and proteoglycans (arabinogalactan-proteins, AGPs) containing of 3-7% of protein.

The existence of lipid components and hydroxyproline and proline amino acids representative of AGPs, have been not documented for MG. This report is focused on the chemical and structural characterization of AGPs of mesquite (Prosopis spp). The obtained data corresponds to a physicochemical composition similar to that reported previously for MG. Lipid composition determination demonstrated the presence of

* Corresponding author: Phone: +52 662 2892400 ext 364, Fax: +52 662 2800421. E-mail address: lopezf@ciad.mx.

C16:0, C18:0 and C22:0 fatty acids. With the EDX analysis, it was possible to detect traces of phosphorus that could be related to the GPI anchor. Additionally, alanine, serine, glycine, threonine and hydroxyproline were detected in MG; these amino acids are part of classical AGPs. The results suggest that the AGP of Sonorean mesquite can be considered classical AGPs.

INTRODUCTION

Arabinogalactan proteins (AGPs) are a family of complex proteoglycans associated with plant gums and exudates. The complete chemical structure of the AGPs is still unknown because of their heterogeneous nature and highly complex structure. The AGPs generally consist mainly of arabinogalactan (AG) of type II and its size varies from 30 to 150 sugar residues (Clark et al., 1979, Fincher et al., 1983, Serpe and Nothnagel, 1999) attached to the hydroxyproline (Hyp) residues. Typically AG molecules have a linear backbone chain of β(1–3) linked D-galactose units substituted at the C (6) with sidechains of β(1–6) linked D-galactose units. These sidechains have terminal residues of L-arabinose, L-rhamnose, D-xylose, or D-glucuronic acid (Nothnagel, 1997). The core protein backbone is rich in hydroxyproline (Hyp), serine (Ser), alanine (Ala), threonine (Thr) and glycine (Gly) (Fincher et al., 1983) but the amino acid content may vary between different species of AGP and tissues. Also, the structural relationship between carbohydrate and protein components of AGPs can vary according to the origin source and the tissue from it was isolated (Komalavilas et al., 1991; Youl et al., 1998; Oxley and Bacic, 1999; Svetek et al., 1999; Majewska-Sawka and Nothnagel, 2000; Schultz et al., 2000, Seifert and Roberts, 2007).

Depending on the core protein the AGPs could be classified in classical and non-classical AGPs. Classical AGPs are defined by their core protein that contain Hyp, Ala, Ser, Thr and Gly as the major amino acid constituents, whereas non-classical AGPs have their carbohydrate moieties attached to core proteins which are different from the classical ones in any number of ways, for example Hyp-poor AGPs, Cys-rich AGPs and Asn-rich AGPs (Baldwin et al., 1993; Chen et al., 1993; Du et al., 1996).

It has been suggested that classical AGPs are glycosylphosphatidylinositol (GPI)-anchored proteins. These anchors are attached to a specific amino acid in the C-terminal portion of classical AGPs concomitant with the elimination of the C-terminal hydrophobic tail from these AGPs. The linker core, of the GPI anchors of mammalian and yeast cells, has the following structure: protein → ethanolamine → PO4 → 6Man → 1→ 2Man → 1 → 6Man → 1 → 4GlcN → 1 → 6myo-inositol1 → PO4 → lipid (Oxley and Bacic, 1999). The oligosaccharide core has been found to carry one or more substituents of Man, ethanolamine-phosphate, Gal, Glc, GalNAc, sialic acid or hexadecanoic acid. The lipid component of GPI anchor is usually a ceramide consisting of tetracosanoic acid and phytosphingosine (Svetek et al., 1999). In a cultured rose cell suspension AGPs with Man, GlcN and inositol, together with tetracosanoic acid and phytosphingosine were detected (Svetek et al., 1999. Similar components of a GPI lipid anchor as Man, GlcN, inositol and lipids were found in gum arabic. These results were consistent with the hypothesis that GPI lipid anchors are present in gum arabic and contribute to the emulsifying action of the gum (Yadav et al., 2007).

A number or biochemical and histochemical studies have been performed to probe AGPs function and interactions. The most commonly approach employed includes the use of a synthetic chemical reagent called β-glucosyl Yariv (β-GlcY). Although the mechanism of interaction between the AGPs and Yariv reagent is not completely known, the β-GlcY has been used to detect, quantify and purify AGPs (Nothnagel 1997).

Based on transmission electron microscopic imaging of AGPs, some AGPs are globular [i.e. Hyp-poor carrot AGP and tobacco transmitting tissue-specific (TTS) proteins] (Baldwin et al., 1993; Cheung et al., 1995), whereas others AGPs are rod-like molecules (i.e. gum arabic) (Qi et al., 1991). Two models of molecular organization are associated with these observed shapes: the "wattle blossom" model (Fincher et al., 1983) and the "twisted hairy rope" model (Qi et al., 1991). In the wattle blossom model, the polysaccharide chains are folded into globular units which decorate the core protein so as to generate an overall spheroidal shape, whereas in the twisted hairy rope model, the polysaccharide chains as well as oligoarabinosides are postulated to wrap around the rod-like core protein. One or both of these models may prove to be correct, as different AGPs may have different molecular shapes.

The AGPs that have been structurally characterized are those that can be obtained in relatively large amounts, for example the AGPs of gum arabic from *Acacia senegal*. Other conventional sources of AGPs were: roses, carrots, blackberries, snuff, rye grass, rice (Nothnagel, 1997) and AGPs from mesquite (López-Franco et al., 2004; 2008).

Mesquite trees are leguminous plant trees that are widespread in arid and semi-arid regions of the world and account for one of the major plant species in such places. It is well documented that the bark of *Prosopis spp* produces an exudate known as mesquite gum (MG) in response to an insect attack, wounding or physiological stresses such as severe drought and hot weather. This gum could be defined as "the dried gummy exudation obtained from the stems and branches of *Prosopis* species". By contrast with commercial gum arabic (*Acacia senegal*), mesquite exudate is not an established hydrocolloid in the world market.

MG is a type II arabinogalactan protein (AGP) to which highly branched polysaccharide chains are attached. The branched polysaccharide is formed by L–Ara and D–Gal as the main carbohydrate residues and traces of D–Glc, D–Man and D–Xyl were also detected (López-Franco et al., 2008). In addition to the polysaccharide component, MG contains a small amount of protein (3-7%) (López-Franco et al., 2004; 2008; 2012) which seems to play an important role on the physical properties in solutions (Goycoolea et al., 1995, Vernon-Carter et al., 1996; Vernon-Carter et al., 1998).

The existence of models that explain the tertiary structure of AGP from mesquite gum has not yet been assessed experimentally. However, light scattering studies and monolayers adsorbed at the air-water interface have shown that mesquite gum *(P. velutina)* (López-Franco et al., 2004), resembles a polydisperse macrocoil in agreement with the "twisted hairy rope" proposal AGP from gum arabic.

The existence of lipid components, hydroxyproline and proline amino acids representative of AGPs have not yet been documented for mesquite gum. This report is focused on the chemical and structural characterization of AGPs from mesquite (*Prosopis spp*).

EXPERIMENTAL

Materials

Mesquite gum (MG) of *Prosopis spp.* was collected in June 2010 in Baviacora, Sonora, Mexico, located between the geographic coordinates 29° 42' north latitude and 110° 09' west longitude, at an average altitude of 620 meters above sea level. The gum was ground using a household coffee mill and stored until its utilization.

All chemical substances and gum arabic were purchased from Sigma-Aldrich Chemicals (St Louis, MO, USA).

Methods

Chemical composition

The moisture content was analyzed gravimetrically drying 3 g of gum in a convection oven (RIOSSA Model HS-41) at 100ºC for 12 h. The protein content was determined as reported elsewhere (López-Franco et al., 2012). The total inorganic matter content was obtained after burning 2 g of the sample in a furnace oven (Thermolyne type 1500 Furnace) at 600°C for 8 h. The fat content was determined according to the AOAC method 920.39 (AOAC, 1990). The total carbohydrate content was determined by the Dubois method (Dubois, 1956).

Hydrophobic Interaction Chromatography

A known volume (100 mL) of mesquite gum solution (10% w/w) in 4.2 M NaCl was fractionated on a phenyl-Sepharose CL–4B column, as described previously (López-Franco et al., 2004). The obtained fractions (FI, FIIa, FIIb, FIII) were dialyzed against water and freeze dried.

Quantification of AGPs by Yariv radial diffusion gel

A 1% (w/w) agarose gel with 0.15 M NaCl, 0.02% (w/v) NaN_3 and 10 μg/mL β-glucosyl Yariv reagent was prepared according to Van Holst and Clarke (1985). Wells of uniform thickness (1.2 mm) were punched in the gel using a Pasteur pipet (Kimble, Toledo, Ohio). The wells were filled with the fractions of mesquite gum (10 μL of 4 mg/L) and incubated for 48 h at room temperature in a box with high relative humidity to avoid drying of the gel. The quantification of AGP was made with standard curve of gum arabic. The β-glucosyl Yariv reagent (1,3,5-tris[4-β-D-glucopyranosyloxyphenylazo]-2,4,6-trihydroxy-benzene) was obtained from Biosupplies Australia Pty Ltd (Australia).

Sugar composition

The samples were hydrolyzed as reported elsewhere (López-Franco et al., 2013). The hydrolysates were analyzed using a HPLC (Varian, Model RI) equipped with a Microsorb 100-3 amino 100 x 4.6 column coupled to a refractive index detector. The analytes were eluted in isocratic condition with acetonitrile:water (80:20). The sample was compared against carbohydrate standards provided by Sigma-Aldrich Chemicals (St Louis, MO).

Amino acids profile

The amino acid profile was obtained by HPLC in accordance with the method reported by (Vazquez-Ortiz et al., 1995). The samples were hydrolyzed using hydrochloric acid 6 M at 150°C for 6 h. After cooling and evaporating, the hydrolysates were buffered to pH 2.2 prior to analysis by HPLC. The instrument used was a HPLC (Hewlett Packard Model 1100) equipped with Microsorb RP-C18 (10 cm x 4.6 mm ID) column and fluorescence detector.

Hydroxyproline and proline were determined by analyzing the mixture of 125 μL of sample with 500 μL of borate buffer and 250 μL de NBD-Cl. The mixture was heated at 60°C for 5 min and stopped with 50 μL of 1 M HCl. The amino acid standard mixture used was Amino Acid Standard H from Thermo Scientific Pierce.

Fatty acids

Samples of AGP fats were dissolved in 1 mL of hexane and transferred to a 10 mL tube with Teflon cap. After adding BF_3 reagent and mixing well, the mixture was heated in a water bath for 10 minutes at 60°C. Next, the mixture was transferred to a separatory funnel and 30 mL of sodium sulphate solution and 30 mL of hexane were added. The tube was mixed carefully by hand for 2 min and left in repose. Finally, the organic layer was removed and evaporates most of the hexane in the water bath at 60°C.

Gas chromatography analyses was performed on a Varian model 3400 gas chromatograph (VARIAN, Milano, Italy) equipped with a flame ionization detector. Analytical separation was achieved on a DB-23 capillary column (30 m × 0.25 mm i.d.) with 0.25 μm film thickness (Supelco, Bellefonte, PA, USA). Helium was used as carrier gas with a constant flow rate of 1 mL/min.

The temperature setting was as follows: injector 250°C and detector 200°C. The oven temperature was held at 70°C for 1 min, programmed to 166°C at 10°C/min, then programmed to 174°C at 1°C/min, and held for 0.5 min, programmed to 194°C at 2°C/min and held for 0.5 min then programmed to 210°C at 1°C/min and held for 1 min. The fatty acids standard was C8 to C24 (FAME mix analytical standard, wt. % varied, Supelco, Bellefonte, PA, USA).

Intrinsic Viscosity [η] and Molecular Weight

The intrinsic viscosity of the MG was measured at 30°C using an OB size Ubbelohde capillary viscometer (Technical Glass Products). MG solutions at 5% (w/w) in 0.1 M NaCl were prepared with magnetic stirring for 24 h at 25°C, followed by filtration with a 0.22 μm membrane. The viscosity average molecular weight (Mν) of the samples was calculated with the obtained intrinsic viscosity values using the Mark Houwink Sakurada equation, $[\eta] = K \cdot M\nu^{\alpha}$, using $K = 1.47 \times 10^{-2}$ and $\alpha = 0.50$ as the constants reported for MG (Rinaudo et al., 2008).

Scanning Electron Microscopy

To examine the morphology, size and electron diffraction pattern of arabinogalactan protein (Fraction IIa and Fraction IIb) samples, these were mounted on metal stubs using double-side adhesive conductive tape and then examined by field emission scanning electron microscope (FESEM, JSM-7600F, JEOL Inc., Japan). Energy Dispersive X-ray microanalysis (EDX) was used to identify the presence of the main elements in the AGP sample.

RESULTS AND DISCUSSION

Chemical Composition

Table 1 shows the results of the chemical analyses of moisture, protein, fat, ashes, and carbohydrate of mesquite gum and gum arabic samples. The moisture content of the gums showed a difference of 0.89% between samples. The moisture of MG is lower than reported previously (López-Franco et al., 2012). Both MG and GA are within the limit of moisture content allowed for this type of additive used for foods (maximum 15%) (Joint FAO/WHO Expert Committee on Food Additives, 1999).

The protein content of the MG (Table 1) was lower than that reported in the literature for gum of *Prosopis* (6.98%) (López-Franco et al., 2012). Conversely, the protein content of gum arabic was similar to that reported in other literature data (Islam et al., 1997, Goycoolea et al., 2000; López-Franco et al., 2012).

Mesquite gum had a noticeably higher content of lipid (5.17%) than GA (1.67%) (Table 1). The value of MG ash was similar to the one previously found (López-Franco et al., 2008; 2012), and lower than that obtained for GA. Total carbohydrate values were lower in both gums compared to those obtained in other investigations (Lopez-Franco et al., 2004; Randall et al., 1989).

Table 1 shows the monosaccharides detected in MG and GA samples after acid hydrolysis. According to the analysis the MG does not contain rhamnose in its structure; this agrees with that previously reported by Rinaudo et al., (2008). Furthermore, the content of arabinose in MG is higher than that of GA. This variation could be due to MG, a higher branched polysaccharide, and its branches are formed mostly by arabinoses.

Quantification of AGPs by Yariv Radial Diffusion Gel

Investigation of the physiological function of AGPs requires a sensitive method for their estimation in plant extracts. The Yariv radial diffusion gel is a good method but cannot be applied to whole plant extracts, because it requires a minimum of 15 μg AGP in the sample. To determine the AGP content, MG was fractionated by hydrophobic interaction chromatography. The AGP concentration in each fraction was determined by a standard curve linear relationship between the amount of AGP in the sample and the area of precipitation measured by square of diameter of the halo formed in the radial diffusion assay. Only in Fraction IIa and Fraction IIb were halos corresponding to AGP given (2.63 and 0.53 mg/mL, respectively). These results, according to that Fraction II, can be regarded as an AGP fraction in MG (López-Franco et al., 2004).

Fatty Acid Composition

Table 2 shows fatty acid composition of mesquite gum and gum arabic. In MG 16 fatty acids and in GA 11 fatty acids were detected.

Table 1. Chemical analyses of mesquite gum and gum arabic

Component (%)	Gum	
	Mesquite [a]	Arabic [a]
Moisture	7.19 ± 0.22	8.08 ± 0.20
Protein *	4.71 ± 0.02	3.73 ± 0.01
Total fat	5.17 ± 0.76	1.67 ± 0.58
Ashes	2.50 ± 0.05	3.77 ± 0.03
Total carbohydrates	63.54 ± 6.34	51.22 ± 0.15
Monosacharide composition (%)		
Arabinose (Ara)	46.31 ± 1.75	13.89 ± 1.40
Galactose (Gal)	4.90 ± 1.62	18.04 ± 1.92
Rhamnose (Rha)	ND	9.79 ± 1.39

All measurements are on a dry weight basis except for moisture.
ND= not detected.
[a] Data are means of triplicate analysis ± SD.
* Protein conversion factor 6.53 for mesquite gum and 6.6 for gum arabic.

Table 2. Fatty acid composition of mesquite gum and gum arabic

Fatty acid (%)	Gum	
	Mesquite*	Arabic♣
C4:0	0.04	0.03
C8:0	0.09	0.04
C10:0	0.79	ND
C11:0	ND	0.12
C12:0	0.94	ND
C13:0	0.59	0.49
C14:0	0.50	ND
C14:1	ND	0.16
C16:0	2.47	7.79
C16:1	ND	1.65
C17:0	4.77	ND
C17:1	5.29	ND
C18:0	5.65	3.29
C18:1 T9	2.38	3.72
C18:1	2.64	3.15
C18:3 Y	1.37	ND
C20:1	11.46	ND
C20:4	50.74	79.56
C22:0	10.29	ND

* % in basis of 200 mg of fat.
♣ % in basis of 60 mg of fat.
ND= not detected.

Essential fatty acids are those which the human body cannot synthesize and therefore must be obtained through the diet. In the MG three essential fatty acids were detected: oleic, linoleic and arachidonic, the last, was found in a higher proportion. However, oleic and arachidonic acid were higher in GA than in MG. These results indicate that both gums could be considered a source of essential fatty acids.

Yadav et al., (2007) determined lipid groups and their contribution to the emulsifying activity of gum arabic. The results showed the presence of short chain fatty acids in major proportion, (C16: 0, C18: 0 and C20: 0), and long chain fatty acids (C22: 0, C23: 0, C24: 0, C25 : 0, C26: 0, C28: 0). Comparing these results with those obtained in this work, in gum arabic only C16: 0 and C18: 0 were detected, and mesquite gum was found in addition to the above, the C22: 0. These differences in the fatty acid profile may be due to the lipid extraction process used.

The results of fatty acid analysis shown in Table 2 are of great relevance to the hypothesis that some AGPs in mesquite gum and gum arabic contain GPI lipid groups that, along with the protein component, contribute to emulsification activity (Yadav et al., 2007). In this sense, this investigation would contribute evidence that support such hypothesis.

The amino acid profile obtained for MG and GA (Table 3) were similar in both samples. The amino acid composition of GA (standard) was very similar to that reported for *Acacia senegal* (Anderson et al., 1985), there were only variation in a higher proportion of alanine, aspartic acid and glutamic acid, glycine, leucine and serine. It is noteworthy that the presence of methionine has been reported in *Acacia fischeri* (30 residues/1000 residues), *Acacia spirocarpa* (2 residues/1000 residues) and *Acacia kamerunensis* (10 residues/1000 residues) (Anderson and Morrison, 1989), however in the commercial sample used in this study it was not detected.

Regarding MG, hydroxyproline was found in greater amount, followed by serine, glycine and valine. In contrast, lysine, arginine, histidine and phenylalanine were present in lower amounts (14.07, 9.50 and 7.79 residues/1000 residues, respectively) than those reported by Anderson et al., (1985) for *Prosopis laevigata*. Previous works (Anderson et al., 1985; López-Franco et al., 2012) have reported the presence of methionine and cystine in mesquite gum, however, these amino acids were not detected in the analyzed MG sample.

Proline and hydroxyproline determination (Table 3) is significant as these amino acids form relevant parts of the protein fraction of the AGPs. Hydroxyproline was detected, reaching considerable amounts in gum arabic (327.59 residues/1000 residues) and mesquite gum (329.79 residues/1000 residues). This amino acid is the higher portion with respect to total amino acids for both gums (24.68% in GA and 27.28% in MG). These results are in accordance with previous reports on the amino acid composition of plant exudates (Anderson et al., 1985, Anderson and Morrison 1989; Osman et al., 1993).

The amino acids present in the classical type AGPs are hydroxyproline, alanine, serine, glycine and threonine. Mesquite gum contains all these amino acids, therefore the Sonoran mesquite AGP could be considered a classical AGP.

The intrinsic viscosity and molecular weight of mesquite and acacia gums are shown in Table 4. Gum arabic showed a higher viscosity than mesquite gum. This tendency was obtained in the molecular weight (Table 4). The results are consistent with the macromolecular species of larger size in GA than in MG, in agreement with López-Franco et al., (2004).

Table 3. Amino acid composition of mesquite gum and gum arabic

Amino acid (residues per 1000 residues)	Gum	
	Mesquite	Arabic
Aspartic acid	96.29	91.98
Glutamic acid	42.28	54.21
Alanine	45.14	72.66
Arginine	9.50	0.07
Phenylalanine	7.79	20.59
Glycine	132.14	197.61
Histidine	21.21	17.51
Isoleucine	41.20	22.00
Leucine	56.89	133.40
Lysine	14.07	15.09
Serine	245.30	230.35
Taurine	ND	0.58
Tyrosine	86.34	13.81
Threonine	58.21	73.20
Valine	112.64	56.95
Hydroxyproline	329.79	327.59
Proline	ND	ND

ND= not detected.

Table 4. Intrinsic viscosity and molecular weight of mesquite gum and gum arabic

	Gum	
	Mesquite	Arabic
Intrinsic viscosity [η] (mL/g)	10.90	21.18
Molecular weight (Mv) (g/mol)	5.5×10^5	1.1×10^6
k^*	1.47×10^{-2}	1.51×10^{-2}
α^*	0.50	0.52

* Data from Rinaudo et al., 2008.

Scanning Electron Microscopy

Morphological and structural studies were investigated using field emission scanning electron microscopy (FESEM). Figure 1A and B display the typical SEM images of Fraction IIa and IIb, respectively. The images exhibit a smooth surface and small particles are present in the matrix of AGP. The left side of the image shows an amorphous shape and the right side image shows some agglomerations which have taken place during the drying process. The development of these forms is expected after the sublimation of ice crystals during the freeze drying process.

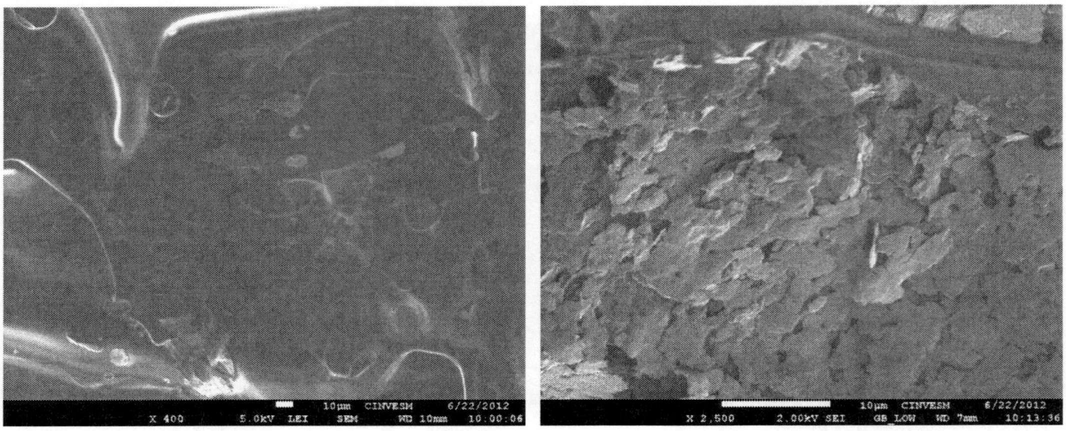

(A) (B)

Figure 1. FESEM images of surface of Fraction IIa at x 400 (A) and Fraction IIb at x 2500 (B).

The energy dispersive x-ray diffraction (EDX) results for AGPs showed peaks for C, O, and N atoms in both fractions. The difference between the fractions was the presence of traces of the phosphorous (P) atom in Fraction IIa at 2.013 eV.

Conclusion

Arabinogalactan proteins (AGPs) were quantified and characterized by physicochemical and microscopic techniques. Results showed that AGPs with a molecular weight of 5.5 x 10^5 g/mol and a protein content of 4.71% were formed mainly by Hyp, Ser, Gly and Val, whereas Ara and Gal were the major carbohydrate constituents.

It was possible to detect palmitic (C16: 0), stearic (C18: 0) and behenic (C22: 0) fatty acids related to the glycosylphosphatidylinositol anchor (GPI). Furthermore, traces of phosphorous (P) were detected in Fraction IIa and were regarded as AGPs in mesquite. The discovery that GPI anchors exist on at least AGPs of exudates represents an important contribution to the field. However, more studies are needed to elucidate the hypothesis of the existence of this molecule in the structure of mesquite gum.

AGPs of mesquite gum possess features of classical AGPs: they contain higher levels of Ara and Gal and abundant quantities of Hyp, Ser, Gly, and are bound to the β-glucosyl Yariv reagent.

Acknowledgments

This work was supported by CONACyT grant SEP-2008-C01-106115, to Y. L. López-Franco, PhD).

The authors would like to thank Dr. P. Quintana Owen from LANBIO at CINVESTAV Unidad Mérida for providing SEM and EDX facilities.

REFERENCES

Anderson, D. M. W., Howlett, J. F. & McNab, C. G. A. (1985). The amino acid composition of gum exudates from *Prosopis* species. *Phytochemistry*, *24*(11), 2718-2720.

Anderson, D. M. W. & Morrison, N. A. (1989). The characterization of four proteinaceous *Acacia* gums which are not permitted food additives. *Food Hydrocolloids*, *3*, 57-63.

AOAC. (1990). *Association of Official Analytical Chemists, Official Methods of Analysis*, 15th ed. AOAC, Washington, USA.

Baldwin, T. C., McCann, M. C. & Roberts, K. (1993). A novel hydroxyproline-deficient arabinogalactan protein secreted by suspension-cultured cells of Daucus carota. Purification and partial characterization. *Plant Physiology*, *103*, 115-123.

Chen, C. G., Mau, S. L. & Clarke, A. E. (1993). Nucleotide sequence and style-specific expression of a novel proline-rich protein gene from *Nicotiana alata*. *Plant Molecular Biology*, *21*, 391-395.

Cheung, A.Y., Wang, H. & Wu, H. M. (1995). A floral transmitting tissue-specific glycoprotein attracts pollen tubes and stimulates their growth. *Cell*, *82*, 383-393.

Clarke, A. E., Anderson, R. L. & Stone, B. A. (1979). Form and function of arabinogalactans and arabinogalactan-proteins. *Phytochemistry*, *18*, 521-540.

Du, H., Clarke, A. E. & Bacic, A. (1996). Arabinogalactan-proteins: a class of extracellular matrix proteoglycans involved in plant growth and development. *Trends in Cell Biology*, *6*, 411-414.

Dubois, M., Gilles, K. A., Hamilton, J. K., Rebers, P. A. & Smith, F. (1956). Colorimetric method for determination of sugars and related substances. *Analytical Chemistry*, *28*, 350-356.

Fincher, G. B., Stone, B. A. & Clarke, A. E. (1983). Arabinogalactan-proteins: structure, biosynthesis and function. *Annual Review of Plant Physiology*. *34*, 47-70.

Goycoolea, F. M., Calderón de la Barca, A. M., Hernández, G., Valenzuela, J. R. & Balderrama, J. R. (2000). Immunochemical, structural and functional properties of mesquite gum compared with gum arabic, 263-276. In E. A. Nothnagel, A. Bacic and A. E. Clarke (eds.), Cell and Developmental Biology of Arabinogalactan Proteins. Klower Academy/Plenum Publishers.

Goycoolea, F. M., Morris, E. R., Richardson, R. K. & Bell, A. E. (1995). Solution rheology of mesquite gum in comparison with gum arabic. *Carbohydrate Polymers*, *27*, 37-45.

Islam, A. M., Phillips, G. O., Sljivo, A., Snowden, M. J. & Williams, P. A. (1997). A review of developments on the regulatory, structural and functional aspects of gum arabic. *Food Hidrocolloids*, *11*(4), 493-505.

Joint FAO/WHO Expert Committee on Food Additives. (1999). Gum Arabic. In Food and Agriculture Organization of United Nations. Ed., Compendium of food additive specifications. 49-50. World Health Organization.

Komalavilas, P., Zhu, J. K. & Nothnagel, E. A. (1991). Arabinogalactan-proteins from the suspension culture medium and plasma membrane of rose cells. *The Journal of Biological Chemistry*, *266*, 15956-15965.

López-Franco, Y. L., Calderón de la Barca, A. M., Valdez, M. A., Peter, M. G., Rinaudo, M., Chambat, G. & Goycoolead, F. M. (2008). Structural characterization of mesquite (*Prosopis velutina*) gum and its fractions. *Macromol. Bioscience*, *8*, 749-757.

López-Franco, Y. L., Cervantes-Montaño, C. I., Martínez-Robinson, K. G., Lizardi-Mendoza, J. & Robles-Ozuna, L. E. (2013). Physicochemical characterization and functional properties of galactomannans from mesquite seeds (*Prosopis* spp). *Food Hydrocolloids*, *30*, 656-660.

López-Franco, Y. L., Córdova-Moreno, R. E., Goycoolea, F. M., Valdez M. A., Juárez-Onofre, J. & Lizardi-Mendoza, J. (2012). Classification and physicochemical characterization of mesquite gum (*Prosopis* spp.). *Food Hydrocolloids*, *26*, 159-166.

López-Franco, Y. L., Valdez, M. A., Hernández, J., Calderón de la Barca, A. M., Rinaudo, M. & Goycoolea, F. M. (2004). Macromolecular dimensions and mechanical properties of monolayer films of sonorean mesquite gum. *Macromolecular Bioscience*, *4*, 865-874.

Majewska-Sawka, A. & Nothnagel, E. A. (2000). The multiple roles of arabinogalactan proteins in plant development. *Plant Physiology*, *122*, 122, 3-9.

Nothnagel, E. A. (1997). Proteoglycans and related components in plant cells. *International Review of Cytology*, *174*, 195-291.

Osman, M. E., Menzies, A. R., Williams, P. A., Phillips, G. O. & Baldwin, T. C. (1993). The molecular characterisation of the polysaccharide gum from *Acacia senegal*. *Carbohydrate Research*, *246*, 303-318.

Oxley, D. & Bacic, A. (1999). Structure of the glycosylphosphatidylinositol anchor of an arabinogalactan protein from *Pyrus communis* suspension-cultured cells. *Proceedings of the National Academy of Sciences of the United States of America*, *96*, 14246-14251.

Qi, W., Fong, C. & Lamport, D. T. A. (1991). Gum arabic glycoprotein is a twisted hairy rope. A new model based on O-Galactosylhydroxyproline as the polysaccharide attachment site. *Plant Physiology*, *96*, 848-855.

Randall, R. C., Phillips, G. O. & Williams, P.A. (1989). Fractionation and characterization of gum arabic from *Acacia senegal*. *Food Hydrocolloids*, *3*(1), 65-75.

Rinaudo, M., Goycoolea, F. M. & Valdez, M. A. (2008). Emulsifying properties of mesquite gum. *Foods & Food Ingredients Journal of Japan*, *213*, 239-248.

Schultz, C. J., Johnson, K. L., Currie, G. & Bacic, A. (2000). The classical arabinogalactan protein gene family of arabidopsis. *The Plant Cell*, *12*, 1751-1767.

Seifert, G. J. & Roberts, K. (2007). The biology of arabinogalactan proteins. *Annual Review of Plant Biology*, *58*, 137-161.

Serpe, M. D. & Nothnagel, E. A. (1999). Arabinogalactan-proteins in the multiple domains of the plant cell surface. *Advances in Botanical Research*, *30*, 207-289.

Svetek, J., Yadav, M. P. & Nothnagel, E. A. (1999). Presence of a glycosylphosphatidylinositol lipid anchor on rose arabinogalactan proteins. *Journal of Biological Chemistry*, *274*, 14724-14733.

Van Holst, G. J. & Clarke, A. E. (1985). Quantification of arabinogalactan-protein in plant extracts by single radial gel-diffusion. *Analytical Biochemistry*, *148*, 446-450.

Vázquez-Ortiz, F. A., Caire, G., Higuera-Ciapara, I. & Hernández, G. (1995). High Performance Liquid Chromatographic Determination of Free Amino Acids in Shrimp. *Journal of Liquid Chromatography*, *18*, 2059-2068.

Vernon-Carter, E. J., Gómez, S. A., Beristain, C. I., Mosqueira, G., Pedroza-Islas, R. & Moreno-Terrazas, R. C. (1996). Color degradation and coalescence kinetics of aztec marigold oleoresin-in-water emulsions stabilized by mesquite or arabic gums and their blends. *Journal of Texture Studies*, *27*, 625-641.

Vernon-Carter, E. J., Pedroza-Islas, R. & Beristain, C. I. (1998). Stability of *Capsicum annum* oleoresin-in-water emulsions containing *Prosopis* and *Acacia* gums. *Journal of Texture Studies*, *29*, 553-567.

Yadav, M. P., Igartuburu, J. M., Yan, Y. & Nothnagel, E. A. (2007). Chemical investigation of the structural basis of the emulsifying activity of gum arabic. *Food Hydrocolloids*, *21*, 297-308.

Youl, J. J., Bacic, A. & Oxley, D. (1998). Arabinogalactan-proteins from *Nicotiana alata* and *Pyrus communis* contain glycosylphosphatidylinositol membrane anchors. *Proceedings of the National Academy of Sciences of the United States of America*, *95*, 7921-7926.

In: Legumes
Editors: Hiroto Satou and Ren Nakamura

ISBN: 978-1-62808-280-7

Chapter 12

COMMON BEAN AS AN EMERGING MODEL GRAIN LEGUME

Virginia Dalla Via, Carolina Rípodas, Joaquín Clúa, Candela Narduzzi, Mauricio Reynoso, Mario Aguilar, María Eugenia Zanetti and Flavio Blanco[1*]

Instituto de Biotecnología y Biología Molecular, Facultad de Ciencias Exactas, Universidad Nacional de La Plata, CCT-La Plata, CONICET, La Plata, Argentina

ABSTRACT

Common bean (*Phaseolus vulgaris*) is a legume plant original from America. It was domesticated independently in the Mesoamerican and Andean geographical centers. From there, it was disseminated to the rest of the world, becoming the most important grain legume in terms of human alimentation. Beans from this plant are the most important source of proteins in the diet of undeveloped and developing countries, representing more than 50% of the grains consumed around the world. Common bean, as well as many other legumes, is able to establish a symbiotic association with soil bacteria, allowing atmospheric nitrogen to be incorporated into ecosystems and soils, having a crucial impact in agriculture. In the last years, many genetic tools have been developed for this species, including sequencing of the genome, developing of transformation and reverse genetic techniques, germoplasm resources and databases of expression libraries. The existence of two genetic pools allowed establishing an experimental model to study genetic affinity between legumes and rhizobia, contributing to understand the genetic bases of nitrogen fixation efficiency.

[*] Corresponding author: fablanco@biol.unlp.edu.ar.

A Grain Legume for Human Consumption

Common bean (*Phaseolus vulgaris*) is considered one of the most important grain legumes worlwide because of the high impact on human alimentation. Also known as green bean, kidney bean, French bean, snap bean or runner bean, it is an herbaceous annual plant whose fruits can be obtained in two months. Common bean can be consumed as grains, pods or leaves. Due to its high percentage of protein content of the grain (20-25%), common bean is considered as the meat of poor people, complementing diets rich in carbohydrates typical from Africa and Latin America. It is also a rich source of vitamins and minerals, such as biotin, iron, phosphorus, magnesium, manganese, zinc, and copper [1].

The Food and Agricultural Organization (FAO) reported an estimation of 23 million tons of grains obtained from approximately 30 million hectares cultivated (http://faostat.fao.org, data from 2010, last updated in August 2012). The main producer is India, followed by Brazil, Myanmar and China. Sustainable agricultural systems have not been implemented in common bean, mainly because production takes place on small farms. Drought, the use of low fertility soils and biotic stresses result in yields that are far below the potential of the crop [1].

Considering the properties of common bean as a crop, its evolution and characteristics of its consumption, many genetic resources have been developed to study this species and to use this knowledge as representative to transfer to other grain legumes of economic importance, such as soybean (*Glycine max*), broad bean (*Vicia faba*) and peanuts (*Arachis hypogaea*). The two species selected as model legumes, *Medicago truncatula* and *Lotus japonicus*, are representative of forage crops and temperate climates, limiting the possibility to extrapolate particular aspects of their biological processes to food crops from tropical regions.

Origin and Domestication

Common bean is a plant original from America. Wild plants are distributed from the north of Argentina to Mexico, including two different genetic pools separated approximately 500 thousand years ago, with partial reproductive barriers [1]. There is still some controversy about the geographical location where the species originated. Molecular studies of seed proteins, as phaseolin [2] led to propose that common bean originated in the region corresponding to the territory that nowadays is occupied by Ecuador. However, a recent work based on molecular markers [3] supported an old hypothesis of the Mesoamerican origin of common bean [4]. The authors concluded that from the original location in Mesoamerica, migration events led to a spread of common bean in South America. Domestication of common bean occurred independently in each of the two genetic diversification centers, but analysis of the process evidenced the existence of a previous genetic bottleneck in the Andean genetic pool.

NITROGEN FIXATION

Legumes are vital to maintain the quality of the soils since they can establish a symbiotic association with bacteria, known as rhizobia, which can convert atmospheric nitrogen (N_2) to nitrogen-containing molecules that can be assimilated by plants into their metabolic pathways. Under low nitrogen conditions in the soil, roots of legumes exudate (iso)flavonoids to the rhizosphere, inducing the synthesis of a signal molecule by rhizobia. This molecule, called Nod Factor, is a lipochitin oligosaccharide with different chemical modifications that are species-specific [5] and primarily responsible for the compatibility of the interaction between certain legumes and particular species of rhizobia [6]. In addition, other molecules secreted by rhizobia or exposed in the cell wall can contribute to plant recognition of bacteria (Figure 1) [7-8].

This signal exchange between both partners of the interaction leads to the activation of two different, but integrated processes: development of a new organ in the root, the nodule, and penetration of bacteria to reach the cortical cells where nitrogen fixation will take place (Figure 2) [9-10]. In most cases, bacteria are attached to the surface of root hairs, which change the direction of the tip growing to entrap the bacterial colony in a curl. From this point, a hollow tube begins to be formed by the plant, inverting the polar growth toward the interior of the cell. This penetration structure, known as the infection thread (Figure 2), growths until the cortex is reached and bacteria are released into cortical cells in a process similar to endocytosis. Concomitantly with the infection process, a developmental program is triggered by cell divisions of the cortex, leading to formation of the nodule primordia. The mature nodule contains differentiated bacteria surrounded by a plant-derive membrane. Inside this structure, called the symbiosome, reduction of atmospheric nitrogen takes place (Figure 2) by the action of the nitrogenase complex under low oxygen conditions.

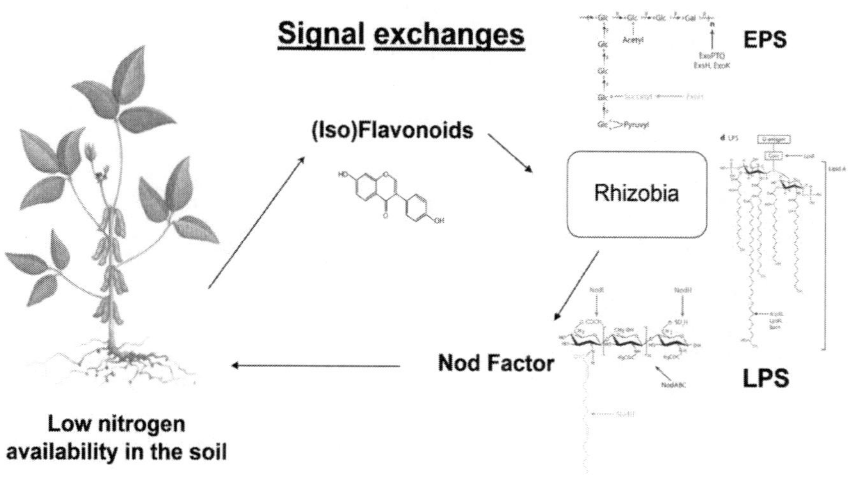

Figure 1. Signal exchanges between roots of legumes and rhizobia. (Iso)flavonoids exudated by the root act as a signal for the synthesis of Nod factor, which together with bacterial molecules as lipopolysaccharides (LPS) and expolysaccharides (EPS) triggers the developmental program of the nodule in the root.

The molecular nature of the plant symbiosis signalling has started to be understood in the last years. Two receptors with LysM extracellular domains and a kinase activity in the cytoplasm would be responsible for Nod Factor perception and initiation of signal transduction [11-12]. One of the earliest responses detected downstream of Nod factor perception is the periodic oscillations of intracellular calcium known as "calcium spiking" [13]. Screening of mutant plants of the two model legumes *L. japonicus* and *M. truncatula* led to the identification of genes whose products are involved in the Nod factor signal transduction pathway. The gene *dmi2* of *M. truncatula* (*Symrk* in Lotus) encodes for a kinase receptor with leucine-rich repeats [14-15], but the putative ligand is still unknown. *dmi1* (*castor* and *pollux* in Lotus) correspond to a cation channel which acts upstream of the calcium spiking [16], whereas *dmi3* encodes for a calcium and calmoduline dependent protein kinase that could be involved in the decoding of the calcium signal [17-18]. Several transcription factors involved in the expression of genes that are required for nodulation have been also genetically identified. These include members of the GRAS [19-20], ERF [21], and NF-Y [22-23] families of transcriptional regulators. All these genetic studies in model legumes have revealed the molecular pathway that is essential to trigger nodule organogenesis and the infection process activated by perception of the Nod factor. However, little is known about the complex ecological process occurring on the rhizosphere, where different species and strains of rhizobia compete to establish the nitrogen-fixing symbiosis with legume plants. The signal exchange between both symbionts allows the plant to recognize and discriminate the best partner before nitrogen fixation reactions take place. Efficiency of biological nitrogen fixation can vary up to ten times for a particular legume species [24], making crucial to understand the molecular networks governing rhizobial strain discrimination to increase the ability to compete of the most efficient strains in agricultural systems.

$$N_2 + 8H^+ + 8e^- + 16\,ATP$$
$$\downarrow$$
$$2NH_3 + H_2 + 16ADP$$

Figure 2. Nodule organogenesis in the root. Bacteria attached to the surface of the root hair and a tube is formed by the plant to guide the infecting bacteria to the cortex (left; IT= infection thread). The mature nodule contains infected cells (right) where nitrogen reduction takes place. The chemical reduction of N_2 is a highly demanding reaction in terms of ATP and reducing power.

MOLECULAR TOOLS TO STUDY SYMBIOSIS WITH RHIZOBIA

Yield of common bean is low compared with other legumes, probably as a result of domestication and selection of traits associated with the stability of the crop [1]. This limitation provides a great opportunity to improve yields using genetic and transformation techniques. Genetic resources for the species have been developed during the last two decades, including recombinant inbred lines (http://www.phaseolus.net/doc/BeanRI.pdf), BAC libraries [25-26], EST [27-28] and microRNA sequences [29]. These individual efforts started to be coordinated by a consortium called Phaseomics, which was launched in Cuernavaca, Mexico, in 2001 (http://www.phaseolus.net/).

Methods for stable transformation of *P. vulgaris* based on ballistic and Agrobacterium transformation are available [30-34], but its usage is limited since this species is very difficult to regenerate from transformed tissues. An attractive alternative to stable transformation was developed few years ago in the laboratory of Federico Sánchez [35-36]. The technique is based on *Agrobacterium rhizogenes* mediated transformation and was successfully adopted by other research groups. Common bean plants are inoculated with *A. rhizogenes* in the stem at the position of cotyledon insertion (Figure 3, left). A callus is formed after transformation and, after a couple of weeks, transgenic roots carrying the R-DNA emerge from the callus (Figure 3, middle). At this point, the wild type root is removed, rendering a composite plant formed by a wild type aerial part and the transgenic root (Figure 3, right). This system is suitable to study the function of genes that have a role in the root biology, including nodule formation.

Taking advantage of this transformation system, we developed an RNA interference (RNAi) based strategy to study gene function by reverse genetics. A construct carrying inverted repeats of the gene that has been selected for silencing is introduced in *A. rhizogenes* (Figure 4). Upon transformation of the root, this DNA is transcribed into an RNA with a stem loop structure that is recognized by the post-transcriptional silencing machinery to produce a small RNAi. This strategy produced an effective reduction of RNA levels of the targeted genes [23, 37-39]. Post-transcriptional gene silencing can vary between different roots since each one is an independent transformation event. Considering the results obtained so far in our laboratory using different genes, we observed a reduction of the transcript levels ranging from 60 to 90% as compared with roots transformed with a control RNAi construct.

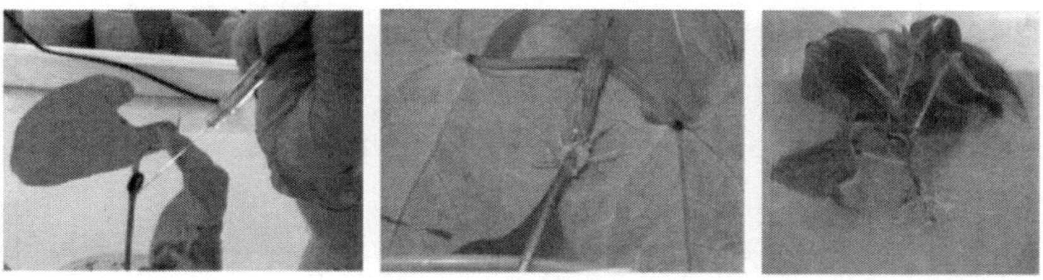

Inoculation with *Agrobacterium rhizogenes* | Callus development and hairy root emergence | Composite plant

Figure 3. *Agrobacterium rhizogenes* mediated transformation in common bean. Common bean plants are inoculated with an *A. rhizogenes* suspension carrying the gene of interest (left). After two weeks, transgenic roots emerge from a callus developed at the inoculation site as a consequence of *A. rhizogenes* infection (middle). The wild type root is cut, resulting in a composite plant whose root is transgenic and the aerial tissue is not transformed (right).

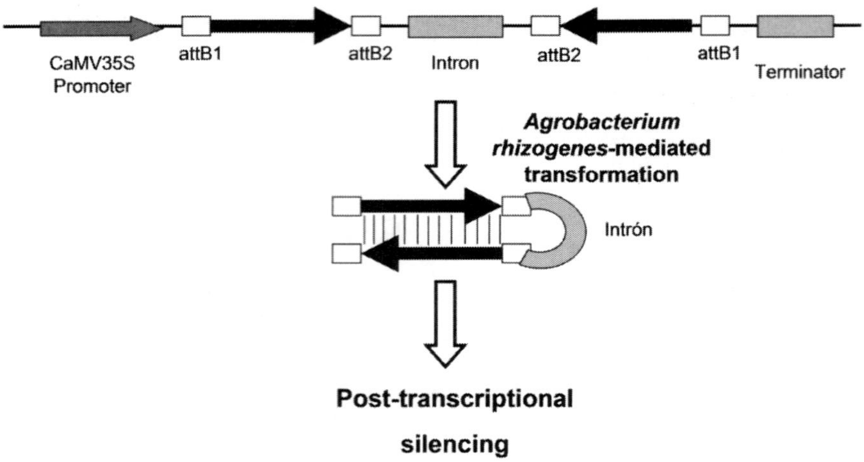

Figure 4. Post-transcriptional gene silencing mediated by RNAi. The construct is composed by the 35S promoter derived from the cauliflower mosaic virus (CaM35S), two copies of a fragment (approximately 200 bp) of the target gene cloned in an inverted orientation (black arrows) and separated by an intron. This construct is made by site-specific recombination using the att recombination sites of the lambda phage and the Gateway technology (Invitrogen). Upon *A. rhizogenes* transformation, this construct is integrated into the plant genome and transcribed to produce an RNA with a stem-loop structure, which is recognized by the post-translational silencing machinery to produce RNAi molecules.

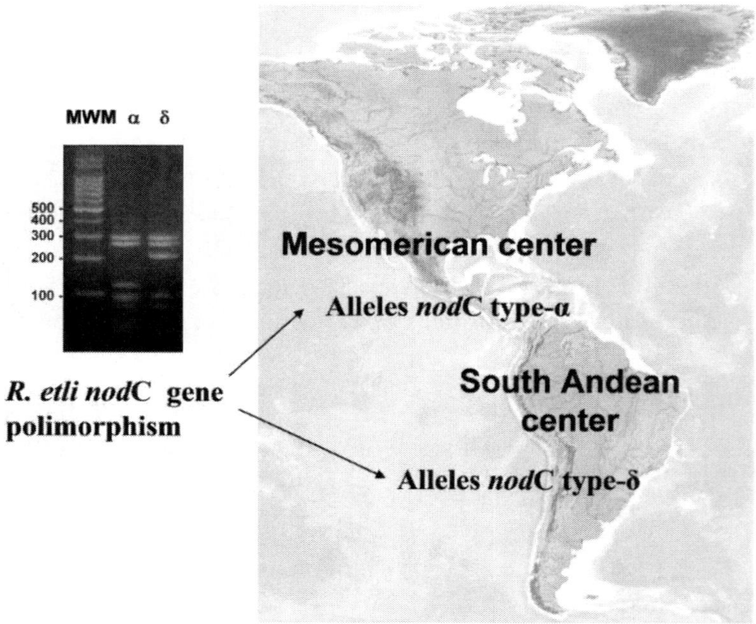

Figure 5. Common bean genetic diversification centers and the correlation with geographical distribution of *R. etli* strains carrying the α and δ alleles of the *nod*C gene. This polymorphism is detected by PCR followed by digestion with *Hin*fI and electrophoresis in agarose gels (left).

MOLECULAR BASIS OF STRAIN PREFERENCE

The origin and evolution of the two genetic pools of *P. vulgaris* constitutes a unique opportunity to study the specificity and evolution of the symbiotic interaction with rhizobia. It has been shown that Mesoamerican plants corresponding to wild or cultivated varieties have an active role in the selection of strains that are predominant in Mesoamerican soils, suggesting the presence of molecular mechanisms that allow the plant to recognize and select rhizobia that are more efficient in the interaction [40]. This study was initially approached by the analysis of the polymorphism of the *nod*C gene of *Rhizobium etli*, the most important microsymbiont of common bean present in the American soil. Strains carrying the α allele of the gene predominated in Mesoamerican soils, whereas the δ alelle was present in most of the strains found in Andean soils (Figure 5). Interestingly, Mesoamerican cultivars were predominantly and more efficiently nodulated by *nod*C α strains in coinoculation experiments, indicating that the molecular mechanisms to discriminate rhizobial strains have evolved in common bean during genetic isolation and possibly during domestication [40].

Taking advantage of this unique scenario, we initiated the screening of genes responsible for this preference. We performed a suppressive subtractive hybridization approach in which a Mesoamerican cultivar was individually inoculated with strains of *R. etli* that differ in the ability to compete and form nodules. The interaction with the more efficient strain SC15, carrying the α alelle of *nod*C gene, was used as the driver and the less competitive strain 55N1, carrying the δ alelle, was used as the tester (Figure 5). Analysis of the sequences obtained identified 41 clones corresponding to common bean genes that are upregulated at the early stage of root interaction with the more efficient *R. etli* strain [41]. Transcripts from a group of these genes accumulate after infection of a Mesoamerican bean roots with the α strain, but not when roots were infected with a strain carrying the δ alelle. Subtle or no differences were observed when an Andean bean cultivar was inoculated with its cognate strain, suggesting involvement of these genes in the cultivar strain-specific response [41].

In order to assess the function of these genes during the symbiotic association, composite plants that express specific RNAi for selected genes were generated by *A. rhizogenes* transformation. Silencing of a monomeric GTPase belonging to the Rab family produced a reduction in the number and length of the root hairs compared with control GUS RNAi plants [37]. This gene, named RabA2, was preferentially accumulated in root hairs and its transcript level increased after infection with a cognate strain of *R. etli*, with a maximum at 12 hs after inoculation. Nodulation was completely abolished in RabA2 RNAi plants, whereas overexpression of this gene did not have any noticeable effect on the phenotype. Bacterial attachment to the root was similar between RabA2 RNAi and control plants, but neither root hair curling nor infection thread formation was observed in roots with reduced levels of RabA2. Furthermore, accumulation of early nodulins after rhizobial infection was severely compromised in RabA2 RNAi plants, suggesting that this Rab protein plays a role in the early response of the root hair that leads to infection thread formation and induction of nodulin genes.

NF-YC1, another gene detected as differential in the subtraction, encodes for a protein highly homologous to the C subunit of NF-Y heterotrimeric transcription factors. Transcripts were found to accumulate at higher levels in *P. vulgaris* roots of the Mesoamerican cultivar at early time points, as well as in young and mature nodules, after inoculation with *R. etli nod*C α strains [23]. RNAi mediated silencing of NF-YC1 dramatically reduced the number of nodules respect to control plants (GUS-RNAi), giving a similar number of nodules when inoculated with either the Mesoamerican or the Andean strain. Overexpression of NF-YC1 did not significantly changed nodulation efficiency of the *nod*C α strain, but resulted in formation of higher number of nodules with *nod*C δ strains. Induction of G2/M transition cell cycle genes in response to rhizobia was reduced in NF-YC1 RNAi roots, suggesting that this transcription factor might promote nodule development by activating cortical cell divisions. Furthermore, overexpression of this gene had a positive impact on nodulation efficiency and selection of strains that are naturally less efficient and bad competitors.

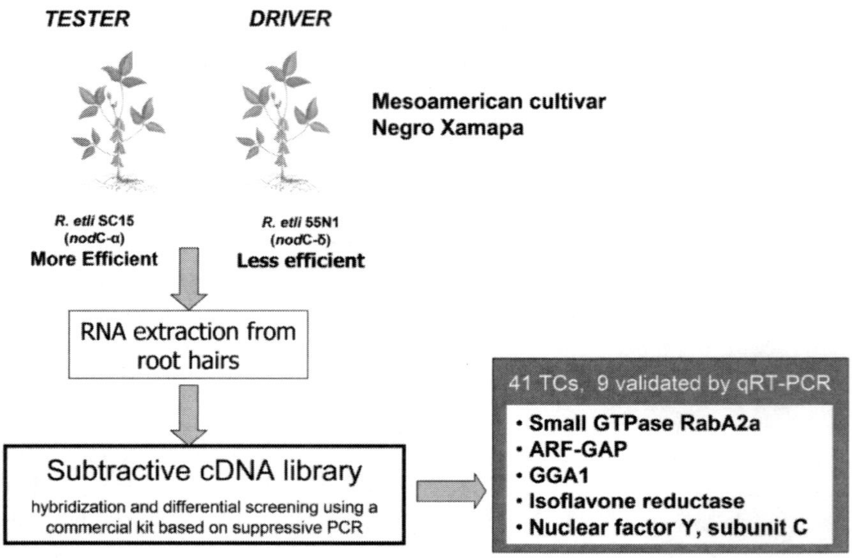

Figure 6. Screening of genes involved in the strain-specific response of a Mesoamerican common bean plant to a more competitive strain of *R. etli*. A subtraction based on suppressive PCR led to the identification of 41 putative genes, whose differential expression was validated by PCR [41].

Discrimination among different partners can be established early during chemical communication in the rhizosphere, through "honest" signals that indicate partner quality [42-43]. Plant genotypes able to select more beneficial strains will be favored by selection. Based on the findings of our studies, we propose a model that provides a link between an early molecular dialogue and the capacity of the plant to select high quality bacteria (Figure 7). In this model, RabA2 is required for NF-YC1 induction [37] in Mesoamerican common beans by their cognate *R. etli* strains, rapidly activating the transcription of cell cycle genes to initiate nodule primordium development. Initial cortical cell divisions precede and are required for infection thread progression, allowing bacteria to penetrate the root and reach the developing nodules. In contrast, *R. etli* strains from the Andean region failed to induce *NF-YC1* at this very early stage, therefore both the infection process and nodule organogenesis are delayed, resulting in a competitive disadvantage. Strains that can initiate the infection process and nodulation faster will out-compete slower ones in terms of nodule occupancy, since early events suppress further nodule organogenesis, probably via a systemic autoregulation of nodulation [44].

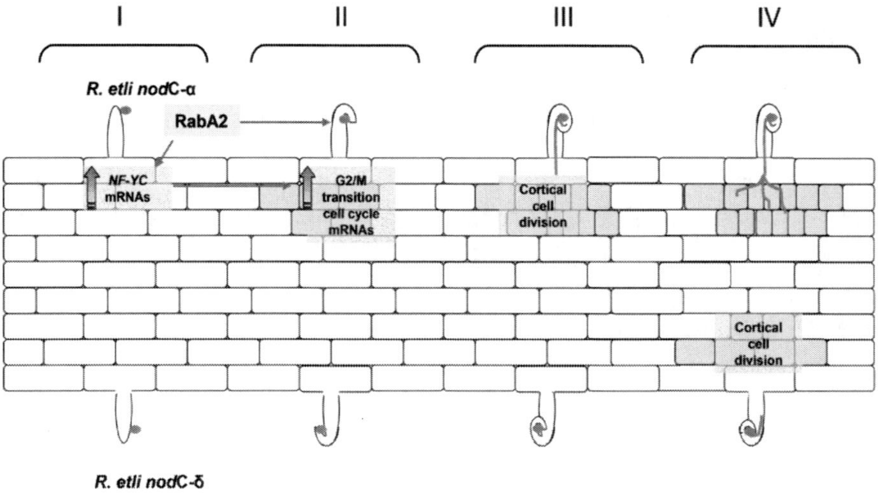

Figure 7. Model for nodule organogenesis and competition in the *P. vulgaris-R. etli* symbiotic association. The more efficient association between a Mesoamerican cultivar and a *R. etli* strain carrying the *nod*C α allele is represented in the upper part of the root, whereas the less efficient strain is in the lower part. In the stage I of the interaction, the recognition of a still unknown molecule of the *nod*C α strain triggers the induction of NF-YC1, a process that requires the presence of RabA2. Higher levels of the NF-YC protein activate expression of cell cycle genes (Stage II), leading to cell division of cortical cells (stage III). The infection process, driven by signals derived from the cortex, reach the nodule primordia (Stage IV) faster than *nod*C δ strains (unable to increase of NF-YC1 levels), resulting in a better competitive ability when both kind of rhizobia are present in the soil.

The evolutionary stability of symbiosis requires constrain of cheating, since partners that accept the benefits of the interaction without reciprocating will displace efficient strains, leading to a transition from symbiosis to parasitism. There is evidence of such stabilizing mechanisms operating in legume-Rhizobium interaction, as partner choice, partner fidelity and host sanctioning [45-50]. Our results suggest that the evolution of common bean and rhizobia at the Mesoamerican center gave origin to a molecular mechanism consistent with the partner fidelity hypothesis [43, 50], in which individuals interact repeatedly according to the "tit for tat" strategy: each player recognizes its partner and adjusts its behaviour according to the previous outcome of the interaction. In this model, fidelity between co-occurring symbionts provides stability to the interaction over the time.

Optimizing functional compatibility between legumes and rhizobia is necessary to improve crop productivity, since availability of nitrogen is one of the most important limitations for plant growth. We believe that the model presented here would help to understand the molecular basis of this functional compatibility in symbiotic interactions, which is the key to manipulate the genetic control of the host to select more efficient strains in detriment of less efficient bacteria that are good competitors to occupy root nodules.

References

[1] Broughton, W.J., et al., Beans (Phaseolus spp.) – model food legumes. *Plant and Soil,* 2003. *252*(1): p. 55-128.

[2] Gepts, P., et al., Phaseolin-protein Variability in Wild Forms and Landraces of the Common Bean (Phaseolus vulgaris): Evidence for Multiple Centers of Domestication. *Economic Botany,* 1986. *40*(4): p. 451-468.

[3] Bitocchi, E., et al., Mesoamerican origin of the common bean (Phaseolus vulgaris L.) is revealed by sequence data. *Proceedings of the National Academy of Sciences,* 2012. *109*(14): p. E788–E796.

[4] Schmit, V., et al., Use of chloroplast DNA polymorphisms for the phylogenetic study of seven Phaseolus taxa including P. vulgaris and P. coccineus. *Theoretical and Applied Genetics,* 1993. *87*(4): p. 506-516.

[5] Dénarié, J. and J. Cullimore, Lipo-oligosaccharide nodulation factors: A new class of signaling molecules mediating recognition and morphogenesis. *Cell,* 1993. *74*: p. 951-954.

[6] Radutoiu, S., et al., LysM domains mediate lipochitin-oligosaccharide recognition and Nfr genes extend the symbiotic host range. *The EMBO Journal,* 2007. *26*(17): p. 3923-35.

[7] Eisenschenk, L., et al., Inhibition of Rhizobium etli Polysaccharide Mutants by Phaseolus vulgaris Root Compounds. *Applied and Environmental Microbiology,* 1994. *60*(9): p. 3315-3322.

[8] Downie, J.A., The roles of extracellular proteins, polysaccharides and signals in the interactions of rhizobia with legume roots. *FEMS Microbiology Reviews,* 2010. *34*(2): p. 150-170.

[9] Oldroyd, G.E. and J.A. Downie, Coordinating nodule morphogenesis with rhizobial infection in legumes. *Annu Rev Plant Biol,* 2008. *59*: p. 519-46.

[10] Oldroyd, G.E., et al., The rules of engagement in the legume-rhizobial symbiosis. *Annu Rev Genet,* 2011. *45*: p. 119-44.

[11] Radutoiu, S., et al., Plant recognition of symbiotic bacteria requires two LysM receptor-like kinases. *Nature,* 2003. *425*(6958): p. 585-92.

[12] Amor, B.B., et al., The NFP locus of Medicago truncatula controls an early step of Nod factor signal transduction upstream of a rapid calcium flux and root hair deformation. *Plant J,* 2003. *34*(4): p. 495-506.

[13] Ehrhardt, D.W., R. Wais, and S.R. Long, Calcium spiking in plant root hairs responding to Rhizobium nodulation signals. *Cell,* 1996. *85*(5): p. 673-81.

[14] Endre, G., et al., A receptor kinase gene regulating symbiotic nodule development. *Nature,* 2002. *417*(6892): p. 962-6.

[15] Stracke, S., et al., A plant receptor-like kinase required for both bacterial and fungal symbiosis. *Nature,* 2002. *417*(6892): p. 959-62.

[16] Ane, J.M., et al., Medicago truncatula DMI1 required for bacterial and fungal symbioses in legumes. *Science,* 2004. *303*(5662): p. 1364-7.

[17] Mitra, R.M., et al., A Ca2+/calmodulin-dependent protein kinase required for symbiotic nodule development: Gene identification by transcript-based cloning. *Proc Natl Acad Sci U S A,* 2004. *101*(13): p. 4701-5.

[18] Lévy, J., et al., A Putative Ca2+ and Calmodulin-Dependent Protein Kinase Required for Bacterial and Fungal Symbioses. *Science,* 2004. *303*(5662): p. 1361-1364.

[19] Kalo, P., et al., Nodulation signaling in legumes requires NSP2, a member of the GRAS family of transcriptional regulators. *Science,* 2005. *308*(5729): p. 1786-9.

[20] Smit, P., et al., NSP1 of the GRAS protein family is essential for rhizobial Nod factor-induced transcription. *Science,* 2005. *308*(5729): p. 1789-91.

[21] Middleton, P.H., et al., An ERF transcription factor in Medicago truncatula that is essential for Nod factor signal transduction. *Plant Cell,* 2007. *19*(4): p. 1221-34.

[22] Combier, J.P., et al., MtHAP2-1 is a key transcriptional regulator of symbiotic nodule development regulated by microRNA169 in Medicago truncatula. *Genes Dev,* 2006. *20*(22): p. 3084-8.

[23] Zanetti, M.E., et al., A C Subunit of the Plant Nuclear Factor NF-Y Required for Rhizobial Infection and Nodule Development Affects Partner Selection in the Common Bean-Rhizobium etli Symbiosis. *Plant Cell,* 2010. *22*(12): p. 4142-57.

[24] Den Herder, G. and M. Parniske, The unbearable naivety of legumes in symbiosis. *Curr Opin Plant Biol,* 2009. *12*(4): p. 491-9.

[25] Vanhouten, W. and S. MacKenzie, Construction and characterization of a common bean bacterial artificial chromosome library. *Plant Molecular Biology,* 1999. *40*(6): p. 977-983.

[26] Kami, J. and P. Gepts, Development of a BAC library in common bean genotype BAT93. *Annu Rept Bean Improv Coop,* 2000. *43*: p. 208-209.

[27] Ramirez, M., et al., Sequencing and analysis of common bean ESTs. Building a foundation for functional genomics. *Plant Physiol,* 2005. *137*(4): p. 1211-27.

[28] Melotto, M., et al., Comparative bioinformatic analysis of genes expressed in common bean (Phaseolus vulgaris L.) seedlings. Genome, 2005. *48*(3): p. 562-570.

[29] Arenas-Huertero, C., et al., Conserved and novel miRNAs in the legume Phaseolus vulgaris in response to stress. *Plant Molecular Biology,* 2009. *70*(4): p. 385-401.

[30] Vianna, G.R., Fragment DNA as vector for genetic transformation of bean (Phaseolus vulgaris L.). *Sci. Hortic.,* 2004. *99*: p. 371-378.

[31] Bonfim, K., et al., RNAi-Mediated Resistance to Bean golden mosaic virus in Genetically Engineered Common Bean (Phaseolus vulgaris). *Molecular Plant-Microbe Interactions,* 2007. *20*(6): p. 717-726.

[32] Rech, E.L., G.R. Vianna, and F.J.L. Aragao, High-efficiency transformation by biolistics of soybean, common bean and cotton transgenic plants. *Nat. Protocols,* 2008. *3*(3): p. 410-418.

[33] Liu, H.K., C. Yang, and Z.M. Wei, Efficient Agrobacterium tumefaciens mediated transformation of soybeans using an embryonic tip regeneration system. *Planta,* 2004. *219*: p. 1042-1049.

[34] Aragao, F.J.L., Inheritance of foreign genes in transgenic bean (Phaseolus vulgaris L.) co-transformed via particle bombardment. *Theor. Appl. Genet.,* 1996. *93*: p. 142-150.

[35] Estrada-Navarrete, G., et al., Agrobacterium rhizogenes transformation of the Phaseolus spp.: a tool for functional genomics. *Mol Plant Microbe Interact,* 2006. *19*(12): p. 1385-93.

[36] Estrada-Navarrete, G., et al., Fast, efficient and reproducible genetic transformation of Phaseolus spp. by Agrobacterium rhizogenes. *Nat Protoc,* 2007. *2*(7): p. 1819-24.

[37] Blanco, F.A., et al., A small GTPase of the Rab family is required for root hair formation and preinfection stages of the common bean-Rhizobium symbiotic association. *Plant Cell,* 2009. *21*(9): p. 2797-810.

[38] Islas-Flores, T., et al., PvRACK1 Loss-of-Function Impairs Cell Expansion and Morphogenesis in Phaseolus vulgaris L. Root Nodules. *Molecular Plant-Microbe Interactions,* 2011. *24*(7): p. 819-826.

[39] Mazziotta, L., et al., Transcriptional and functional variation of NF-YC1 in genetically diverse accessions of Phaseolus vulgaris during the symbiotic association with Rhizobium etli. *Plant Biology,* 2012: p. DOI: 10.1111/j.1438-8677.2012.00683.x.

[40] Aguilar, O.M., O. Riva, and E. Peltzer, Analysis of Rhizobium etli and of its symbiosis with wild Phaseolus vulgaris supports coevolution in centers of host diversification. *Proc Natl Acad Sci U S A,* 2004. *101*(37): p. 13548-53.

[41] Peltzer Meschini, E.P., et al., Host genes involved in nodulation preference in common bean (Phaseolus vulgaris)-rhizobium etli symbiosis revealed by suppressive subtractive hybridization. *Mol Plant Microbe Interact,* 2008. *21*(4): p. 459-68.

[42] Simms, E.L. and D.L. Taylor, Partner Choice in Nitrogen-Fixation Mutualisms of Legumes and Rhizobia. Integrative and Comparative Biology, 2002. *42*(2): p. 369-380.

[43] Simms, E.L. and D.L. Taylor, Partner Choice in Nitrogen-Fixation Mutualisms of Legumes and Rhizobia. *Integr. Comp. Biol.,* 2002. *42*(2): p. 369-380.

[44] Oka-Kira, E. and M. Kawaguchi, Long-distance signaling to control root nodule number. Curr Opin Plant Biol, 2006. *9*(5): p. 496-502.

[45] Parker, M.P., Plant Fitness Variation Caused by Different Mutualist Genotypes. *Ecology,* 1995. *76*(5): p. 1525-1535.

[46] Heath, K.D. and P. Tiffin, Stabilizing mechanisms in a legume-rhizobium mutualism. *Evolution,* 2009. *63*(3): p. 652-62.

[47] Kiers, E.T., et al., Host sanctions and the legume-rhizobium mutualism. *Nature,* 2003. *425*(6953): p. 78-81.

[48] Bull, J.J. and W.R. Rice, Distinguishing mechanisms for the evolution of co-operation. *J Theor Biol,* 1991. *149*(1): p. 63-74.

[49] Marco, D.E., et al., An experimental and modelling exploration of the host-sanction hypothesis in legume-rhizobia mutualism. *J Theor Biol,* 2009. *259*(3): p. 423-33.

[50] Axelrod, R. and W.D. Hamilton, The evolution of cooperation. *Science,* 1981. *211*(4489): p. 1390-6.

In: Legumes
Editors: Hiroto Satou and Ren Nakamura

ISBN: 978-1-62808-280-7

Chapter 13

HEALTH BENEFITS OF LEGUMES

T. B. Ng[1,*] and C. C. W. Ng[1]
[1]School of Biomedical Sciences, Faculty of Medicine, The Chinese University of Hong Kong, Shatin, New Territories, Hong Kong, China

ABSTRACT

Legumes exhibit a variety of health benefits. Fenugreek seeds lower blood cholesterol and glucose levels. They are rich in fiber and saponins, and thus suppress accumulation of fat and upregulate the low density lipoprotein receptor. They also manifest anticancer and antidiabetic activities.Peanuts have beneficial effects on cholesterol level, hypertension, cancer, inflammation, oxidative stress, and vascular reactivity. Faba beans protect against DNA damage induced by hydrogen peroxide, inhibit the actions of pancreatic lipase and angiotensin converting enzyme, and suppress the proliferation of a variety of cancer cells besides exerting alpha-glucosidase activity. Adzuki bean demonstrate hydrogen peroxide, 2,2'-diphenyl-1-picrylhydrazyl, and superoxide anion scavenging activities. Kidney beans exhibit antimutagenic activity. Garden pea manifests angiotensin converting enzyme inhibitory and antioxidant activities. Mung beans have antidiabetic activity. Chickpea has an abundance of the essential fatty acids linoleic and oleic acids, vitamins and plant sterols. Bowman inhibitors possess anticarcinogenic properties.Soybean protein exerts a cardioprotective action and soybean phospholipids demonstrate hypolipidemic activity. Soybean is characterized by an abundance of phytoestrogens which alleviate postmenopausal symptoms and prevent osteoporosis and cardiovascular and malignant diseases. Although controversies regarding some of the aforementioned health benefits exist such as the anticancer action of isoflavones, in general consumption of legumes will promote health.

INTRODUCTION

Beans and peas are good to health.Intake of kidney beans results in significant decline in the levels of both total and low density lipoprotein-cholesterol, while intake of groundnuts

* E-mail: b021770@mailserv.cuhk.edu.hk.

lowers total cholesterol level. Mineral availability from legumes varies depending on the legume and may be due to their mineral content, mineral-mineral interaction and their tannic acid and phytic acid content. The above findings provide a scientific basis for choosing legumes as functional foods (Trinidad et al., 2012). The prevailing guidelines recommend copious consumption of fiber in the form of fruits and vegetables, whole grains, and dried beans and peas. Dried beans are good sources of soluble fiber. High fiber consumption is associated with reductions in serum cholesterol levels, risk of coronary heart disease, blood pressure and risk of certain forms of cancer. It is also related to enhanced weight control, and improved control of blood glucose level and gastrointestinal function (Anderson et al., 1994).The aim of the present article is to review the health promoting action of the common beans and peas.

(i) **Peanuts (*Arachis Hypogaea*)**

Peanuts are abundant in unsaturated fatty acids, fiber, high-quality vegetable protein, minerals, tocopherols, phytosterols, and phenolic compounds. Nut consumption has been associated with a lowered incidence of coronary heart disease and gallstones in both men and women and diabetes in women, and has beneficial effects on cholesterol level, hypertension, cancer, inflammation, oxidative stress, and vascular reactivity. Blood pressure, visceral adiposity and the metabolic syndrome also appear to be positively influenced by nut consumption (Ros et al., 2010).

(ii) **Chickpea (*Cicer Arietinum*)**

The effects of consumption of a chickpea-supplemented diet on serum lipid level, glucose tolerance, satiety and bowel function for five weeks have been described. There was an improvement of perceived general bowel health, and a small but significant reduction in serum levels of triglyceride and low density lipoprotein-cholesterol . Greater satiety has been reported by some participants (Pittaway et al., 2007).

Chickpea (*Cicer arietinum* L.) has an abundance of carbohydrates and protein, and the quality of proteins is superior to that of other pulses. All of the essential amino acids with the exception of sulphur-containing amino acids, as well as the important unsaturated fatty acids linoleic and oleic acids, the important sterols campesterol, â-sitosterol, and stigmasterol, are found in chickpea. Ca, Mg, P and, especially, K are also present in chickpea seeds. Chickpea is rich in â-carotene, folate, niacin, riboflavin, and thiamin. In conjunction with other pulses and cereals, chickpea could have beneficial effects on cardiovascular disease, type 2 diabetes, digestive diseases and some cancers (Jukanti et al., 2012).

Bowman-Birk inhibitors from chickpea, lentil, pea, and soybean are capable of withstanding acidic conditions and proteolytic degradation, and pass through the gastrointestinal tract intact.This allows sufficient active inhibitor to reach the large intestine and exert their anti-inflammatory and anti-cancer actions by acting on trypsin- and chymotrypsin-like proteases involved in carcinogenesis (Clemente et al., 2011). The possible utility of chickpeas to the management and/or prevention of degenerative diseases associated with free radical damage has been suggested (Heiras-Palazuelos_et al., 2013).

Legumes (including alfalfa, broad beans, chickpeas, clover, dry beans, dry peas, green beans and peas, lentils,lupins, peanuts, soybeans)represent an important component of the human diet in several areas of the world, especially in the developing countries, where they complement the lack of proteins from cereals, roots,. The majority of legumes contain phytochemicals: bioactive compounds, including enzyme inhibitors, phytohemagglutinins (lectins), phytoestrogens, oligosaccharides, saponins, and phenolic compounds, which play metabolic roles in humans who frequently consume these foods. Dietary intake of phytochemicals may provide health benefits, protecting against numerous diseases or disorders, such as coronary heart disease, diabetes, high blood pressure and inflammation (Bouchenak et al., 2013).

(iii) Soybean (*Glycine Max*)

Genistein, daidzein and glycitein are soybean isoflavones with potent estrogenic activity. These phytoestrogens, which structurally resemble 17-â-estradiol, can bind to both alpha and beta estrogen receptors, and mimic estrogenic action on target organs. Genistein and daidzein may be useful in chemoprevention of breast and prostate cancer, cardiovascular disease, osteoporosis, and alleviation of postmenopausal symptoms. The chemical form in which isoflavones occur affects bioavailability and hence biological activity. Glucose-conjugated isoflavones, which are highly polar and water-soluble, are less biologically active than the aglycone counterparts and are poorly absorbed by the intestinal epithelium. Colonic microorganisms can convert glycosylated isoflavones into aglycones. Clinical studies rvealed significant differences between the aglycone and conjugated forms of genistein and daidzein. In order to enhance bioavailability, lipid-based formulations including drug incorporation into oils, emulsions and self-microemulsifying formulations, and complexation with cyclodextrin have been made available (Vitale et al., 2012).

Soya protein lowers circulating levels of low density lipoprotein-cholesterol. Evidence that soya foods lowers the risk of coronary heart disease independent of their effects on lipid levels has also been presented. Recently published clinical and epidemiological data do not lend support to the results of rodent experiments showing that soyabean isoflavones elevate the risk of breast cancer. In postmenopausal women, exposure to isoflavone does not exert a deleterious effect on breast tissue density or breast cell proliferation. In both US and Chinese prospective epidemiological studies, soya consumption after diagnosis is associated with a better prognosis. Thus soyafoods should be regarded as healthy foods to be included in the diet with an intent to lower the risk of coronary heart disease irrespective of their breast cancer status (Messina et al., 2012).

Saponins are hypocholesterolemic steroid or triterpene glycosides present in soy beans and chickpeas. Some saponins complex with cholesterol thus preventing small intestinal cholesterol absorption. Other saponins cause an increase of bile acid excretion in the feces, and thus removal of cholesterol since bile acids are derived from cholesterol. Soybean saponins reduce the rate of bile salt absorption due the formation of large mixed micelles with a molecular weights exceeding one million daltons (Sidhu et al., 1986).

Equol (7-hydroxy-3-(4'-hydroxyphenyl)-chroman) derived from the soya isoflavone daidzein is formed by human intestinal bacteria. Inter-individual variations in daidzein conversion to equol are attributed to dietary differences and dissimilarities in intestinal microflora composition. Equol exhibits estrogenic activity and antagonizes dihydrotestosterone *in vivo*. Compared with daidzein, equol exhibits a higher affinity for the estrogen receptor and antioxidant potency.The ability to produce equol is associated with lowered risk of breast and prostate cancer, improved bone health and reduced incidence of hot flushes, but negative and contrary findings exist (Magee, 2011).

Chinese adult equol excretors with higher daily consumption of soy isoflavones had significantly reduced carotid artery intima-media thickness and higher HDL-cholesterol levels compared with those with lower isoflavone consumption, while no association between isoflavone consumption and serum lipid level or carotid artery intima-media thickness was discernible in equol non-excretors.Hence, the cardiovascular benefits of soya isoflavones may be apparent only among equol excretors (Cai et al., 2012).

Glyceollins are phytoalexins produced from daidzein in the soybean upon exposure to some fungal species. The efficiency of its production depends on the species of fungus, the soybean variety, and the extent of physical damage to the soybean. Glyceollins have potent anti-inflammatory and antioxidant activities, and suppress the proliferation and migration of human aortic smooth muscle cells, signifying a potential for inhibition of atherosclerosis. Glyceollins suppress the growth of prostate and breast cancer cells in xenograft animal models, probably due to their anti-estrogenic activity (Kim et al., 2012).

Soy products may affect skin pigmentation, photoaging, photodamage, overall skin health, and progression of skin cancer. Soybean serine protease inhibitors reduce skin pigmentation by preventing protease-activated receptor-2-mediated phagocytosis of melanosomes by keratinocytes. Recent clinical evidence suggests that the application of soy products for the therapy of hyperpigmentation is safe and effective (Leyden et al., 2011).

Niwa et al. (1998) have reported the surge in lipid peroxide levels and reduction in superoxide dismutase inducibility in severe atopic dermatitis patients. After attachment to the stratum corneum, lipid peroxides enhance loss of skin moisturization thus aggravating atopic dermatitis. Cataracts occurring in severe atopic dermatitis are produced by the diffusion of lipid peroxides through the posterior lens. The decline in sperm motility in the seminal plasma of infertile men can be reversed by antioxidant enzymes and vitamins including catalase, glutathione-peroxidase and ascorbate. Niwa et al. (1998) have reported the development of an oral antioxidant, designated as AOA, from seeds and plants (e.g., soybean, sesame, wheat germ), heated by exposure to far infrared rays (4-14 microns wavelength), brewed in the presence of *Aspergillus oryzae*, and lipophilized with similarly heated sesame oil. As a consequence, low-molecular-weight antioxidants initially in an inactive form of repeating subunits of polymers, are released to produce free, activated forms of antioxidants. AOA, has been employed in the treatment of both cataracts complicated with atopic dermatitis and male infertility with half the patients demonstrating much improvement (Niwa et al., 1998).

Pretreatment with three major anthocyanins (cyanidin-3-O-glucoside, delphinidin-3-O-glucoside, and petunidin-3-O-glucoside) isolated from the black soybean (*Glycine max* L.) cv. Cheongja 3 seed coat attenuated the toxicity of H_2O_2 on human brain neuroblastoma SK-N-SH cells, decreased intracellular ROS level and inactivated apoptosis signal-regulating kinase (ASK1, Thr845), p38, and c-Jun N-terminal kinase (JNK) proteins. The HO-1 and Neu1 mRNA levels were elevated by H_2O_2, and further augmented by anthocyanin pretreatment. Sialic acids attenuated H_2O_2cytotoxicity and curtailed intracellular reactive oxygen species level. The data indicate that black soybean seed coat anthocyanins exert neuroprotective effects against H_2O_2-induced oxidative stress by suppressing the activation of ASK1-JNK/p38 pathways, scavenging ROS, upregulating HO-1expression and, recruiting cellular free sialic acids by enhancing Neu1 sialidase gene expression (Kim et al., 2012).

(iv) Kidney Bean (*Phaseolus Vulgaris*)

Polyphenols in fruits, vegetables (e.g., flavonols like quercetin) and tea (e.g., catechins such as epigallocatechin gallate) are good antioxidants with antimutagenic and anticarcinogenic properties. de Mejía et al. (1999) used *Salmonella typhimurium* tester strain YG1024 in the plate-incorporation test to study the inhibitory action of phenolic compounds extracted from common beans (*Phaseolus vulgaris*), on mutagenicity of beno[a] pyrene.Water extract,water/methanol extract and methanol extract of the beans exhibited antimutagenic activity (de Mejía et al., 1999).

Transcript levels from a number of stress- and toxicity-related genes were examined in female rats fed beans at doses relevant to human intake worldwide. Following intake of cooked red bean incorporated into a purified diet formulation at 0, 7·5, 15, 30 or 60 % w/w, the expression of Cyp3a11, Cyp7a1, Fmo1, Gstm1, Mif and Ugt1a6 was elevated. On the other hand, the expression of Hspa8 was down-regulated. Hepatic gene expression was not modulated in a manner indicative of an adverse response. Only the expression of genes encoding cholesterol 7á hydoxylase and UDP-glucuronosyltransferase increased in a dose-dependent manner at nutritionally relevant dietary bean concentrations. These candidate genes may contribute to the health benefits caused by increased bean consumption (Daniell et al., 2012).

(v) Pea (*Pisum Sativum*)

The health benefits of peas are attributed to the concentration and properties of fiber, minerals, protein, starch, phytochemicals and vitamins in them. Seed coat fiber and cotyledon cell walls play a beneficial role in gastrointestinal function and health, and they reduce the digestibility of pea starch. The intermediate amylose content of pea starch also accounts for the lowered starch digestibility and glycemic index. Pea protein, after hydrolysis, may give rise to peptides with antioxidant activity and inhibitory activity on angiotensin I-converting enzyme. The pea vitamin and mineral contents may prevent diseases related to deficiencies of folic acid and selenium. Phytochemicals in peas comprise polyphenolics with anticarcinogenic and antioxidant activities, saponins with anticarcinogenic and hypocholesterolemic activities, and galactose oligosaccharides with beneficial prebiotic effects in the large intestine (Dahl et al., 2012).

The proliferation of HT29 colon cancer cells was inhibited by the recombinant pea Bowman Birk protease inhibitor rTI1B, whereas the mutant, with amino acid substitutions at the P1 positions in the two inhibitory domains and inactive against both trypsin and chymotrypsin, was devoid of a similar action. Neither rTI1B nor its mutant affected the proliferation of non-malignant colonic fibroblast CCD-18Co cells, indicating that serine proteases are targets of BBI in their inhibitory action on colorectal carcinogenesis (Clemente et al., 2012).

(vi) Fenugreek (*Trigonella foenum* Graecum) Seeds

Lipid Lowering and Antioxidative Activities

Plasma cholesterol and triglycerides in hyperlipidemic rats were lowered by an alcoholic extract of fenugreek seeds at a dose of 200mg/kg body weight by activating lecithin-acyl-cholesterol-transferase, post-heparin lipolytic activity, triglyceride lipase, and lipoprotein lipase and promoting fecal bile acid excretion. The seed extract exerted an inhibitory action on generation of superoxide anions and hydroxyl free radicals in both enzymatic and non-enzymatic systems, and restored the activities of the antioxidative enzymes catalase and superoxide dismutase as well as suppressed plasma lipid peroxidation, and hepatic levels of 4-hydroxynonenal and isoprostanes (Chaturvedi et al., 2013). The ethyl acetate extract of fenugreek seeds, but not extracts prepared using dichloromethane, hexane, methanol, and water, displayed in rats fed a cholesterol-rich diet, a suppressive action on total cholesterol, low density lipoprotein cholesterol, and triglycerides, and elevated high density lipoprotein-cholesterol. The extract also lowered the renal, hepatic, and cardiac content of thiobarbituric acid reactive substances, and the activities of catalase and superoxide dismutase (Belguith-Hadriche et al., 2013).

Improvement of Parkinson's Disease

Nathan et al. (2013) carried out a double-blind placebo-controlled clinical investigation in which Parkinson's disease patients stabilized on L-Dopa therapy received twice daily IBHB capsules containing 300 mg standardized extract of Fenugreek for 6 months. Total scores on the Unified Parkinson's Disease Rating Scale exhibited a much more gradual increase in IBHB-treated patients compared to the placebo group. Improvement as indicated by Hoehn and Yahr (HandY) staging was observed. Safety assessment comprising hematology, urinalysis and adverse event monitoring did not reveal any undesirable consequences. Hence IBHB may be useful as an adjuvant with L-Dopa in the therapy of patients with Parkinson's disease.

Anticancer Activity

Several anticancer compounds, including cedrene, eugenol, gingerol, vanillin and zingerone are present in fenugreek extract. Incubation of human T lymphoma Jurkat cells with an alcoholic extract of dry fenugreek seeds in the range of concentrations from 30 to 1500 μg/mL for up to 72 h brought about death of the cells via autophagy which was preceded by the appearance of multiple large vacuoles. (Al- Daghri et al., 2013).

Antidiabetic Activity

Alterations in serum levels of glucose and insulin, activities of serum glutathione-S-transferase and catalase as well as muscle and hepatic glycogen content due to streptozotocin-induced diabetes were attenuated following oral administration of an aqueous suspension of Fenugreek seeds powder (Marzouk et al., 2013).

The compound GII purified from the water extract of fenugreek seeds by Murthy and his colleagues (patented in India and USA) at a daily dose of 50 mg/kg body weight suppressed the fasting blood glucose levels in subdiabetic, moderately diabetic and severely diabetic rabbits to normal after 12, 15 and 28 days of treatment. Serum levels of HbA1C and insulin were improved.

Intermittent therapy with GII at the same dose once weekly for 6 weeks lowered the fasting blood glucose levels in the subdiabetic and in the moderately diabetic rabbits (Puri et al., 2012).

Treatment with GII (50 mg/kg/day) for 15 days in the subdiabetic and moderately diabetic rabbits and for 30 days in the severely diabetic rabbits (i) suppressed the heightened serum levels of total cholesterol, triglycerides, low density lipoprotein cholesterol and very low density lipoprotein cholesterol, and increased the decreased high density lipoprotein cholesterol, (ii) suppressed the heightened hepatic and cardiac levels of total cholesterol and triglycerides, (iii) elevated the suppressed hepatic and muscle glycogen, (iv) elevated the suppressed glycolytic enzymes hexokinase, glucokinase, and pyruvate kinase, malic enzyme, glucose-6-phosphate dehydrogenase, and the antioxidant enzymes superoxide dismutase and glutathione peroxidase, (v) suppressed the heightened gluconeogenic enzyme glucose- 6-phosphatase, and the polyol pathway enzymes sorbitol dehydrogenase and aldose reductase. Histopathological abnormalities (fatty infiltration and other cellular changes) in the heart, kidneys, liver, and pancreas were rectified. In fact partially damaged pancreas was repaired. Hepatic and renal function test results were normal indicating the safetyof the treatment (Puri et al., 20113).

The alkaloids (contained in alkaline chloroform fraction) and flavonoids (contained in aqueous and acidified chloroform fractions) of fenugreek seeds were responsible for antinociceptive and anti-inflammatory (inhibition of carrageenan-induced paw edema) activities of the plant, respectively (Kumar et al., 2012).

Amelioration of Adriamycin-Induced Cytotoxicity and Testicular Alterations

The aqueous extract of fenugreek seeds ameliorated the cytotoxicity and testicular aberrations elicited by adriamycin in rats and this may be mediated by its potent antioxidant effects as evidenced by deminished lipid peroxidation and augmented activities of the antioxidative enymes superoxide dismutase and catalase (Sakr et al., 2012).

(vii) Faba Bean (*Vicia Faba*)

Faba bean extracts displayed a protective action against H_2O_2-induced DNA damage in RAW264.7 cells, and an antiproliferative action on AGS, BL13, Hep G2 and HT-29 human cancer cells, but a negligible effect on the non-transformed human CCD-18Co cells. The extracts induced apoptosis in acute promyelocytic leukaemia HL-60 cells, and demonstrated angiotensin converting enzyme, á-glucosidase and pancreatic lipase inhibitory activities.

Nura (buff-colored) and Rossa (red-colored) extracts were comparable, whereas TF(Ic*As)*483/13(white-colored) extract displayed the lowest phenolic content and antioxidant and enzyme inhibitory activities. (Siah et al., 2012).

(viii) Black Gram (*Vigna Mungo*)

The phenolic acids in the black gram husk extract comprise ferulic, gallic, gentisic and protocatechuic acids. The IC(50) values for DPPH radical scavenging and α-glucosidase inhibitory activities were 3.92 μg gallic acid equivalents (GAE) and 2.78 μg GAE,respectively. The IC(50) values for the extract for H_2O_2 induced hemolysis of rat erythrocyte were 11.5 μg of GAE. The extract protected erythrocyte membrane from morphological changes induced by H_2O_2 (Girish et al., 2012).

Adzuki Bean (Vigna Angularis), Moth Bean (Vigna Aconitifolius), and Mung Bean (Vigna Radiata)

Phenolic extracts of the aforementioned Vigna species have been examined for phenolic contents, antioxidant activities, and inhibitory properties against α-glucosidase and pancreatic lipase. Adzuki bean contained the highest phenolic index, metal chelating, and 2,2'-diphenyl-1-picrylhydrazyl, hydrogen peroxide and superoxide anion scavenging activities. Mung bean had the lowest antioxidant activity.

All of the extracts displayed higher lipase inhibitory activity than α-glucosidase inhibitory activity. Vigna species are thus sources of antioxidant phenolics and for α-glucosidase and lipase inhibitors (Sreerama et al., 2012). The antioxidant flavonoids vitexin and isovitexin accounted for the most of total DPPH radicals scavenging, ferric reducing and ABTS(+) reducing scavenging activities in soup in all the three cultivars of mung bean examined (Li et al., 2012).

Kaempferol (3,5,7-trihydroxy-2-(4-hydroxyphenyl)-4H-1-benzopyran-4-one) is a flavonoid found in many edible plants including beans. Kaempferol and some of its glycosides display a diversity of actions, including analgesic, antiallergic, anticancer, antidiabetic, anti-inflammatory, antimicrobial, antioxidant, antiosteoporotic, anxiolytic, cardioprotective, estrogenic/antiestrogenic, and neuroprotective activities (Calderón-Montaño et al., 2011).

(ix) *Lanchocarpus Neuroscapha*

4-Hydroxycordoin, a natural isopentenyloxychalcone, is a plant secondary metabolite isolated from the Fabaceae plant *Lanchocarpus neuroscapha.* It exerted pronounced antibacterial activity against the three major periodontal pathogens, *Fusobacterium nucleatum, Porphyromonas gingivalis,* and *Prevotella intermedia.* It prevented attachment of *P. gingivalis* to oral epithelial cells,and suppressed the *Aggregatibacter actinomycetemcomitans* lipopolysaccharide-induced secretion of inflammatory mediators which was associated with reduced activation of the nuclear factor-êB (NF-êB) p65 and activator protein-1 (AP-1) pathway (Feldman et al., 2011).

Conclusion

The above review disclosed that a variety of legumes display health promoting effects including antihyperlipidemic, antidiabetic, antioxidant, and anticancer activities. Because beans and peas are relatively inexpensive and are very rich in proteins,they should form an important part of the diet.Continuing research on legumes will uncover the mechanisms of many of their health promoting effects and also additional beneficial actions.

References

Al-Daghri N. M, Alokail M. S, Alkharfy K. M, Mohammed A. K, Abd-Alrahman S. H, Yakout S. M., Amer O. E., Krishnaswamy S. Fenugreek extract as an inducer *BMC Complement Altern. Med.* 2012;12:202.

Anderson J. W., Smith B. M., Gustafson N. J. Health benefits *Am. J. Clin. Nutr.*, 1994;59(5 Suppl):1242S-1247S.

Belguith-Hadriche O., Bouaziz M., Jamoussi K., Simmonds M. S., El Feki A., Makni-Ayedi F. Comparative study on hypocholesterolemic and antioxidant *Food Chem.*, 2013;138:1448-53.

Bouchenak M., Lamri-Senhadji M. Nutritional quality of legumes, and their role in cardiometabolic risk *J. Med. Food,* 2013;16:185-98.

Cai Y., Guo K., Chen C., Wang P., Zhang B., Zhou Q., Mei F., Su Y. Soya isoflavone *Br. J. Nutr.*, 2012;108:1698-704.

Calderón-Montaño J. M., Burgos-Morón E., Pérez-Guerrero C., López-Lázaro M. A review on the dietary flavonoid kaempferol *Mini Rev. Med. Chem.*, 2011;11:298-344.

Chaturvedi U., Shrivastava A., Bhadauria S., Saxena J. K., Bhatia G. A mechanism based pharmacological evaluation of efficacy of Trigonella-foenom graecum (fenugreek) seeds in regulation of dyslipidemia *J. Cardiovasc. Pharmacol.*, 2013 Feb 20. [Epub ahead of print]

Clemente A., Carmen Marín-Manzano M., Jiménez E., Carmen Arques M., Domoney C. The anti-proliferative effect of TI1B, a major Bowman-Birk isoinhibitor from pea (*Pisum sativum* L.), on HT29 colon cancer cells is mediated through protease inhibition. *Br. J. Nutr.*, 2012;108 Suppl 1:S135-44.

Clemente A., Sonnante G., Domoney C. Bowman-Birk inhibitors from legumes and human *Curr. Protein. Pept. Sci.,* 2011;12:358-73.

Dahl W. J., Foster L. M., Tyler R. T. Review of the health *Br. J. Nutr.,* 2012;108 Suppl 1:S3-10.

Daniell E. L., Ryan E. P., Brick M. A., Thompson H. J. Dietary dry bean effects on hepatic expression of stress *Br. J. Nutr.*, 2012;108 Suppl 1:S37-45.

de Mejía E. G., Castaño-Tostado E., Loarca-Piña G. Antimutagenic effects of natural phenolic compounds in beans. *Mutat. Res.*, 1999;441:1-9.

Girish T. K., Vasudevaraju P., Prasada Rao U. J.Protection of DNA *Food Chem. Toxicol.*, 2012;50:1690-6.

Heiras-Palazuelos M. J., Ochoa-Lugo M. I., Gutiérrez-Dorado R, López-Valenzuela J. A., Mora-Rochín S, Milán-Carrillo J, Garzón-Tiznado J. A., Reyes-Moreno C. Technological properties, antioxidant *Int. J. Food Sci. Nutr.,* 2013;64:69-76.

Jukanti A. K., Gaur P. M., Gowda C. L., Chibbar R. N. Nutritional quality and health *Br. J. Nutr.*, 2012;108 Suppl 1:S11-26.

Kim S. M., Chung M. J., Ha T. J., Choi H. N., Jang S. J., Kim S. O., Chun M. H., Do S. I., Choo Y. K., Park YI.Neuroprotective effects of black soybean anthocyanins via inactivation of ASK1-JNK/p38 pathways *Life Sci.*, 2012;90:874-82.

Kim H. J., Lim J. S., Kim W. K., Kim J. S. Soyabean glyceollins: biological effects and relevance *Proc. Nutr. Soc.*, 2012;71:166-74.

Kumar P, Kale R. K., Baquer N. Z. Antihyperglycemic and protective effects of Trigonella foenum graecum seed *Eur. Rev. Med. Pharmacol. Sci.,* 2012 16 Suppl 3:18-27.

Leyden J., Wallo W. The mechanism of action and clinical benefits *Int. J. Dermatol.*, 2011 Apr;50(4):470-7.

Li H, Cao D., Yi J., Cao J., Jiang W. Identification of the flavonoids *Food Chem.*, 2012;135:2942-6.

Magee P. J. Is equol production beneficial to health *Proc. Nutr. Soc.*, 2011;70:10-8.

Marzouk M., Soliman A. M., Omar T. Y. Hypoglycemic and antioxidative effects of fenugreek and termis seeds powder in streptozotocin-diabetic rats. *Eur. Rev. Med. Pharmacol. Sci.,* 2013 Feb;17(4):559-65.

Messina M., Messina V., Jenkins D. J. Can breast cancer *Br. J. Nutr.*, 2012;108:810-9.

Nathan J., Panjwani S., Mohan V., Joshi V., Thakurdesai P. A. Efficacy and Safety of Standardized Extract of Trigonella foenum-graecum L Seeds as an Adjuvant to L-Dopa in the Management of Patients with Parkinson's Disease. *Phytother. Res.*, 2013 Mar 20. [Epub ahead of print].

Niwa Y., Tominaga K., Yoshida K. Successful treatment of severe atopic dermatitis-complicated cataract and male infertility with a natural product antioxidant. *Int. J. Tissue React.,* 1998;20:63-9.

Pittaway J. K., Ahuja K. D., Robertson I. K., Ball M. J. Effects of a controlled diet supplemented with chickpeas on serum lipids, glucose tolerance, satiety and bowel function *J. Am. Coll. Nutr.,* 2007;26:334-40.

Puri D., Prabhu K. M., Murthy P. S. Antidiabetic Effect of GII Compound Purified from Fenugreek (Trigonella foenum graecum Linn) Seeds in Diabetic Rabbits. *Indian J. Clin. Biochem.*, 2012;27:21-7.

Puri D., Prabhu K. M., Dev G., Agarwal S., Murthy P. S. Mechanism of Antidiabetic Action of Compound GII Purified from Fenugreek (Trigonellafoenum graecum) Seeds. *Indian J. Clin. Biochem.,* 2011;26:335-46.

Ros E. Health benefits of nut consumption. *Nutrients,* 2010;2:652-82.

Sakr S. A., El-Shenawy S. M., Al-Shabka AMAqueous fenugreek seed *Reprod. Sci.,* 2012;19:70-80.

Siah S. D., Konczak I., Agboola S., Wood J. A., Blanchard C. L. In vitro investigations of the potential health *Br. J. Nutr.*, 2012;108 Suppl 1:S123-34.

Sidhu G. S., Oakenfull D. G. A mechanism for the hypocholesterolaemic activity of saponins. *Br. J. Nutr.,* 1986;55:643-9.

Sreerama Y. N., Takahashi Y., Yamaki K. Phenolic antioxidants in some Vigna species of legumes and their distinct inhibitory effects on α-glucosidase and pancreatic lipase activities. *J. Food Sci.,* 2012;77:C927-33.

Vitale D. C., Piazza C., Melilli B., Drago F., Salomone S. Isoflavones: estrogenic activity, biological effect and bioavailability *Eur. J. Drug Metab. Pharmacokinet.*, 2012 Nov 17. [Epub ahead of print]

In: Legumes
Editors: Hiroto Satou and Ren Nakamura
ISBN: 978-1-62808-280-7

Chapter 14

Current Update in Methodologies for Extraction and Analysis of Proteins and Isoflavones

***Savithiry S. Natarajan*[1*] *and Devanand L. Luthria*[2]**
[1]USDA-ARS, Soybean Genomics and Improvement Laboratory,
[2]Food Composition and Methods Development Laboratory, Beltsville, MD, US

Abstract

Legumes such as soybeans are staple diet for a significant proportion of the Asian population. In the past several decades, soybeans and its derived food products have become popular globally because of its protein, oil, and other bioactive phytochemicals content. This short communication summarizes recent advances in analytical methodologies used for the extraction and analysis of proteins and isoflavones from soybean and soy products.

1. Introduction

Soybean is an important crop and a major source of nutritional plant proteins for human consumption and animal husbandry. It has unique physicochemical properties suitable for various food ingredients. Soybean is one of the widely used raw materials for manufacturing of large numbers of industrial and food products. The soy meal products are used for livestock feed, fish and pet foods, and soy protein products are used as meat imitations for vegetarian foods, protein drinks, protein enriched baked foods, soup bases, and gravies. In addition, it is being used for pharmaceutical applications. Several reviews were available about the study of soybean proteins and their functions (Adachi et al. 2003; Fehr et al. 2003;

[*] Corresponding author: Dr. Savithiry S. Natarajan, Soybean Genomics and Improvement Laboratory, ARS-USDA, 10300 Baltimore Avenue, Beltsville, MD 20705 Phone Number: 301-504-5258, Fax Number: 301-504-5728, Email: savi.natarajan@ars.usda.gov.

Poysa et al. 2006). In addition to the protein, the oil extracted from soybeans is used in various forms such as cooking oil, margarine, and shortening.

2. Protein Extraction and Analysis

In order to investigate the roles of different proteins, it is important to extract, separate, and isolate proteins. Extraction of proteins from plant cells is difficult due to the presence of salts, organic acids, phenolics, pigments, terpenes, lipids, large quantities of polysaccharides, proteases, oxidative enzymes, and other secondary metabolites which interfere with some of the proteomic techniques such as two-dimensional polyacrylamide gel electrophoresis (2D-PAGE) (Carpentier et al. 2005; Wang et al. 2008). Proteomics techniques are effective in separating and visualizing proteins from a complex mixture, and are applied in several fields including biomedicine, agronomy, genotyping, and food science (Gorg et al. 2004; Gonzalez-Buitrago et al. 2007; Salekdeh and Kamatsu, 2007; Alm et al. 2007; Saz and Marina, 2007). Protein analysis by 2D-PAGE systems are used to separate proteins from developing and dry mature soybean seeds in different studies including study of temperature effects treatments and heavy metal effect (Agrawal et al. 2008; Danchenko et al. 2009; Ren et al. 2009; Klubicova et al. 2012; Natarajan et al. 2012 and 2013).

The quality and discrete distribution of protein spots depends primarily on the choice of sample preparation methodology used for protein extraction. In addition, the application of proteomics in plants is difficult due to the lack of effective protein extraction methods and it is necessary to optimize extraction method for sample selection. The plant cell consists of a cell wall, which requires effective mechanical devices for rupturing, namely sonication, crushing with beads and sand (Isaacson et al. 2006). In addition, several detergents such as SDS, CHAPS, cholate, Triton X-114, and tween are used to remove impurities and contaminants from the protein extract including lipids, cellulose, and nucleic acids (Gorg et al. 2004; Wang et al. 2008). The detergents help to prevent protein loss by preventing hydrophobic interactions between protein domains, which result in aggregation and precipitation, and help to solubilize membrane proteins. To prevent protein oxidation and to reduce disulfide bonds that achieve complete protein unfolding during extraction, several reductants including dithiothreital (DTT), β-mercaptoethanol are generally used. Denaturing agents such as urea, thiourea, and acids are also used to disrupt protein-protein interactions and the protein's secondary and tertiary structure by altering solution ionic strength. Enzymes (e.g., DNAse, RNAse) may be utilized to digest contaminating nucleic acids, carbohydrates, and lipids. In this chapter, we will mostly focus on some of the recent extraction methodologies used to extract plant proteins. In general, TCA/acetone precipitation is used to remove various contaminants from plant tissues.

Since different sample tissues need specific extraction methodologies, we describe several recently published methods (Table 1). Zhang et al. (2012) compared three methods of protein extraction from banana peels including TCA-acetone precipitation, phenol extraction, and TCA precipitation. These authors reported that the modified phenol method was the most optimal in producing high quality protein spots in 2D-PAGE. Chatterjee et al. (2012) initially evaluated two methods, TCA/acetone and phenol, for extracting protein from chickpea roots and reported that the latter method was more efficient. Furthermore, these authors used SDS in phenol with/without sonication and concluded that SDS with phenol and sonication method successfully isolated high quality proteins for MS analysis from Chickpea roots. Unlike seed materials, root protein extraction is difficult due to the low quantity of protein and the high quantity of proteases and other interfering substances. Rodrigues et al. (2012) used modified phenol/SDS extraction method for solubilizing soybean root tissues and their method allowed high quality identification of proteins for MS analysis. Natarajan et al. (2009) used various concentrations of isopropanol with acetone to enhance the extraction of low abundant proteins and deplete abundant proteins in soybeans. The authors reported that the method effectively removes 87% of the highly abundant proteins thereby enabling detection of 541 less abundant proteins. Ammonium acetate with methanol has proven useful with plant samples containing high levels of interfering substances like phenolics, fats and carbohydrates. Natarajan et al. (2013) and Boonmee et al. (2011) used phenol based extraction followed by ammonium acetate/methanol precipitation to isolate soybean embryonic axis and *Curcuma comosa* (a flowering plant of *Zingiberaceae* family) proteins respectively. Both authors used the protein for 2D separation and identified 335 proteins from embryonic axis and 183 proteins from *Curcuma comosa* using mass spectrometry (MS) analysis. Barbosa et al. (2013) also used ammonium acetate/methanol to precipitate proteins from sunflower seed embryos using extraction buffer containing different concentrations of solvents. Wu et al. (2012) used sequential extraction to isolate the tuber protein of *Pinellia ternate*, a member of the Araceae family and a well-known Chinese traditional medicine. Since the tuber of *Pinellia ternate* has few high abundant proteins (HAP), mainly mannose-binding lectin, the authors used acetic acid extraction to extract acid soluble HAPs and used SDS containing buffer to extract the remaining acid-insoluble proteins. The protein content was further analyzed for 2D-PAGE and MS analysis.

3. Isoflavones Extraction and Analysis

Isoflavones, also known as phytoestrogens, are non-steroidal secondary plant metabolites that exhibit biological responses similar to endogenous estrogens due to close structural resemblance to 17β-estradiol. Isoflavones are known to occur in free aglycon (daidzein, glycitein and genistein) forms as well as conjugated forms with sugars at position 7 that are often esterified with either acetyl of malonyl group at 6' position of the sugar group (Faraj and Vasanthan, 2004; Luthria and Natarajan, 2009; Rostagna et al. 2009).

In our earlier communications (Luthria et al. 2007; Luthria and Natarajan, 2009; Natarajan and Luthria, 2010 and 2011), we had carried out a systematic comparative study using wide array of commonly used extraction solvents and techniques for the extraction of isoflavones from soybeans. The results showed that optimum extraction yield of isoflavones was achieved with dimethyl sulphoxide:ethanol:water (5:70:25, v/v/v) solvent mixture using a pressurized liquid extractor at 1000 psi at 100 °C, with a 5 min equilibration time, a 7 min static time, a 90 s purge time and three extraction cycles (Luthria et al. 2007). We had summarized the extraction solvent composition (aqueous-organic solvents mixtures at different acidification levels), physical techniques (Soxhlet, stirring, ultrasonic, microwave, pressurized, supercritical-fluid, high-speed counter-current chromatography), and the analysis parameters (temperature, pressure, number of cycles, solid-solvent ratio) that influence quantitative extraction of isoflavones from different matrices. Furthermore, a brief overview of structures, sources, bioactivities, separation, and detection used for isoflavones analysis was also presented (Luthria and Natarajan, 2009).

Table 1. Extraction methods used for plant proteins

Serial#	Extraction Procedure	Method	Reference
1	Chickpea root tissue was extracted by several methods, TCA/acetone and phenol extraction and phenol with SDS. The phenol with/without SDS extraction was further evaluated with and without sonication method. The protein was analyzed by 2D and MS analysis.	TCA/acetone; Phenol; phenol with sonication; phenol with SDS; phenol with SDS and sonication	Chatterjee et al.2012
2	Soybean roots washed with TCA in acetone , methanol and acetone followed by a vacuum-drying and extracted using Phenol/SDS method and the protein was used for 2D and MS analysi.	Phenol/SDS	Rodrigues et al. 2012
3	Soybean seed powder was extracted using isopropanol and precipated with acetone. The pellet was resuspended in lysis buffer (9 M urea, 1%, CHAPS, 0.5 M IPG buffer, and 1% DTT) and the protein was used for 2D and MS analysis.	Isopropanol /acetone	Natarajan et al. 2009
4	Protein was extracted from peel of the banana by three methods: TCA/Acetone precipitation; TCA precipitation; Phenol based extraction. The extracted protein from each procedure was analyzed by 2D-PAGE and mass spectrometry.	TCA / Acetone; TCA; Phenol	Zhang et al. 2012
5	Sunflower seed embryos were ground and washed with petroleum ether for oil removal. The powder was then exposed to protein extraction buffer (KCl, DTT, PMSF, and SDS) with varying solvent (de-ionized water or Tris-HCl) to test their efficacy in protein extraction. The extracted protein was washed with ammonium acetate-methanol solution, and subsequently by cold acetone, dried, after which the protein content was resolubilized and analyzed by 2D-PAGE and MS.	KCl, DTT, PMSF, and SDS in varying solvent (ddH_20 OR Tris-HCL)	Barbosa et al. 2013
6	Soybean embryonic axis were ground and the powder was incubated on ice in a sucrose based extraction buffer (0.9M Sucrose, 100mM Tris-HCl pH 8.8, 0.4% 2-mercaptoethanol, protease inhibitor solution) followed by addition of equal volume of Tris-equilibrated phenol. The mixture was further incubated, spun down, and the upper phenol phase containing the protein was obtained. Protein within solution was then precipitated using ammonium acetate, dried, resolubilized, and analyzed via 2D-PAGE and MS.	Sucrose / Phenol based extraction followed by ammonium acetate-methanol precipitation	Natarajan et al. 2013
8	Soybean seed tissue was ground and protein was extracted using Trichloroacetic (TCA) / Acetone precipitation method, followed by cold Acetone washes. The precipitate was then dried, resolubilized in a Urea / CHAPS based lysis	Acid extraction: TCA/Acetone precipitation	Natarajan et al. 2012

Serial#	Extraction Procedure	Method	Reference
	buffer, and subjected to 2D-PAGE analysis as well as MS for protein identification.		
9	Pinellia kernata, tuber was sequentially extracted using 10% acetic acid, followed by SDS-containing buffer. The SDS extract was further missed with equal volume of phenol The phenol phase was precipitated with methanol containing 0.1M ammonium acetate and further dissolved in rehydration buffer for 2D analysis	Acid extraction /acetone precipitation / phenol extraction	Wu et al. 2012
10	Curcuma comosa plant tissue powder was extracted using buffer (0.5M Tris, 30mM HCl, 0.1M KCl, 0.7M sucrose, 1% 2-mercaptoethanol). The precipitate was extracted by phenol, precipitated by ammonium acetate in methanol overnight. The resulting protein pellet was precipitated with acetone and used for further 2D analysis.	Phenol method/ precipitated with ammonium acetate-methanol /acetone	Boonmee et al. 2011

It is well documented in the literature that a wide range of analysis procedures have been used for qualitative and quantitative assay of isoflavones from soybean and its processed food products. These include classical thin layer chromatography that can be used for qualitative and semi-quantitative analysis of hydrolyzed aglycons. Gas chromatography coupled to flame ionization and/or mass spectrometry detection methods have been used for quantification of isoflavones but these methods require derivatization which is often laborious and time consuming. The most commonly used procedures for separation and quantification of simple and conjugated forms of isoflavones have been liquid chromatography coupled with diode-array and/or mass spectrometry detectors (Rostagno et al. 2009; Luthria and Natarajan, 2009; Huang and Liu, 2012).

Some of the recent updates on isoflavones extraction and analysis from soybeans are briefly described in the current communication. Proximate composition of flavonoids content and *in vitro* antioxidant activity of ten legume seeds grown in China was recently reported by Ren et al. (2012). In a separate study, Lee et al. (2008) carried out analysis of 322 Korean soybean varieties collected from six different cultivated sites that were classified into three subgroups based on 100 seed weight as small, medium, and large to provide basic understanding of soybean germplasm with higher isoflavonoids concentration (Lee at al. 2008). Development and validation of an HPLC-DAD method for analysis of six isoflavones (three conjugated -daidzin, genistin, glycitin and their respective three aglycons daidzein, genistein, and glycitein) in the extracts from different soy processing stages is recently reported by Gasparetto et al. 2012. The method validation parameter indicated that the linear range for all six isoflavonoids was between 1.5 to 150 μg/mL, recovery efficiency between 95-102%, and intraday and interday relative standard deviation of <5.0%. The robustness study indicated that flow rate was the only limiting factor (Gasparetto et al. 2012).

Lojza et al. (2012) developed direct analysis in real time (DART) ionization-high resolution mass spectrometry for analysis of isoflavones from soybeans. The results showed that higher concentration of free aglycons reported in their analysis may be attributed to the cleavage of conjugated forms of isoflavones in the DART ion source. The effect of high power ultrasonic–assisted and toasting on the extraction of isoflavones from defatted soybeans was investigated by Pananun et al. (2012). The authors observed higher recovery (1.2 to 1.5 times) of isoflavones as compared to classical methods due to approximately 10 fold reduction in particle size. Longer toasting of defatted soy flakes at 150 °C resulted in the increase concentration of aglycons. In another study, Benedetti et al. (2013) showed that it was possible to concentrate malonyl glucosides and β-glucosides by nanofilteration of defatted soy-flour extract. The influence of thermal treatment on two forms varied, malonyl glycosides concentration was impacted by temperature whereas the concentration of β-glucosides was dependent on both time and temperature.

A new simplified chip based nanoflow high performance liquid chromatography coupled with mass spectrometry was developed, validated, and applied for the profiling of flavonoids from soybeans by Chang et al. (2012). This method can be used for metabolic profiling of soy extracts because of satisfactory, reproducibility, linearity, and recovery. Shen et al. (2012) developed high performance liquid chromatography coupled to diode-array and mass spectral detections for the separation of 19 isoflavones form tofu-type soybeans. A range of total isoflavones contents for tofu type soybeans was established to evaluate the applicability of chromatographic fingerprint approach for quality control use (Shen et al. (2012). The authors determined that the seeds size, composition, flavor, and nutritional content of tofu-type soybeans are different from conventional soybean varieties.

In conclusion, this chapter briefly summarizes some of the recent updates on the extraction and analysis of proteins and isoflavones from soybeans and its products.

ACKNOWLEDGMENTS

The authors thank Dr. M.A. Talo Pastor-Corrales for critical review of this manuscript. Funding for this research was provided by ARS project 1275-21000-224-00D.

REFERENCES

Adachi, M., Kanamori, J., Masuda, T., Yagasaki, K., Kitamura K. & Mikami, B. (2003). crystal structure of soybean 11S globulin glycinin A3B4 homohexamer, *Proc Natl Acad Sci*, *100*, 7395-7400).

Agrawal, G. K., Hajduch, M., Graham, K. & Thelen, J. J. (2008). In-depth investigation of the soybean seed-filling proteome and comparison with a parallel study of rapeseed, *Plant Physiol*, *148*, 504-518.

Alm, R., Ekefjard, A., Krogh, M., Hakkinen, J. & Emanuelsson, C. (2007). Proteomic variation is as large within as between strawberry varieties, *J Proteome Res*, *6*, 3011-3020.

Barbosa, H. S., de Souza, D. L., Koolen, H. H. F., Gozzo, F. C. & Arruda, M. A. Z. (2013). Sample preparation focusing on plant proteomics: extraction, evaluation and identification of proteins from sunflower seeds, *Anal Methods*, *5*, 116-123.

Benedetti, S., Prudencio, E. S., Mandarino, J. M. G., Rezzadori, K., Petrus, J. C. C. (2013). Concentration of soybean isoflavones by nanofiltration and the effects of thermal treatments on the concentrate, *Food Res Int*, *50*, 625-632.

Boonmee, A., Srisomsap, C., Chokchaicharmnankit, D., Karnchanatat, A. & Sangvanich, P., (2011). A proteomic analysis of Curcuma comosa Roxb. rhizomes, *Proteome Sci*, *9*, 43.

Chang, H., Lin, H. C. & Chang, Y. A. (2012). Development of the research in isoflavones, *Asian J Chem*, *24*, 3309-3313.

Carpentier, S. C., Witters, E., Laukens, E., Deckers, P., Swennen, R. & Panis, B. (2005). Preparation of protein extracts from recalcitrant plant tissues: an evaluation of different methods for two-dimensional gel electrophoresis analysis, *Proteomics*, *5*, 2497-2507.

Chatterjee, M., Gupta, S., Bhar, A. & Das, S. (2012). Optimization of an efficient protein extraction protocol compatible with two-dimensional electrophoresis and mass spectrometry from recalcitrant phenolic rich roots of chickpea (Cicer arietinum L.), *Int J Proteomics*, 536963.

Danchenko, M., Skultety, L., Rashydov, N. M., Berezhna, V. V., Matel, L., Salaj, T., Pretova, A. & Hajduch, M. (2009). Proteomic analysis of mature soybean seeds from the Chernobyl area suggests plant adaptation to the contaminated environment, *J Proteome Res*, *8*, 2915-2922.

Faraj, A. & Vasanthan, T., (2004). Soybean isoflavones: effects of processing and health benefits, *Food Rev Int*, *20*, 51-75.

Fehr, W. R., Hoeck, J. A., Johnson, S. L., Murphy, J. D., Nott, P. A. & Padilla, G. I. (2003). Genotype and environmental influence on protein components of soybean, *Crop Sci*, *43*, 511-514.

Gasparetto, J. C., Smolarek, F. S. F., De Francisco, T. M. G., Miranda, L. C., Pontarolo, R. & Siqueira, P. F. (2012). Development and validation of an HPLC-DAD method for analysis of the six major isoflavones in extracts from soybean processing, *J Amer Oil Chem Soc*, *89*, 1211-1222.

Gorg, A., Weiss, W. & Dunn, M. J. (2004). Current two-dimensional electrophoresis technology for proteomics, *Proteomics*, *4*, 3665-3685.

Gonzalez-Buitrago, J. M., Ferrerira, L., Isidoro-Garcia, M., Sanz, C., Lorente, F. & Davila, I., (2007). Proteomic approaches for indentifying new allergens and diagnosing allergic diseases, *Clin Chim Acta*, *385*, 21-27.

Huang M. & Liu, D. (2012). Thin layer chromatography densitometric determination of soybean isoflavones in wild soybean (*Glycine soja*) seeds, *Asian J Chem*, *24*, 1322-1324.

Isaacson, T., Damasceno, C. M., Saravanan, R. S., He, Y., Saladieand, M. & Rose, J. K. (2006). Sample extraction techniques for enhanced proteomic analysis of plant tissues, *Nat Protoc*, *1*, 769-774.

Klubicova, K., Danchenko, M., Skultety, L., Berezhna, V. V., Uvackova, L., Rashydov, N. M. & Hadjuch, M. (2012). Soybeans grown in the Chernobyl area produce fertile seeds that have increased heavy metal resistance and modified carbon metabolism, *PLoS One*, *7*, e48169.

Lee, S. J., Kim, J. J., Moon, H. I., Ahn, J. K., Chun, S. C., Jung, W. S., Lee, O. K. & Chung, I. M. (2008). Analysis of isoflavones and phenolic compounds in Korean soybean [*Glycine max* (L.) Merrill] seeds of different seed weights, *J Agric Food Chem*, *56*, 2751-2758.

Lojza, J., Cajka, T., Schulzova, V., Riddellova, K. & Hajslova, J., (2012). Analysis of isoflavones in soybeans employing direct analysis in real-time ionization-high-resolution mass spectrometry, *J Sep Sci*, *35*, 476-481.

Luthria, D. L., Biswas, R. & Natarajan, S. (2007). Comparison of extraction solvents and techniques used for the assay of isoflavones for soybeans, *Food Chem*, *105*, 325-333.

Luthria, D. L. & Natarajan, S. S. (2009). Influence of sample preparation on the assay of isoflavones, *Planta Med*, *74*, 704-710.

Natarajan, S. S., Krishnan, H. B., Lakshman, S. & Garrett, W. M. (2009). An efficient extraction method to enhance the analysis of low abundant proteins from soybean seed. *Anal Biochem*, *394*, 259-268.

Natarajan, S. S., Krishnan, H. B., Khan, F., Chen, X., Garrett, W. M. & Lakshman, D. (2013) Analysis of soybean embryonic axis proteins by two-dimensional gel electrophoresis and mass spectrometry, *J Basic Appl Sci*, *9*, 309-332.

Natarajan, S. S. & Luthria, D. L. (2010). Extraction and characterization of soybean proteins, In (Eds) J. E. Maxwell, *Soybean: Cultivation, Uses and Nutrition*, Nova Science Publishers, Inc. Hauppage, NY, 465-470.

Natarajan, S. S. & Luthria, D. L. (2011). Dietary sources of isoflavones and the methodology used for the analysis, In (Eds) K. Yamane and Y. Kato, Flovonoids: *Dietary Sources, Properties and Health Benefits*, Nova Science Publishers, Inc. Hauppage, NY, 527-532.

Natarajan, S. S., Xu, C., Garrett, W. M., Lakshman, D. & Bae, H. (2012). Assessment of the natural variation of low abundant metabolic proteins in soybean seeds using proteomics, *J Plant Biochem Biotech*, *21*, 30-37.

Pananun, T., Montalbo-Lomboy, M., Noomhorm, A., Grewell, D. & Lamsal, B. (2012). High-power ultrasonication-assisted extraction of soybean isoflavones and effect of toasting, *LWT-Food Sci Tech*, *47*, 199-207.

Poysa., V., Woodrow, L. & Yu, K. (2006). Effect of soy protein subunit composition on tofu quality, *Food Res Int*, *39*, 309-317.

Ren, S. C, Ze-Long, L. & Wang, P. (2012). Proximate composition and flavanoids content and in vitro antioxidant activity of 10 varieties legume seeds grown in China, *J Med Plant Res*, *62*, 301-308.

Ren, N., Xie, T. & Xing, D., (2009). Composition of extracellular polymeric substances influences the autoaggregation capability of hydrogen-producing bacterium Ethanoligenens harbinense, *Bioresour Technol*, *100*, 5109-5113.

Rostagno, M. A., Villares, A., Guillamon, E., Garcia-Lafuente, A. & Martinez, J. A. (2009). Sample preparation for the analysis of isoflavones from soybeans and soy foods, *J Chromatogr*, *1216*, 2-29.

Rodrigues, E. P., Torres, A. R., da Silva Batista, J. S., Hergo, L. & Hungria, M. (2012). A simple, economical and reproducible protein extraction protocol for proteomics studies of soybean roots, *Genet Mol Biol*, *35*, 348-352.

Salekdeh, G. H. & Komatsu, S., (2007). Crop proteomics: aim at sustainable agriculture of tomorrow, *Proteomics*, *7*, 2976-2996.

Saz, J. M. & Marina, M. L. (2007). High performance liquid chromatography and capillary electrophoresis in the analysis of soybean proteins and peptides in foodstuffs, *J Sep Sci*, *30*, 431-451.

Shen, D., Wu, Q., Sciarappa, W. J. & Simon, J. E. (2012). Chromatographic fingerprints and quantitative analysis of isoflavones in tofu-type soybeans, *Food Chem*, *130*, 1003-1009.

Wang, W., Tai, F. & Chen, S., (2008). Optimizing protein extraction from plant tissues for enhanced proteomics analysis, *J Sep Sci*, *21*, 2032-2039.

Wu, X., Xiong, E., An, S., Gong, F. & Wang, W. (2012). Sequential extraction results in improved proteome profiling of medicinal plant Pinellia ternate tubers, which contain large amounts of high-abundance proteins, *PLoS One*, *7*, e50497.

Zhang, L. L., Feng, R. J. & Zhang, Y. D. (2012). Evaluation of different methods of protein extraction and identification of differentially expressed proteins upon ethylene-induced early-ripening in banana peels, *J Sci Food Agric*, *92*, 2106-2115

INDEX

#

20th century, xi, 179, 194
21st century, 138, 151

A

abiotic stresses, viii, 2, 4, 21, 24, 26, 31
access, 217, 230
accessions, vii, 2, 8, 9, 12, 22, 23, 25, 26, 27, 28, 36, 184, 186, 187, 199, 282
accommodation, 220
accounting, 132, 194
acetic acid, 28, 297, 299
acetone, 105, 296, 298, 299
acetonitrile, 229, 260
acetylcholine, 131
acidic, 133, 143, 147, 284
acidity, 169
active compound, 14, 134
AD, 4, 13, 34, 36, 38, 120, 152
ADA, 118
adaptation(s), x, 3, 11, 22, 24, 32, 179, 230, 234, 301
additives, 28, 116
adenocarcinoma, 35
adhesion, 137
adipocyte, 102, 103, 152
adipose, 55, 94, 99, 100, 111, 113, 115
adipose tissue, 55, 94, 99, 100, 111, 113, 115
adiposity, 102, 114, 115, 139
adjustment, 30
adsorption, 202, 252, 255
adults, 86, 89, 94, 101, 105, 110, 118, 129, 138, 141, 227, 238
adverse effects, 135
adverse event, 288
Africa, x, 6, 9, 84, 161, 162, 164, 165, 175, 176, 179, 182, 194, 197, 226, 272
age, xi, 28, 87, 168, 175, 193, 194, 234
ageing population, ix, 93
aggregation, 296
aging population, 227
agonist, 96, 143
agriculture, xi, xiii, 10, 32, 36, 149, 179, 182, 191, 195, 203, 207, 210, 242, 271, 302
Agrobacterium, 211, 219, 223, 275, 281
air pollutants, 21
alanine, xiii, 70, 77, 78, 190, 258, 264
albumin, 105, 121
alcohol consumption, 136, 227
aldosterone, xii, 225
alfalfa, xi, 3, 10, 12, 209, 210, 215, 285
Algeria, 10
alimentation, xiii, 271, 272
alkaloids, 226, 289
allele, 277, 279
allergens, 3, 149, 301
allergy, 130
alternative treatments, 237
amine, 242, 248
amino acids, ix, xii, 2, 30, 42, 44, 45, 66, 67, 68, 71, 72, 77, 78, 79, 82, 83, 87, 91, 92, 106, 114, 129, 140, 160, 162, 165, 181, 183, 184, 190, 195, 197, 210, 226, 257, 259, 264, 284
amino groups, 229
ammonia, 210
ammonium, 160, 297, 298, 299
amylase, 97, 101, 109, 111, 119, 122, 140, 153, 182, 203
anaerobic bacteria, 136
analgesic, 290
anatomy, 27
angiogenesis, 141
angiotensin converting enzyme, xiii, 283, 289
angiotensin II, xii, 225, 227
animal husbandry, 195, 295
animal nutrition, viii, 41, 60

animal welfare, 233
anorexia, 101
ANOVA, 231
antagonism, 243
anther, 28
anthocyanin, 116, 286
anti-cancer, 149, 284
anticancer activity, 141
antidiabetic activities, xiii, 140, 283
antigen, 150
antioxidant, viii, xiii, 41, 42, 44, 56, 57, 58, 59, 60, 62, 63, 64, 65, 66, 83, 86, 105, 111, 112, 115, 117, 123, 135, 137, 186, 203, 227, 283, 286, 287, 289, 290, 291, 292, 299, 302
antioxidant activity, viii, 41, 42, 44, 57, 58, 59, 60, 62, 63, 64, 65, 66, 83, 86, 111, 112, 186, 287, 290, 299, 302
antioxidative activity, 59, 60, 64
antipyretic, 18
antisocial behavior, 43
anxiety, 142
apoptosis, 35, 44, 86, 134, 141, 286, 289
appetite, 95, 99, 103, 106, 117, 120, 121, 123
aqueous suspension, 288
arabinogalactan, vii, xii, 257, 258, 259, 261, 267, 268
Argentina, 85, 148, 209, 226, 271, 272
arginine, 67, 70, 77, 78, 139, 160, 162, 264
aromatic compounds, 242
aromatic rings, 133
arrest, 134, 141
ARS, 149, 295, 300
arsenic, 201, 253
arterial hypertension, xi, 225, 236
artery(s), 227, 238, 286
arthritis, 43, 49
arthropods, 21
ascorbic acid, 201
Asia, x, 5, 114, 141, 151, 153, 179, 180, 182, 191, 194, 197, 199
Asian countries, 126, 134, 141, 155
aspartic acid, 44, 67, 78, 160, 264
assessment, 71, 110, 152, 177, 288
asthma, 49
atherogenesis, 137, 148, 152
atherosclerosis, 136, 145, 234, 236, 286
atherosclerotic plaque, 137
atmosphere, 210
atmospheric deposition, 241
atomic emission spectrometry, 244
atoms, 266
atopic dermatitis, 286, 292
ATP, 274
atrophy, 143
attachment, 28, 212, 217, 268, 277, 286, 290
Austria, 39
autoimmune disease(s), 20, 88, 135
auxins, 214
avoidance, 95
awareness, 161, 241

B

BAC, 275, 281
Bacillus subtilis, x, 160, 172
backcross, 191
bacteria, xi, xiii, 22, 114, 115, 131, 133, 136, 137, 160, 168, 202, 209, 210, 212, 213, 215, 271, 273, 274, 278, 279, 280, 285
bacterial artificial chromosome, 281
bacterial fermentation, 169
bacterial infection, 211, 215
bacterium, 212, 302
ban, 181, 188, 189
Bangladesh, 66, 186, 197, 199, 200
barriers, 161, 272
base, 3, 37, 197, 211
baths, 19
BD, 9, 119, 120, 123, 156
beef, 11
beer, 169
Belgium, 147
beneficial effect, viii, xiii, 41, 43, 44, 97, 99, 103, 106, 117, 131, 135, 138, 139, 140, 204, 283, 284
benefits, vii, xii, xiii, 1, 3, 43, 57, 86, 98, 101, 105, 118, 121, 122, 130, 131, 133, 134, 136, 137, 139, 142, 146, 180, 195, 199, 202, 203, 204, 225, 227, 279, 283, 285, 286, 287, 291, 292, 301
benzene, 133, 260
beta-carotene, xi, 193, 194, 195, 201
beverages, 169
bias, 238
bile, 136, 137, 285, 288
bile acids, 136, 137, 285
bile duct, 136
bioavailability, xi, 132, 134, 149, 168, 183, 191, 193, 196, 201, 203, 204, 205, 206, 242, 243, 244, 253, 255, 256, 285, 293
bioavailable mineral micronutrients, xi, 193
biochemistry, 27
biodiversity, 42, 43, 45, 84, 204
biogeography, 190
biological activity(s), xii, 21, 31, 32, 134, 225, 227, 285
biological consequences, 188
biological control, 166

biological processes, 210, 218, 242, 272
biological responses, 143, 297
biologically active compounds, 111
biomarkers, 145, 153
biomass, 3, 11, 12, 27, 186, 247, 248, 252
biomolecules, 20
biosynthesis, viii, 2, 4, 28, 33, 47, 49, 141, 181, 184, 190, 210, 212, 217, 218, 267
biosynthetic pathways, 12
biotechnology, 27, 177, 195
biotic, viii, 2, 4, 21, 22, 30, 218, 272
biotic factor, 30
biotin, 197, 272
birds, 22
blends, 268
blood, ix, xi, xiii, 3, 18, 43, 94, 98, 99, 103, 105, 106, 109, 110, 115, 116, 117, 121, 123, 136, 137, 143, 152, 157, 225, 227, 230, 233, 234, 236, 238, 283, 284, 289
blood flow, 143
blood pressure, xi, 43, 109, 136, 137, 152, 225, 227, 230, 233, 236, 238, 284
body composition, 102, 111, 119, 138, 139, 152
body fat, 101, 136, 138
body fluid, xii, 225
body mass index (BMI), ix, 94, 98, 101, 105, 138
body size, 21
body weight, ix, 94, 95, 101, 102, 105, 111, 139, 230, 288, 289
bonds, 135, 229, 242, 296
bone, 89, 144, 145, 150, 156, 157, 286
bone mass, 144
bone resorption, 144, 157
borderline personality disorder, 85
bowel, 122, 284, 292
brain, 227
Brazil, 145, 272
breast cancer, 44, 86, 141, 145, 150, 154, 285, 286, 292
breeding, vii, viii, 1, 2, 4, 9, 22, 24, 25, 26, 27, 28, 30, 31, 149, 181, 184, 186, 187, 188, 190, 195, 196, 204
browsing, 38
burn, 10
by-products, 86, 147

C

Ca^{2+}, 221, 280
cabbage, 254
cadmium, 27, 201, 253, 254, 255
caecum, 133
caffeine, 96
calcium, x, 132, 144, 159, 162, 166, 173, 203, 213, 214, 217, 221, 222, 226, 245, 274, 280
calcium carbonate, 245
calibration, 244
caloric restriction, 138
calorie, 166, 194
calyx, 6
CAM, 38
Canary Islands, vii, 2, 4, 6, 8, 9, 10, 11, 22, 23, 26, 33, 36
cancer, xiii, 3, 44, 86, 89, 127, 129, 133, 135, 136, 138, 141, 142, 144, 145, 146, 148, 150, 151, 183, 184, 201, 283, 284, 285, 289
cancer cells, xiii, 142, 283, 289
cancer therapy, 183
candidates, 21, 71, 79
capillary, 261, 302
carbohydrate(s), x, xi, xii, 2, 43, 85, 98, 100, 101, 103, 110, 114, 115, 116, 119, 123, 130, 131, 140, 160, 162, 166, 168, 172, 173, 193, 195, 199, 202, 203, 217, 257, 258, 259, 260, 262, 263, 266, 272, 284, 296, 297
carbohydrate metabolism, 103
carbon, 47, 133, 136, 210, 222, 242, 301
carbon atoms, 133
carbon dioxide, 136
carboxyl, 228
carcinogen, 141
carcinogenesis, 119, 284, 288
carcinogenicity, 20
cardiac arrhythmia, 148
cardiovascular disease(s), ix, xii, 49, 88, 95, 97, 125, 126, 127, 134, 136, 137, 140, 151, 152, 225, 226, 227, 233, 234, 284, 285
cardiovascular disorders, 47
cardiovascular morbidity, 227
cardiovascular risk, 120, 145, 238
cardiovascular system, 234
carob, x, 159, 161
carotene, 62, 201, 284
carotenoids, 85
casein, 105, 121, 153
cataract, 292
cation, 132, 214, 218, 274
cattle, 4, 11
Caucasus, 7, 38
CDC, 116
cDNA, 216
cell biology, 27
cell cycle, 134, 141, 278, 279
cell differentiation, 141
cell division, 212, 214, 215, 273, 278, 279
cell invasiveness, 141

cell line(s), 35, 142
cell membranes, xii, 89, 257
cell signaling, 141, 203
cell surface, 268
cellulose, 105, 130, 296
central nervous system, 181, 191
ceramide, 258
challenges, 195, 204
cheese, 12, 169
chelates, 242
chemical(s), vii, xii, 1, 2, 27, 79, 85, 86, 91, 92, 175, 203, 241, 242, 243, 244, 245, 252, 253, 255, 257, 258, 259, 260, 262, 273, 274, 278, 285
chemical properties, 85
chemokines, 137
chemoprevention, 89, 126, 149, 285
chemopreventive agents, 133, 154
chemotherapy, 149
chicken, 161, 254
child development, xi, 193
childhood, 194
children, xi, 129, 138, 193, 194, 206
Chile, 3
China, 37, 126, 142, 197, 226, 253, 272, 283, 299, 302
chitin, 221
chloroform, 289
chloroplast, 187, 280
cholesterol, xiii, 3, 86, 103, 128, 130, 136, 137, 139, 151, 283, 284, 285, 286, 287, 288, 289
choline, 131
chromatographic fingerprint, 300
chromatography, 261, 262, 297, 299, 300, 301
chronic diseases, ix, xi, 89, 95, 97, 111, 117, 125, 127, 129, 193, 194
chymotrypsin, 135, 166, 168, 226, 284, 287
cirrhosis, 132
City, 239
classes, 85, 96, 97, 115, 133, 138, 227
classical methods, 300
classification, 188, 190
cleavage, 299
climate(s), xi, 3, 4, 35, 38, 126, 132, 179, 272
climate change, xi, 179
clinical application, 20
clinical depression, 95
clinical trials, 14, 19, 143, 145, 152
cloning, 218, 221, 280
CO_2, 11, 131
coatings, 2
cobalamin, 197
codominant, 188
coffee, x, 159, 161, 226, 260
cognitive function, 134
cognitive impairment, 194
collaboration, 4, 30
colon, 35, 43, 44, 106, 111, 119, 133, 136, 202, 204, 287, 291
colon cancer, 43, 44, 111, 119, 204, 287, 291
colonisation, 10
colonization, 115, 203, 218, 223
color, 127, 130, 165, 185, 226
commercial, vii, viii, ix, xii, 2, 3, 22, 27, 42, 57, 59, 60, 61, 62, 66, 68, 73, 78, 79, 83, 115, 146, 225, 226, 241, 243, 248, 250, 259, 264
commodity, 146
communication, xiii, 278, 295, 299
community(s), 9, 42, 137, 210
comorbidity, 102
compatibility, 273, 279
compensation, 106
competition, 279
competitors, 278, 279
complement, 3, 199, 285
complex carbohydrates, 111, 130
complex interactions, 252
compliance, 142
complications, 95, 98, 103, 111, 118, 227
compounds, ix, 3, 4, 19, 20, 21, 22, 24, 28, 38, 44, 55, 56, 57, 60, 62, 86, 94, 95, 106, 111, 114, 117, 121, 130, 131, 133, 134, 135, 136, 137, 140, 141, 142, 149, 160, 176, 180, 183, 184, 188, 203, 242, 243, 285, 288
computer, 27
Concise, 147
conductivity, 27, 30
Congress, 37
consensus, 38, 120, 148
conservation, vii, viii, 1, 2, 4, 26, 27, 28, 33, 186
constipation, 131
constituents, x, 20, 130, 132, 160, 258, 266
consumer education, 204
consumers, 101, 102, 180, 181
contaminated soil(s), xii, 241, 243, 247, 250, 251, 254, 255
contamination, 176, 253
control condition, 27
control group, 107, 230
controlled trials, 157
controversial, 145
controversies, xiii, 283
Convention on Biological Diversity, 42
cooking, x, 55, 86, 89, 90, 135, 159, 161, 169, 170, 171, 172, 174, 181, 183, 191, 296
cooling, 261
cooperation, 31, 42, 282

coordination, 242
copper, 27, 132, 164, 196, 226, 243, 253, 254, 255, 272
cornea, 194
corolla, 14
coronary heart disease, 89, 95, 98, 123, 128, 146, 148, 152, 227, 238, 284, 285
correlation(s), 30, 44, 47, 49, 56, 58, 61, 64, 78, 86, 110, 116, 276
corruption, 17
cortex, 212, 214, 217, 273, 274, 279
cosmetics, vii, 1, 2
cost, 94, 168, 187
cotton, 281
cotyledon, 115, 116, 123, 199, 275, 287
coumarins, 20
counseling, 98
counterbalance, 181
covering, 211
crop(s), vii, viii, x, xi, 2, 3, 4, 9, 19, 32, 36, 38, 41, 43, 44, 45, 56, 89, 109, 126, 129, 132, 162, 165, 166, 177, 179, 180, 188, 190, 193, 194, 195, 196, 197, 199, 200, 201, 204, 205, 206, 207, 209, 210, 218, 219, 239, 272, 275, 279, 295
crop production, 194, 195
crude oil, 127
cryopreservation, 30
crystal structure, 300
crystals, 30, 265
CT, 120, 147
cues, 222
cultivars, viii, 2, 4, 22, 25, 31, 41, 84, 85, 90, 112, 114, 116, 149, 183, 199, 200, 226, 277, 290
cultivation, ix, xi, 24, 27, 30, 42, 43, 60, 71, 79, 126, 164, 179, 180, 184, 185, 189, 190, 199, 254
culture, 4, 27, 28, 29, 210, 234, 267
culture medium, 28, 267
curcumin, 133
cure, 14
current balance, 55
customers, 189
cutin, 217, 218
CV, 151, 184
CVD, 136, 137, 227
cyanide, x, 159, 173, 174
cycles, 297
cyclooxygenase, 117
cysteine, 44, 67, 78, 83, 129
cystine, 160, 264
cytochromes, 20
cytokines, 137, 214
cytology, 143
cytoplasm, 274
cytotoxicity, 286, 289

D

DART, 299
database, 187, 211, 220
DBP, 230, 233, 234, 235, 236, 237
deaths, xi, 193, 194
decoding, 274
decomposition, xii, 225
decontamination, 253
decortication, 86
defects, 222
defence, 21, 183
deficiency(s), 132, 144, 194, 195, 201, 204, 205, 243, 287
deformation, 213, 280
degradation, 3, 62, 102, 135, 150, 254, 268, 284
dehydration, 3
demography, 153
denaturation, 183
Denmark, 228
Department of Agriculture, 10, 118, 122, 126
Department of Health and Human Services, 118
deposition, 99, 100, 111, 119, 241
depression, 135, 142, 248
deprivation, 27
depth, 300
derivatives, 49, 221
dermatitis, 286
destruction, 30
detection, 191, 244, 297, 298, 299
detergents, 296
detoxification, 141, 181, 183, 186
developed countries, 136
developing countries, x, xiii, 160, 161, 179, 194, 199, 201, 204, 271, 285
diabetes, vii, ix, 43, 44, 93, 94, 95, 96, 97, 98, 99, 102, 105, 110, 111, 114, 117, 118, 119, 120, 122, 125, 129, 136, 138, 139, 140, 152, 284, 285, 288
diabetic patients, 103, 118, 121, 140
diarrhea, 136
diastolic blood pressure, 227
dietary fat, 55, 88
dietary fiber, ix, 103, 121, 125, 129, 130, 136, 137, 152, 199, 202
Dietary Guidelines, 97, 118, 120
Dietary Guidelines for Americans, 118
dietary habits, 130
dietary intake, 242
dietary supplementation, 195
diffraction, 266
diffusion, 260, 262, 268, 286

digestibility, ix, 42, 68, 70, 71, 72, 78, 79, 80, 82, 83, 87, 90, 91, 116, 123, 134, 166, 168, 182, 287
digestion, 101, 106, 111, 114, 117, 135, 136, 140, 168, 169, 202, 276
digestive enzymes, 68
dipeptides, 228, 232
diploid, xi, 209, 210
discomfort, 130, 136
discrimination, 274
diseases, ix, xi, xiii, 3, 14, 18, 19, 44, 49, 93, 114, 118, 125, 130, 134, 136, 138, 140, 141, 186, 194, 195, 197, 202, 203, 225, 283, 284, 285, 287, 301
disorder, x, 20, 179
dispersion, 228, 234
dissociation, 135
distilled water, 30, 228
distribution, 2, 5, 6, 8, 52, 86, 95, 137, 226, 245, 276, 296
diversification, viii, 41, 42, 182, 195, 196, 272, 276, 282
diversity, 3, 4, 6, 7, 8, 14, 42, 122, 187, 188, 190, 191, 290
DNA, xiii, 9, 20, 37, 39, 86, 141, 142, 187, 188, 210, 211, 215, 220, 275, 280, 281, 283, 289, 291
DNA damage, xiii, 86, 283, 289
DNA sequencing, 9, 37
docosahexaenoic acid, 49, 131
dogs, 9
DOI, 37, 123, 149, 253, 282
domestication, 275, 277
dosage, 243
double helix, 20
dough, 176
down-regulation, 141
drainage, 3
drawing, 5, 185
drinking water, 230
drought, vii, x, 2, 3, 4, 11, 22, 23, 25, 26, 27, 35, 162, 179, 180, 184, 185, 259
drug metabolism, 20
drug treatment, 95
drugs, 4, 20, 95, 96, 97, 142, 236, 237
dry matter, 126, 132, 135, 182
drying, 170, 171, 260, 265, 298
dyslipidemia, 291

E

East Asia, 7
ecology, 17, 195
economic change, 38
economic development, 195
ecosystem, 42, 114, 168
Ecuador, 272
edema, 289
editors, 32, 33, 34, 35, 36, 38
education, 96
effluent, 241
egg, 105, 121
Egypt, 5, 6, 7
eicosapentaenoic acid, 131
electrical conductivity, 30
electrolyte, 30
electron, 242, 259, 261
electron diffraction, 261
electrophoresis, 276, 296, 301, 302
embryogenesis, 4, 28, 29, 211
emission, 255, 261
EMS, 211
emulsions, 268, 269, 285
encapsulation, 4
encoding, 215, 217, 287
endocrine, 131, 137, 138
endocrine disorders, 138
endocrine glands, 131
endosperm, 132
endothelium, 137, 236
energy, ix, xi, 2, 12, 87, 88, 94, 98, 99, 101, 105, 110, 111, 113, 119, 122, 138, 140, 166, 168, 169, 193, 194, 203, 210, 266
energy consumption, 111
energy expenditure, 99, 119
energy input, 169
England, 253
environment(s), vii, 2, 3, 9, 11, 23, 31, 36, 37, 142, 148, 152, 184, 191, 203, 206, 242, 255, 301
environmental aspects, 152
environmental conditions, 3, 21, 66, 67, 116, 252
environmental effects, 39
environmental factors, 22, 29
environmental influences, 200
environmental services, 239
environmental stress(s), 26, 39
enzymatic activity, 230
enzyme(s), x, xii, xiii, 84, 101, 123, 134, 141, 153, 168, 169, 179, 180, 211, 217, 218, 225, 227, 228, 229, 231, 238, 239, 283, 285, 286, 287, 288, 289, 296
enzyme inhibitors, x, 179, 180, 285
EPA, 131
epidemic, vii, 94
epidemiologic, 121, 137, 146
epidemiologic studies, 146
epidemiology, 192
epidermis, 212, 214, 217
epigenetics, 149

epilepsy, 14, 18
epithelia, 143
epithelial cells, 135, 290
epithelium, 143, 154, 285
EPS, 273
equilibrium, 180, 181
Eritrea, 197
essential fatty acids, xiii, 49, 85, 195, 264, 283
EST, 187, 191, 192, 211, 275
ester, 19, 230, 236
estrogen, 139, 141, 142, 143, 144, 154, 156, 285, 286
ethanol, 184, 297
ethyl acetate, 288
ethylene, 303
etiology, 94, 143
EU, 126
Europe, x, 6, 7, 14, 32, 36, 98, 126, 131, 147, 179, 180, 182
European Parliament, 148
European Union, 126, 148
evidence, ix, 9, 94, 97, 98, 99, 102, 103, 110, 111, 117, 118, 126, 138, 139, 142, 146, 147, 152, 153, 157, 204, 213, 264, 279, 286
evolution, 204, 223, 272, 277, 279, 282
excavations, 10
excitation, 180, 181, 184
excretion, 26, 132, 137, 144, 157, 285, 288
exercise, 103, 104, 120, 136, 238
experimental condition, 164, 234
exploitation, 42, 184, 186
exports, 190
exposure, 20, 21, 27, 29, 108, 109, 143, 189, 285, 286
expressed sequence tag, 187
extracellular matrix, 267
extraction, vii, xii, xiii, 22, 127, 225, 227, 228, 241, 244, 245, 251, 252, 254, 255, 256, 264, 295, 296, 297, 298, 299, 300, 301, 302, 303
extracts, viii, 42, 56, 59, 63, 64, 66, 86, 90, 101, 105, 123, 140, 145, 153, 184, 228, 262, 268, 288, 289, 290, 299, 300, 301
extrusion, 110, 135, 183
exudate, 259, 273

F

families, 20, 26, 135, 210, 274
famine, 180
farmers, 42, 180, 188, 189, 190, 194, 200
farms, 272
FAS, 99, 100
fasting, 99, 103, 104, 105, 106, 110, 119, 289
fat, x, xiii, 43, 45, 55, 94, 95, 97, 99, 101, 102, 103, 105, 107, 111, 113, 114, 115, 119, 121, 128, 131, 132, 136, 138, 139, 147, 159, 162, 166, 172, 173, 226, 260, 262, 263, 283
fat intake, 55
fatty acid composition, viii, 41, 43, 46, 47, 48, 50, 51, 53, 55, 84, 88, 89, 182, 262
fatty acids, viii, xiii, 41, 43, 44, 46, 47, 48, 49, 50, 51, 52, 54, 55, 83, 84, 85, 88, 89, 99, 111, 115, 131, 137, 140, 148, 197, 203, 258, 261, 262, 264, 266, 284
feces, 285
feedstock, x, 179
feelings, 103, 113
female rat, 287
fermentable carbohydrates, 114, 203
fermentation, vii, x, 85, 114, 115, 119, 127, 133, 137, 140, 147, 153, 159, 161, 168, 169, 171, 172, 173, 174, 175, 176, 177, 178, 183, 191
fertility, 160, 162, 180, 272
fertilization, 196
fertilizers, 201
fiber(s), ix, xiii, 43, 94, 97, 101, 103, 105, 114, 116, 122, 125, 130, 136, 137, 141, 152, 226, 228, 283, 284, 287
fiber content, 103, 116, 226
fibroblasts, 234
fidelity, 279
field emission scanning electron microscopy, 265
film thickness, 261
films, 268
filtration, 261
fingerprints, 302
Finland, 196, 201, 206
fish, 148, 162, 295
fish oil, 148
fisheries, 149
fitness, 183
fixation, 180, 190, 274
flame, 261, 299
flatulence, x, 130, 134, 136, 160, 168, 174, 199, 226
flavonoids, 116, 133, 212, 221, 273, 289, 290, 292, 299, 300
flavonol, 86
flavor, 127, 168, 169, 170, 237, 300
floods, 180
flora, 3, 6, 10, 143
flour, 50, 55, 56, 57, 58, 59, 60, 61, 62, 63, 64, 99, 103, 105, 110, 121, 122, 127, 162, 170, 172, 174, 176, 178, 228, 300
flowers, 5, 7, 18, 19, 24, 38, 127, 199, 211
fluctuations, 99
fluid, 14, 143, 144, 297

fluorescence, 261
folate, 284
folic acid, 195, 197, 287
food additive(s), 267
Food and Drug Administration (FDA), 128, 146
food chain, xii, 241, 242
food industry, 126, 145, 231
food intake, ix, 94, 103, 106, 117, 121, 141
food production, 194
food products, xiii, 130, 295, 299
food safety, 241, 255
food security, 42, 162, 165, 166, 195
forage crops, 272
Ford, 191
formation, 20, 28, 30, 136, 168, 204, 212, 213, 214, 215, 217, 218, 221, 273, 275, 277, 278, 281, 285
fractures, 144
fragments, 187
France, 31, 33, 38, 126, 149, 200, 226, 241, 243, 255
free radicals, 111, 203, 288
freezing, 4, 30, 39, 91
frost, 24, 30
fructose, 130
fruits, 7, 12, 18, 19, 21, 44, 90, 127, 195, 203, 272, 284, 287
functional food, xii, 116, 117, 121, 126, 130, 145, 195, 226, 237, 284
fungi, xi, 140, 209, 210, 211, 216, 217, 220
fungus, 21, 177, 217, 286
furan, 20

G

galacto-oligosaccharides, ix, 125, 133, 144
gallstones, 284
gastric enzymes, xii, 225
gastric mucosa, 228
gastrointestinal tract, 106, 114, 135, 181, 284
gel, 260, 262, 268, 296, 301, 302
gene expression, 141, 142, 211, 215, 219, 220, 287
gene regulation, 141
gene silencing, 218, 275, 276
genes, 21, 99, 105, 138, 141, 201, 211, 213, 214, 215, 218, 219, 274, 275, 277, 278, 279, 280, 281, 282, 287
genetic diversity, 4, 42, 189, 191
genetic engineering, 4
genetic factors, 137
genetic marker, 37
genetic predisposition, 49
genetic screening, 218
genetics, 214, 218, 275
genome, xi, xiii, 187, 199, 209, 210, 211, 220, 223, 271, 276
genomics, 218, 220, 281
genotype, 132, 175, 185, 191, 206, 250, 281
genotyping, 296
genus, 5, 13, 21, 43, 44, 45, 46, 50, 52, 56, 57, 61, 62, 68, 71, 73, 78, 79, 88, 126, 180, 187, 188, 189, 199, 210, 226
Georgia, 5, 226
Germany, 228
germination, 22, 25, 26, 32, 161, 203
germplasm resources, viii, 2
ginger, 123
glucagon, 107, 108, 115, 137, 139
glucose, ix, xiii, 94, 95, 96, 98, 99, 103, 104, 105, 106, 109, 110, 116, 117, 121, 122, 123, 130, 137, 139, 140, 153, 283, 284, 288, 289, 292
glucose regulation, 140
glucose tolerance, 95, 103, 106, 109, 284, 292
glucose tolerance test, 106
glucosidases, 140
glucoside, 140, 166, 174, 176, 286
glutamic acid, 45, 78, 226, 264
glutamine, 70, 77
glutathione, 286, 288, 289
glycerol, 131, 218
glycine, xiii, 70, 77, 78, 139, 242, 258, 264
glycogen, 99, 288, 289
glycol, 27
glycolysis, 99
glycoproteins, ix, 94, 135
goat milk, 35
gout, 18
GPS, 242, 243, 248, 252, 253
GRAS, 213, 215, 217, 218, 220, 222, 274, 280, 281
grass, vii, x, 84, 85, 86, 87, 91, 179, 180, 181, 182, 183, 184, 185, 186, 187, 188, 189, 190, 191, 192, 259
grasslands, 38
grazing, vii, 3, 4, 9, 10, 11, 12, 25, 26, 35, 38
Greece, 24, 31, 35
green revolution, viii, 41, 194
greenhouse, 29
grounding, 243
growth, 23, 24, 25, 27, 88, 123, 133, 135, 136, 137, 141, 160, 168, 183, 194, 197, 202, 203, 213, 242, 244, 248, 267, 273, 286
growth factor, 141
growth rate, 183
Guatemala, 226
guidance, 204
guidelines, 89, 99, 102, 120, 202, 284
Guinea, 10, 33

H

habitat(s), 6, 8, 21, 23, 42
hair, 212, 213, 277
halos, 262
HAP, 297
haploid, viii, 2, 28, 199
hardness, 171
harvesting, 129, 186
HDAC, 142
health care, 194
health condition, 126
health effects, vii, ix, 89, 105, 125, 126, 128, 140, 145, 150, 157, 204
health problems, 136, 138, 227
health services, ix, 93
heart attack, 43, 98
heart disease, 55, 128, 129, 136, 138, 151, 204, 227, 285
heart failure, 234
heart rate, 98, 102
heat loss, 143
heavy metals, 4, 22, 27, 35, 39, 201, 241, 247, 253, 254, 255, 256
height, 127, 138, 194
hematology, 288
hemicellulose, 99, 105
hemostasis, 18
herbicide, vii, xii, 241, 242, 243, 244, 245, 246, 248, 249, 251, 252, 253
heritability, 200, 201
heterogeneity, 99, 247
hexane, 261, 288
HHS, 118
high blood pressure, 227, 285
high density lipoprotein, 288, 289
high fat, 106, 111, 139
high fiber foods, 130
highlands, 180
Hispanic population, 10
histidine, 69, 76, 264
histone(s), 141, 142, 154, 184
histone deacetylase, 142
history, 14, 19, 180
HIV, 201
HM, 123, 146, 147, 155
HO-1, 286
homeostasis, 106, 110, 114, 140, 153, 205
Honduras, 226
Hong Kong, 283
hormone(s), xii, 106, 117, 121, 123, 133, 136, 137, 141, 142, 144, 145, 156, 225
host, 20, 38, 114, 115, 130, 134, 137, 160, 202, 217, 279, 280, 282
human body, 49, 138, 264
human brain, 286
human exposure, xii, 241
human health, vii, 49, 90, 131, 135, 183, 194, 203, 204, 205, 242
human subjects, 88, 119, 120, 139
humidity, 11, 230, 260
hybrid, viii, 2, 24, 25, 26, 213
hybridization, 37, 277, 282
hydrogen, xiii, 131, 136, 226, 283, 290, 302
hydrogen cyanide, 226
hydrogen peroxide, xiii, 283, 290
hydrolysates, vii, xii, 225, 227, 228, 230, 231, 232, 233, 234, 235, 236, 237, 238, 239, 260, 261
hydrolysis, 103, 116, 140, 150, 229, 231, 232, 239, 262, 287
hydroxyl, 133, 288
hydroxyl groups, 133
hypercholesterolemia, ix, 102, 125
hyperglycemia, 111, 117, 139, 140
hyperinsulinemia, 138, 139
hyperlipidemia, ix, 125, 138, 139
hypermethylation, 141, 154
hyperplasia, 135
hypertension, xiii, 95, 105, 129, 136, 153, 227, 234, 236, 237, 238, 239, 283, 284
hypertrophy, 135, 234, 238
hypolipidemic activity, xiii, 283
hypotensive, xii, 226, 228, 230, 233, 234, 236
hypothalamus, 143
hypothesis, 10, 103, 114, 137, 141, 143, 144, 181, 258, 264, 266, 272, 279, 282
hypoxia, 27
hysteria, 14, 18

I

ICE, 32
ID, 191, 261
ideal, 22, 195, 233
identification, 22, 166, 213, 215, 218, 221, 274, 278, 280, 297, 299, 300, 303
ileum, 101
image(s), 265, 266
immigrants, 145
immobilization, xii, 241, 254
immune system, 131, 134, 194
immunomodulatory, 147
immunoprecipitation, 218
improvements, 98, 110, 122, 139
impurities, 296

in vitro, viii, 2, 4, 12, 22, 28, 29, 71, 78, 83, 90, 91, 106, 116, 119, 133, 140, 141, 142, 147, 153, 182, 183, 188, 191, 231, 299, 302
in vivo, 12, 30, 78, 133, 141, 183, 286
inbreeding, 24
incidence, 44, 111, 126, 137, 139, 141, 153, 169, 183, 189, 227, 284, 286
income, 199
incubation period, 108
incubator, 170
India, 14, 57, 66, 114, 186, 191, 197, 199, 226, 272, 289
Indians, 120
individuals, 12, 26, 31, 95, 105, 110, 111, 118, 130, 145, 152, 187, 202, 279
inducer, 291
induction, 28, 44, 134, 213, 221, 277, 278, 279
industry(s), vii, 2, 4, 19, 27, 126, 194, 210
infarction, 227
infection, 27, 131, 140, 212, 213, 214, 215, 220, 221, 222, 273, 274, 275, 277, 278, 279, 280
infertility, 286, 292
inflammation, xiii, 88, 114, 117, 123, 283, 284, 285
inflammatory mediators, 290
ingestion, 21, 26, 110, 131, 139
ingredients, ix, xii, 106, 125, 130, 131, 147, 168, 202, 226, 239, 248, 252, 253, 295
inheritance, 187
inhibition, xii, 101, 103, 117, 135, 141, 142, 221, 225, 227, 230, 236, 238, 254, 286, 289, 291
inhibitor, 96, 101, 111, 119, 122, 135, 166, 168, 169, 173, 174, 176, 182, 188, 191, 203, 230, 236, 284, 287, 298
initiation, 214, 215, 274
injury, 30
inoculation, 212, 214, 275, 277
inositol, 135, 183, 203, 204, 258
insects, x, 135, 179, 181
insecurity, 194
insertion, 211, 218, 222, 275
institutions, 22, 186
insulin, 3, 94, 95, 97, 98, 103, 105, 106, 107, 109, 110, 116, 117, 119, 122, 137, 138, 139, 152, 153, 205, 288, 289
insulin resistance, 98, 103, 105, 106, 119, 138, 139
insulin sensitivity, 98, 109, 110, 137
insulin signaling, 106
integration, 195
integrin, 137
integrity, 145
interface, 11, 259
interference, 213, 275
intervention, 99, 103, 137, 139, 144, 156, 157, 233
intestine, 115, 135, 136
intima, 286
intracellular calcium, 274
intrinsic viscosity, 261, 264
intron, 276
iodine, 195
ionization, 261, 299, 301
ions, 242
Iowa, 254
Iran, 109, 197
Ireland, 118
iron, xi, 114, 132, 164, 193, 195, 203, 205, 206, 226, 253, 254, 272
irrigation, 11, 12, 35
Islam, 91, 191, 206, 262, 267
islands, 6, 8, 9, 10
isoflavone, ix, 111, 125, 134, 140, 141, 144, 145, 150, 152, 157, 285, 286, 291
isoflavonoid(s), 140, 144, 153, 157, 212, 299
isolation, 120, 227, 277
isoleucine, 69, 76, 160, 162, 197
Israel, 4, 7, 25, 31
issues, xi, 193, 194, 195
Italy, 3, 5, 7, 31, 36, 38, 126, 179, 180, 184, 186, 187, 189, 192, 226, 261

J

Japan, 131, 155, 190, 261, 268
Japanese women, 157
juvenile rheumatoid arthritis, 206

K

K^+, 213
kaempferol, 133, 291
keratinocytes, 286
kidney(s), x, xii, 68, 71, 84, 99, 101, 102, 103, 111, 119, 131, 159, 161, 163, 181, 225, 272, 283, 289
kinase activity, 274
kinetics, 106, 268

L

lactic acid, 133, 136, 168
Lactobacillus, 172, 174
languages, 14, 15
large intestine, 131, 136, 284, 287
L-arginine, 230, 236
Latin America, xi, 179, 272
LDL, 117, 130, 136, 137, 139, 140
leaching, 247, 255

lead, 99, 181, 202, 253, 254, 255
leakage, 30
lecithin, 131, 148, 288
legislation, 234
legs, 180
lens, 114, 199, 286
leptin, 102, 111, 113
lesions, 117
leucine, 69, 76, 160, 162, 197, 229, 264, 274
leukotrienes, 49
liberation, 231
life expectancy, 138
lifetime, 227, 239
ligand, 144, 156, 274
light, 20, 38, 142, 186, 230, 259
light scattering, 259
lignans, 155
lignin, 105, 202
linoleic acid, 48, 49, 50, 54, 62, 83, 131
lipid metabolism, 102, 103, 119, 138, 139, 152
lipid peroxidation, 139, 288, 289
lipid peroxides, 286
lipids, viii, 20, 41, 43, 84, 85, 98, 103, 111, 131, 136, 146, 148, 152, 157, 180, 258, 292, 296
lipolysis, 115, 139
lipoproteins, 88, 136
liquid chromatography, 190, 229, 238, 299, 300, 302
liver, 3, 111, 132, 136, 139, 149, 184, 205, 289
liver cancer, 149
livestock, 3, 4, 9, 10, 11, 160, 295
localization, 156
loci, 187, 188, 190
locus, 175, 187, 280
logging, x, 179, 180
loss of libido, 142
low temperatures, 26
low-density lipoprotein, 47, 117, 152
lumbar spine, 145
lumen, 140
Luo, 150
lymphoma, 20, 38, 288
lysine, 45, 68, 78, 114, 129, 160, 162, 184, 197, 264
lysis, 298

M

machinery, 215, 275, 276
macromolecules, xii, 257
magnesium, x, 132, 160, 162, 164, 173, 203, 204, 226, 272
majority, 11, 211, 213, 216, 285
Malaysia, 226
malnutrition, xi, 162, 181, 189, 193, 194, 195, 204, 205, 206
mammals, 130
man, 181
management, 31, 95, 97, 105, 111, 114, 117, 118, 164, 239, 284
manganese, 226, 253, 254, 272
manifolds, 197
manipulation, 94, 95, 255
mannitol, 116
manufacturing, x, 179, 295
manure, 3, 226, 254
mapping, 188
marketing, 195
mass, 4, 99, 102, 103, 104, 111, 138, 139, 140, 194, 297, 298, 299, 300, 301, 302
mass spectrometry, 297, 298, 299, 300, 301, 302
materials, 168, 200, 244, 297
matrix, 144, 201, 265
matter, 11, 12, 126, 145, 260
Mauritius, 226
MB, 34, 35, 118, 119, 145, 146, 147, 152
measurement(s), 30, 138, 233, 244, 263
meat, 162, 163, 272, 295
mechanical properties, 268
media, 90, 286
medical, vii, 1, 14, 43, 138, 231
medication, 227
medicine, 4, 14, 19, 20, 297
Mediterranean, vii, xi, 1, 2, 3, 4, 5, 6, 7, 8, 9, 12, 13, 19, 22, 23, 24, 25, 26, 28, 31, 32, 33, 35, 37, 38, 71, 89, 151, 209, 210
Mediterranean climate, 12
Mediterranean countries, 6
mellitus, ix, 20, 93, 102, 114, 140
membranes, 30, 49, 136
menopause, 139, 143, 144, 155
menstruation, 14, 18
mercury, 201
meristem, 30, 222
messenger RNA, 154
meta-analysis, 98, 119, 143, 150, 152, 155, 157, 238
Metabolic, 122, 152, 205
metabolic disorder(s), 114, 138
metabolic pathways, 3, 273
metabolic responses, 115
metabolic syndrome, 117, 118, 120, 138, 284
metabolism, 26, 94, 95, 99, 122, 132, 137, 140, 141, 145, 149, 151, 152, 155, 157, 242, 248, 252, 253, 301
metabolites, vii, 1, 2, 25, 27, 38, 57, 102, 137, 140, 141, 185, 186, 252, 296, 297
metal ion(s), 135, 242, 254

metals, vii, xii, 27, 241, 242, 243, 245, 248, 250, 252, 253, 254, 255
metastasis, 141
meter, 30
metformin, 95
methanol, 287, 288, 297, 298, 299
methodology, 229, 296, 302
methylation, 141, 142
Mexico, 101, 109, 164, 228, 237, 239, 257, 260, 272, 275
mice, 86, 102, 115, 119, 120, 121, 139, 153, 154, 205
microbial communities, 254
microbiota, 114, 115, 122, 131, 147, 202, 205
micrograms, 112
micronutrients, xi, 140, 193, 195, 196, 199, 200, 201, 203, 204, 206, 207, 254
microorganism(s), x, 21, 135, 140, 160, 168, 169, 172, 177, 181, 183, 210, 242, 252, 285
microRNA, 275
microsatellites, 187
microscope, 261
microscopy, 212
microspore, 28
migration, 137, 272, 286
milking goats, vii, 2, 4, 11, 12, 23
mimicry, 234
mixing, 171, 244, 261
model system, 177, 183
modelling, 282
models, 140, 144, 228, 233, 234, 238, 259, 286
moderate activity, 14
modifications, 117, 273
moisture, 4, 168, 197, 226, 260, 262, 263
moisture content, 260, 262
molecular biology, 27
molecular tools, xi, 22, 209
molecular weight, xii, 135, 226, 261, 264, 265, 266, 285
molecules, xii, 9, 20, 133, 137, 140, 220, 257, 258, 259, 273, 276, 280
monolayer, 268
monomers, 130, 217, 218
monounsaturated fatty acids, 48, 54, 88, 131
Montenegro, 190
Moon, 301
morbidity, 227
Morocco, 6, 7, 193, 197, 201
morphogenesis, 220, 280
morphology, 186, 261
mortality, 21, 95, 98, 153, 227, 238
mosaic, 216, 276, 281
Moses, 176
motif, 213, 219
motor neurons, 181
MR, 32, 35, 38, 39, 119
mRNA(s), 9, 111, 113, 141, 215, 216, 222, 286
mucosa, 106, 130, 135, 136, 143, 168
multiple sclerosis, 20
multiplication, 28
mung bean, 99, 103, 232, 290
mung bean starch, 103
mutagenesis, 223
mutant, xi, 209, 211, 213, 215, 218, 222, 274, 287
mutation, 223
Myanmar, 109, 272
mycorrhiza, xi, 209, 217, 218, 220, 221, 222

N

NaCl, 27, 260, 261
National Academy of Sciences, 205, 268, 269, 280
National Health and Nutrition Examination Survey, 119, 227
natural compound, 188
natural food, 94, 125
natural resources, 42
nausea, 144
necrosis, 30, 248
negative consequences, xi, 193
negative effects, 49, 204
Nepal, 197, 199, 200, 201
nephropathy, 95
nervous system, 131
neuroblastoma, 286
neurodegenerative diseases, 86
neuropathy, 95
neurotransmitter, 132
neutral, 47, 130, 143
New England, 38, 146
New South Wales, 38
New Zealand, 196
NHANES, 101, 227
niacin, 197, 284
nickel, 196
Nigeria, 88, 159, 162, 164, 165, 167, 175, 176
nitrates, 160
nitric oxide, 184, 236
nitrogen, x, xi, xiii, 66, 87, 91, 160, 176, 179, 180, 195, 209, 210, 212, 215, 219, 222, 226, 229, 271, 273, 274, 279
nitrogen compounds, 91
nitrogen fixation, x, xiii, 179, 222, 271, 273, 274
nitrogenase, 210, 273
NMR, 153

nodules, 85, 210, 211, 213, 214, 215, 218, 277, 278, 279
North Africa, 191, 197
North America, 5, 197, 201
Northern blot, 113
Norway, 89
nuclear receptors, 141
nuclei, 213, 218
nucleic acid, 111, 296
nucleotide sequence, 187
nucleotides, 210
nucleus, 133, 213
nutraceutical, vii, xi, 1, 85, 117, 119, 179, 181, 183
nutrient(s), viii, x, xi, 27, 39, 41, 85, 88, 94, 95, 97, 101, 105, 116, 126, 132, 134, 135, 136, 148, 150, 159, 166, 168, 169, 172, 173, 174, 180, 182, 191, 190, 191, 193, 194, 195, 197, 200, 203, 204, 217, 218, 222, 226
nutrient composition, x, 85, 159
nutrition, vii, viii, xi, 41, 47, 60, 84, 86, 88, 89, 95, 99, 114, 121, 122, 135, 147, 148, 149, 151, 175, 180, 193, 194, 195, 196, 199, 204, 205, 206, 207, 209, 223, 254
nutritional deficiencies, 181, 194

O

obesity, vii, ix, 93, 94, 95, 98, 99, 102, 111, 114, 115, 117, 118, 119, 120, 121, 122, 123, 125, 135, 136, 138, 140, 150, 152, 153, 195, 202, 227
obstacles, 95
obstructive sleep apnea, 95, 138
OECD, 148
oil, viii, x, xiii, 18, 20, 41, 46, 47, 48, 50, 51, 53, 89, 119, 127, 131, 132, 148, 159, 168, 210, 286, 295, 296, 298
oil production, 210
oilseed, 2, 126
oleic acid, xiii, 43, 48, 49, 50, 54, 83, 111, 131, 283, 284
oligosaccharide, 202, 217, 258, 273, 280
olive oil, 19
omega-3, 89
operations, 183
opportunities, 195, 203, 206
optical microscopy, 212
optimism, 127
optimization, 218
organ(s), 20, 28, 34, 165, 194, 210, 211, 227, 273
organelle, 212
organic matter, 242
organic solvents, 297
organism, vii, xi, 71, 131, 209
osteoarthritis, 95, 138
osteoporosis, xiii, 126, 134, 136, 144, 145, 150, 156, 157, 283, 285
overweight, ix, 94, 95, 105, 118, 119, 138, 227
ox, 204
oxidation, xii, 99, 100, 105, 119, 132, 134, 137, 140, 196, 225
oxidative stress, xii, xiii, 30, 111, 123, 141, 146, 181, 226, 283, 284, 287
oxygen, 155, 273

P

p53, 35
Pacific, 93
pain, 19
paints, 2
Pakistan, 85, 88, 109
pancreas, 135, 139, 229, 289
pancreatic cancer, 142
pantothenic acid, 197
parallel, 300
paralysis, 18
paresis, 180
participants, 98, 99, 110, 111, 284
particle bombardment, 281
pasta, 117, 127
pasture, vii, 1, 2, 4, 11, 25, 27, 32, 34, 35, 37, 38, 160, 210
pastures, 3, 10, 27, 219
pathogenesis, 233
pathogens, 3, 21, 203, 290
pathology, 181
pathophysiological, 234
pathways, 9, 102, 114, 117, 139, 140, 223, 287, 292
PCR, 8, 187, 216, 276, 278
pea starch, 287
pedigree, 26
pepsin, xii, 203, 225, 228, 229, 231, 232
peptide(s), viii, xii, 41, 96, 103, 108, 115, 140, 189, 225, 227, 228, 229, 230, 231, 236, 237, 239, 287, 302
periodontal, 290
permeability, 115
permission, 196
permit, 27
peroxide, xiii, 283, 286
personality, 43
personality disorder, 43
Peru, 164
pests, x, 3, 166, 179
petroleum, 298

pH, 11, 143, 170, 181, 228, 229, 242, 245, 252, 261, 298
phage, 276
phagocytosis, 286
pharmaceutical(s), vii, viii, 1, 2, 4, 14, 19, 22, 24, 25, 27, 31, 34, 295
pharmacological treatment, 227
phenol, 44, 57, 296, 298, 299
phenolic compounds, ix, 19, 20, 42, 86, 111, 116, 123, 140, 212, 284, 285, 287, 291, 301
phenotype, 205, 218, 277
phenylalanine, 69, 76, 160, 162, 197, 264
phloem, 196
phosphate(s), 131, 132, 135, 183, 203, 204, 217, 218, 229, 258, 289
phosphatidylcholine, 131
phospholipids, xiii, 131, 283
phosphorous, 164, 173, 176, 203, 226, 266
phosphorus, x, xiii, 114, 132, 149, 159, 176, 223, 242, 258, 272
photomicrographs, 107
photosynthesis, 217, 242
phototoxicity, 14, 19
phylum, 216
physical activity, 136, 138, 151, 227
physical characteristics, 148, 168
physical properties, 86, 259
physical treatments, 151
physicians, 14
physicochemical properties, 116, 295
Physiological, 35, 144
physiology, 14, 27, 35, 183, 184, 205
phytodiversity, viii, 41
phytoremediation, 243
phytosterols, 140, 284
pigmentation, 286
pigs, 9, 89
pilot study, 85
pioglitazone, 95
placebo, 85, 143, 288
plant breeding programmes, viii, 2, 31
plant growth, 27, 28, 30, 219, 243, 253, 267, 279
plant sterols, xiii, 283
plaque, 136
plasma membrane, 30, 153, 213, 214, 267
platelet aggregation, 236
playing, 228
pleurisy, 18
point mutation, 211
Poland, 189, 190
polar, 215, 222, 273, 285
pollen, viii, 2, 267
pollen tube, 267
pollination, 9
pollinators, 22, 24
pollutants, 242
polyacrylamide, 296
polymerase chain reaction, 216
polymers, 286
polymorphism(s), 187, 188, 190, 276, 277, 280
polyphenols, viii, 41, 42, 43, 44, 56, 57, 59, 60, 61, 62, 64, 66, 83, 90, 101, 110, 116, 133, 136, 142, 149, 166, 168, 180, 186, 190, 201
polypropylene, 244
polysaccharide chains, 259
polysaccharide(s), 101, 103, 114, 115, 119, 130, 147, 202, 259, 262, 268, 280, 296
polyunsaturated fatty acids, 48, 49, 52, 54, 131, 148
pools, xiii, 255, 271, 272, 277
population, xiii, 2, 9, 10, 27, 28, 59, 98, 116, 131, 142, 145, 150, 153, 180, 191, 199, 211, 215, 216, 218, 227, 295
population growth, 199
Portugal, 6, 8, 32, 200
positive correlation, 185
potassium, x, 114, 132, 159, 164, 173, 226, 229, 254
potato, 203
potential benefits, 95
poultry, 88, 151
precipitation, 11, 242, 262, 296, 298, 299
predators, 135
preparation, 140, 145, 155, 161, 296, 300, 302
preschool, 194
preschool children, 194
preservation, 84, 144
preservative, 169
prevention, ix, 44, 89, 93, 94, 95, 97, 120, 125, 129, 130, 133, 134, 141, 145, 149, 150, 152, 154, 189, 205, 238, 284
primary function, 203
primary sector, 42
principal component analysis, 253
prisoners, 85
probe, 258
probiotic(s), 115, 122, 123, 205
processing stages, 299
producers, 126, 145, 197
production costs, 242
professionals, 94
prognosis, 285
project, 10, 237, 300
proliferation, xiii, 134, 141, 154, 283, 285, 286, 287
proline, xii, 4, 27, 70, 77, 78, 257, 259, 261, 267
promoter, 154, 213, 214, 216, 276
propagation, vii, viii, 2, 4, 22, 27, 28
prostaglandins, 49, 131

prostate cancer, 126, 134, 141, 142, 150, 153, 154, 285, 286
protease inhibitors, 134, 141, 203, 226, 286
protection, 21, 117, 123, 149, 201, 203, 255
protective role, 141
protein components, 137, 258, 301
protein family, 222, 281
protein hydrolysates, vii, 228, 231, 232, 233, 234, 235, 236, 237, 238, 239
protein oxidation, 296
protein synthesis, 242
protein-protein interactions, 296
proteoglycans, xii, 257, 258, 267
proteolysis, 135
proteolytic enzyme, 183, 239
proteome, 300, 303
proteomics, 3, 296, 300, 301, 302
psoriasis, 20, 38
psychiatric illness, 138
puberty, 141
public awareness, 195
public health, 138
Puerto Rico, 226
pulp, 127
pure line, 92
purification, 150, 216
pyrimidine, 20

Q

quality control, 244, 300
quantification, 260, 299
quartile, 98
quercetin, 116, 133, 287

R

Rab, 277, 281
radiation, 20
radicals, 290
rain forest, 164
rainfall, vii, 2, 3, 4, 11, 23, 37, 184, 185
raw materials, 126, 295
reaction time, 231
reactions, 274
reactive oxygen, 20, 86, 286
reactivity, xiii, 283, 284
reagents, 244
real time, 299
recall, 95
receptors, 135, 137, 139, 143, 144, 184, 213, 215, 217, 274, 285
recognition, 213, 220, 273, 279, 280
recombination, 24, 276
recommendations, ix, 42, 49, 66, 77, 83, 97, 102, 120, 152
reconstruction, 222
recovery, 11, 12, 34, 168, 299, 300
recruiting, 287
red blood cells, 135, 168
redistribution, 196
redundancy, 215
refractive index, 260
regenerate, 275
regeneration, viii, 2, 4, 28, 29, 37, 210, 211, 281
regions of the world, viii, 41, 226, 259
regression, 230, 238
regression analysis, 230
rehydration, 299
reintroduction, xi, 179
rejection, 20
relatives, 88
relevance, vii, 1, 4, 43, 180, 238, 264, 292
renal failure, 120
repressor, 184
requirements, 50, 69, 76, 83, 91, 126, 169, 196
researchers, 4, 98, 111
reserves, 101
residues, 213, 258, 259, 264, 265
resins, 2
resistance, x, 34, 105, 117, 131, 139, 153, 179, 183, 190, 191, 200, 238, 301
resolution, 299, 301
resources, viii, xiii, 2, 8, 9, 31, 42, 175, 189, 194, 210, 211, 271, 272, 275
response, 4, 27, 35, 103, 105, 106, 110, 116, 117, 121, 123, 141, 153, 213, 214, 215, 222, 229, 234, 259, 277, 278, 281, 287
resveratrol, 133
retail, 150, 153
reticulum, 217
retinopathy, 95
RH, 33, 123, 149
rheology, 267
Rhizopus, 85, 170, 171, 172, 174, 177, 178
riboflavin, 169, 197, 284
ribosome, 222
rings, 133
risk(s), viii, ix, xii, 41, 43, 49, 55, 86, 88, 89, 93, 94, 95, 97, 98, 102, 111, 115, 116, 118, 123, 128, 130, 135, 136, 137, 141, 144, 145, 146, 150, 151, 154, 156, 201, 204, 225, 227, 234, 239, 241, 255, 284, 285, 286, 291
risk assessment, xii, 241, 255
risk factors, 55, 94, 98, 118, 136

RNA, 113, 210, 213, 216, 219, 275, 276
RNAi, 213, 275, 276, 277, 278, 281
room temperature, 181, 228, 244, 260
root growth, 247
root hair, 212, 213, 214, 219, 273, 274, 277, 280, 281
root system, 127, 210, 211
rosiglitazone, 95
rotations, 3
Rouleau, 116
routes, xii, 226, 241
rules, 222, 280
rural areas, 243

S

safety, 131, 155, 183
salinity, 27
saliva, 111
Salmonella, 287
salts, 132, 296
sanctions, 282
saponin, 136, 166
saturated fat, ix, 43, 46, 47, 48, 49, 51, 52, 54, 110, 125, 128, 136
saturated fatty acids, 43, 46, 47, 48, 49, 51, 52, 54
school, 126
science, 36, 88, 122, 145, 147, 151, 176, 296
scientific knowledge, 42
sea level, 9, 11, 24, 180, 260
seasonal changes, 21
secondary metabolites, vii, 1, 2, 27, 38, 57, 185, 186, 296
secretion, 99, 100, 103, 106, 139, 152, 212, 290
security, 191, 195
sediment(s), 170, 228, 254
seeding, 10, 11
seedlings, 88, 181, 190, 210, 223, 281
selective estrogen receptor modulator, 142
selenium, xi, 193, 194, 195, 201, 205, 206, 287
sensations, 103, 106, 117
sensitivity, 110, 143
sequencing, xi, xiii, 209, 215, 216, 271
serine, xiii, 70, 77, 78, 203, 258, 264, 286, 287
serotonin, 96
serum, 86, 103, 104, 115, 122, 129, 146, 147, 152, 157, 284, 286, 288, 289, 292
SES, 226
sewage, 242
sex steroid, 143
shape, 127, 259, 265
shear, 110
sheep, 9, 11, 19, 26, 32, 38
shoot(s), 24, 27, 28, 29, 30, 211, 216, 243, 244, 247, 248, 249, 250, 251
short supply, 199
showing, viii, 2, 26, 28, 30, 31, 83, 95, 164, 285
shrubs, 39
sialic acid, 258, 287
side chain, 130
side effects, 144, 237, 243
signal peptide, 103
signal transduction, 221, 274, 280, 281
signaling pathway, 149, 154, 215, 217, 218
signalling, 220, 274
signals, 212, 218, 221, 278, 279, 280
Singapore, 142
skeletal muscle, 106
skin, 20, 21, 143, 286
skin cancer, 286
skin diseases, 20
sleep disturbance, 142
sludge, 242
small intestine, 168
smoking, 136
smooth muscle, 110, 286
smooth muscle cells, 286
society, 126, 227
sodium, x, xii, 139, 159, 173, 201, 225, 226, 229, 230, 261
software, 231
soil erosion, 10
soil particles, 244
soil type, x, 179
solid matrix, 168
solid state, 168, 176
solubility, 183, 242
solution, xi, 27, 103, 117, 193, 195, 204, 206, 230, 234, 235, 242, 243, 244, 245, 260, 261, 296, 298
solvents, 127, 297, 302
sorption, 242, 254
South Africa, 162
South America, 197, 272
South Asia, 180, 194, 197, 199
Southeast Asia, 199
soybean(s), vii, ix, xiii, 45, 71, 84, 125, 126, 127, 129, 132, 133, 134, 135, 136, 137, 140, 142, 145, 146, 148, 150, 151, 153, 157, 162, 177, 281, 285, 295, 296, 297, 299, 300, 301, 302
soymilk, 126, 150
Spain, viii, 1, 3, 4, 6, 8, 9, 12, 14, 22, 24, 25, 26, 30, 31, 32, 33, 34, 35, 36, 37, 38, 41, 43, 44, 45, 46, 50, 60, 62, 66, 89, 90, 91, 92, 125, 126, 145, 146, 148, 151, 190, 200, 226
spastic, 180
speciation, 27, 245, 252, 253, 255

specifications, 267
sperm, 286
sphygmomanometer, 230
Sprague-Dawley rats, 102
SS, 151, 153
stability, 147, 186, 252, 275, 279
standard deviation, 48, 53, 55, 61, 68, 69, 70, 72, 74, 75, 77, 80, 81, 112, 113, 299
standard error, 107, 116
starch, 90, 99, 101, 103, 109, 110, 114, 115, 116, 117, 130, 135, 202, 228, 238, 287
starvation, 194
state(s), 85, 102, 168, 176, 189, 228, 234
steel, 243
steroids, 136
sterols, 131, 284
stimulant, 18
stimulation, 137, 141
stock, 226
stomach, 95
storage, 87, 99, 115, 129, 132, 150, 183, 185, 186, 187, 203, 243
stress, 4, 23, 35, 106, 122, 155, 181, 185, 200, 218, 281, 287, 291
stroke, 95, 152, 227, 238
structural changes, 140
structure, 20, 35, 130, 133, 144, 212, 217, 242, 258, 259, 262, 266, 267, 273, 275, 276, 296
style, 195, 267
subgroups, 299
subsistence, x, 179, 180
subsistence farming, x, 179
substitutes, 96, 162
substitution(s), 130, 137, 213, 287
substrate(s), 47, 49, 100, 119, 131, 137, 152, 153, 169, 177, 229, 231
subtraction, 277, 278
sucrose, 106, 109, 130, 298, 299
sulfur, 45, 77, 78, 129, 132
sulphur, ix, 42, 44, 67, 78, 83, 135, 183, 284
Sun, 187, 192, 220, 221, 255
supplementation, 111, 117, 122, 143, 154, 206
suppression, 106, 215
surfactant(s), 243 248
survival, 21, 28, 29
susceptibility, 138
sustainability, 219, 222
sustainable development, xi, 193
Sweden, 31, 35, 228
sweeteners, 106, 108
swelling, 213
symbiosis, xi, 11, 209, 210, 211, 212, 213, 214, 215, 216, 217, 218, 220, 221, 222, 274, 279, 280, 281, 282
symptoms, xiii, 30, 126, 134, 142, 143, 155, 156, 189, 233, 234, 247, 283, 285
syndrome, 152
synergistic effect, 111, 140
synthesis, 21, 102, 131, 136, 137, 215, 273
Syria, 186, 189, 197, 199, 200, 201
systemic lupus erythematosus, 20
systolic blood pressure, 98, 227

T

tachycardia, 143
tannins, 26, 38, 84, 86, 89, 90, 133, 180, 182, 188, 226
tar, 13
target, 99, 141, 195, 218, 227, 276, 285
target organs, 285
Task Force, 238
taxa, viii, 6, 7, 8, 11, 12, 41, 43, 45, 50, 51, 52, 53, 55, 58, 59, 61, 62, 63, 64, 66, 73, 74, 75, 77, 78, 79, 80, 83, 84, 226, 280
taxonomy, 88
teams, 22, 24
techniques, viii, xi, xiii, 2, 196, 209, 233, 266, 271, 275, 296, 297, 301, 302
technology(s), 37, 147, 211, 215, 218, 276, 301
telephone, 93
temperature, 4, 27, 28, 30, 127, 143, 162, 172, 181, 261, 296, 297, 300
terpenes, 296
territorial, 37
territory, 272
testing, 199
testosterone, 136
texture, 169
therapeutic agents, 85, 95, 120, 237
therapeutic interventions, 234
therapeutic use, 14, 113
therapy, 20, 38, 94, 95, 138, 142, 144, 145, 150, 156, 189, 286, 288, 289
thermal treatment, 135, 300
thiamin, 164, 197, 284
threonine, xiii, 69, 76, 129, 162, 197, 258, 264
thromboxanes, 49
thymine, 20
thyroid, 137, 151
thyroxin, 137
tissue, 27, 28, 106, 111, 115, 119, 131, 136, 141, 143, 244, 249, 250, 255, 258, 259, 267, 275, 285, 298, 299

tobacco, 136, 220, 222, 259
tocopherols, 85, 132, 284
tofu, 126, 127, 300, 302
tones, 126
total cholesterol, 110, 137, 139, 284, 288, 289
toxic effect, 252
toxic metals, xii, 241, 242, 255
toxicity, xi, 20, 26, 149, 179, 181, 183, 184, 189, 191, 242, 243, 252, 255, 286, 287
toxin, 181, 183, 185, 186, 188, 190
trace elements, 132, 195, 255
trade, 2, 128, 222
trafficking, 100
training, 238
traits, vii, viii, x, 1, 2, 4, 9, 22, 24, 25, 26, 31, 36, 115, 179, 186, 189, 190, 191, 200, 275
transcription, 100, 105, 141, 154, 213, 214, 215, 216, 217, 218, 219, 220, 221, 222, 274, 277, 278, 281
transcription factors, 213, 214, 215, 218, 219, 274, 277
transcripts, 215
transduction, 274
transformation, xi, xiii, 209, 211, 219, 271, 275, 276, 277, 281
translation, 13, 215, 216
translocation, 154, 196, 243, 251, 252
transmission, 259
transpiration, 25
transplant, 20
transport, 136, 140, 141, 215, 221, 222, 253
treaties, 13
treatment, ix, 14, 20, 38, 85, 86, 87, 94, 95, 103, 105, 106, 110, 117, 125, 129, 135, 140, 142, 143, 144, 145, 151, 153, 155, 186, 201, 211, 213, 215, 227, 230, 231, 238, 243, 244, 245, 246, 247, 248, 252, 254, 286, 289, 292
trial, 12, 24, 25, 85, 98, 99, 105, 145, 154, 155, 156, 233, 236
triggers, 223, 273, 279
triglycerides, 103, 131, 137, 139, 288, 289
Trinidad, 284
trypsin, x, 3, 100, 129, 135, 140, 149, 159, 166, 168, 169, 173, 174, 183, 188, 191, 203, 226, 284, 287
tryptophan, 67, 69, 76, 78, 83, 160, 162, 197
tumor(s), 18, 141, 142, 144, 149, 154
tumor progression, 141
tumorigenesis, 154
Turkey, 5, 31, 85, 88, 109, 114, 186, 197, 201
turnover, 145
type 1 diabetes, 20
type 2 diabetes, ix, 93, 94, 95, 102, 105, 114, 116, 117, 120, 123, 138, 140, 153, 284
Tyrosine, 67, 78, 129, 265

U

undernutrition, 138
United Kingdom (UK), 34, 121, 122, 147, 175, 176, 200, 205
United Nations, 42, 84, 88, 205, 267
United States (USA), 10, 33, 35, 109, 116, 118, 122, 126, 131, 145, 147, 148, 149, 164, 177, 193, 200, 202, 226, 227, 228, 260, 261, 267, 268, 269, 289, 295
updating, 147
urban, 241, 242, 245, 255
urbanization, ix, 93
urea, 296, 298
urinalysis, 288
urine, 18, 105, 150, 181
USDA, 11, 105, 118, 122, 146, 149, 295
uterus, 144
UV light, 20

V

vacuum, 298
vagina, 143
Valencia, 37
validation, 299, 301
valine, 69, 76, 160, 162, 197, 264
vanadium, 196
variations, 8, 67, 143, 181, 245, 285
varieties, vii, 2, 3, 6, 8, 9, 11, 12, 22, 24, 25, 26, 28, 33, 42, 85, 89, 116, 132, 148, 176, 181, 186, 187, 188, 199, 200, 226, 277, 299, 300, 302
vascular system, 18
vasoconstriction, xii, 225
vasodilation, 143
vasodilator, 236
vasomotor, 143
vector, 281
vegetable oil, 89
vegetables, xii, 44, 90, 98, 136, 160, 161, 163, 195, 203, 241, 255, 284, 287
vegetation, 7
vegetative stages, 184
velvet, x, 86, 159, 161
Vermeer, 156
vertebrates, 181
very low density lipoprotein, 289
victims, 19
viral infection, 201
visceral adiposity, 119, 284
viscosity, 261, 264, 265
vitamin A, 194

vitamin B1, 169
vitamin B12, 169
vitamin D, 136
vitamin E, 132
vitamins, xiii, 43, 85, 114, 169, 180, 195, 203, 272, 283, 286, 287
vitiligo, 20

W

waist circumference, ix, 94, 98, 101, 102, 105
war, vii, ix, 94
Washington, 122, 148, 207, 267
waste disposal, 241
waste incineration, 241
water, viii, x, 3, 4, 11, 12, 27, 30, 35, 38, 39, 41, 105, 106, 114, 126, 127, 168, 170, 171, 172, 179, 180, 183, 185, 195, 202, 228, 230, 242, 243, 244, 259, 260, 261, 268, 269, 285, 287, 288, 289, 297, 298
wavelengths, 21
web, 211, 220
weight control, 94, 97, 284
weight gain, ix, 93, 95, 99, 103, 144
weight loss, ix, 94, 95, 96, 101, 106, 111, 113, 116, 117, 118, 137, 153, 202
weight management, 105, 111, 118
weight reduction, 95, 227, 248
well-being, xi, 130, 193
wells, 260
West Africa, 165, 175
Western Australia, 1, 11, 30, 32, 36, 37
Western countries, 134, 141, 227, 234
wheat germ, 286
World Health Organization (WHO), 69, 76, 83, 88, 89, 91, 94, 118, 138, 145, 152, 176, 194, 205, 207, 227, 262, 267
wild type, 211, 216, 275
wood, 10, 171
worldwide, ix, xi, 42, 93, 94, 138, 179, 194, 210, 225, 227, 238, 287

X

xerophthalmia, 194

Y

yeast, 168, 213, 258
yield, 11, 12, 22, 23, 26, 31, 35, 59, 87, 90, 160, 161, 164, 168, 176, 186, 188, 191, 197, 200, 229, 297

Z

Zimbabwe, 162, 226
zinc, xi, 27, 114, 132, 193, 195, 203, 204, 206, 226, 228, 254, 255, 272

